OHIO STATE UNIVERSITY BIOSCIENCES COLLOQUIA

Genetics and Biogenesis
Of Mitochondria and Chloroplasts

THEMES, DATES, AND ORGANIZERS

Genetics and Biogenesis of Mitochondria and Chloroplasts
5–7 September 1974
C. W. Birky, Jr., P. S. Perlman, and T. J. Byers

Biological Control Mechanisms
4–6 September 1975
J. C. Copeland and G. A. Marzluf

Analysis of Ecological Systems
29 April–1 May 1976
D. J. Horn, G. R. Stairs, and R. D. Mitchell

EDITED BY C. WILLIAM BIRKY, JR.
PHILIP S. PERLMAN
AND THOMAS J. BYERS

Genetics and Biogenesis of Mitochondria and Chloroplasts

OHIO STATE UNIVERSITY PRESS : COLUMBUS

Manufactured in the United States of America

Library of Congress Cataloging in Publication Data

Biosciences Colloquium, 1st, Ohio State University, 1974.
Genetics and biogenesis of mitochondria and chloroplasts.

(Ohio State University biosciences colloquia)
Bibliography: p.
Includes index.
1. Mitochondria—Congresses. 2. Chloroplasts—Congresses. 3. Genetics—Congresses. I. Birky, C. William. II. Perlman, Philip S., 1945- III. Byers, Thomas J. IV. Title. V. Series: Ohio State University, Columbus. Ohio State University biosciences colloquia. [DNLM: 1. Chloroplasts—Congresses. 2. Mitochondria—Congresses. QH603 G328]
QH603.M5B53 1974 574.8'734 75-20271
ISBN 0-8142-0236-5

Contents

Preface

The annual Biosciences Colloquium of the College of Biological Sciences, Ohio State University, was developed to provide a forum in which research workers from the college and from other institutions would present their work to each other and to interested third parties in a direct and personal way.

The program was instituted by the Research Committee of the college with the support and encouragement of Dean Richard H. Böhning. The program is supported by funds allotted by the college for encouragement of scholarly activities of the faculty, and this commitment has shaped the nature of the series. The individual colloquia are produced by one or more of the college faculty, and the topics are drawn from their actual research interests. The topics are selected by the College Research Committee from proposals submitted by the faculty in an annual competition, and continuity in format is provided by a standing Colloquium Series Committee, which solicits the proposals and presents them to the Research Committee for their disposition. Thus the series will provide a record of research in biological sciences at the Ohio State University and will help to relate this research to activities at other institutions.

In addition to our debt to Dean Böhning, without whose support this program could not exist, we wish to thank Weldon A. Kefauver, the director of the Ohio State University Press, for his support in bringing these proceedings to publication.

Julius P. Kreier, Professor
Richard O. Moore, Associate Dean
College of Biological Sciences
Colloquium Series Committee

First Annual Biosciences Colloquium
College of Biological Sciences
Ohio State University
5–7 September 1974

GENETICS AND BIOGENESIS OF MITOCHONDRIA AND CHLOROPLASTS

Organizers

C. William Birky, Jr., Associate Professor of Genetics and Developmental Biology Program, Ohio State University

Philip S. Perlman, Assistant Professor of Genetics and Developmental Biology Program, Ohio State University

Thomas J. Byers, Associate Professor of Microbiology and Developmental Biology Program, Ohio State University

Speakers

Giuseppe Attardi, Division of Biology, California Institute of Technology

C. William Birky, Jr., Department of Genetics and Developmental Biology Program, Ohio State University

David E. Griffiths, Department of Molecular Sciences, University of Warwick, England

J. Kenneth Hoober, Department of Biochemistry, Temple University School of Medicine

John R. Laughnan, Provisional Department of Genetics and Development, University of Illinois

Henry R. Mahler, Chemical Laboratories, Indiana University

Philip S. Perlman, Department of Genetics and Developmental Biology Program, Ohio State University

Ruth Sager, Department of Biological Sciences, Hunter College of the City University of New York

R. A. E. Tilney-Bassett, Department of Genetics, University College of Swansea, Wales

S. G. Wildman, Department of Biology, University of California, Los Angeles

Introduction

It is appropriate that the Ohio State University Biosciences Colloquium Series begin with a field that has blossomed within the past ten years into one of the most active topics of genetics and cell biology. Although the genetics of chloroplasts has been studied since 1908, and mitochondrial inheritance since 1949, it was not until the early 1960s that the discovery of DNA in these organelles transformed the investigation of mitochondrial and chloroplast biogenesis into a popular pastime for molecular and cell biologists and gave new impetus to investigations of organelle genetics. The importance, both scientific and practical, of acquiring more knowledge about the energy-producing centers of the cell cannot be overemphasized.

In selecting the speakers and topics for the colloquium, the editors tried to cover a wide variety of topics of special interest. In the area of mitochondrial biogenesis most of our knowledge about the location of genes coding for mitochondrial components, and sites of synthesis of those components, has come from studies on yeast and HeLa cells. These areas are reviewed by Mahler and Attardi; their papers bring out the striking similarities in conclusions obtained with the two experimental systems. Griffiths describes the extensive work of his group on nuclear and mitochondrial mutants affecting oxidative phosphorylation in yeast, which are providing the tools for an analysis of the mechanism of oxidative phosphorylation. Hoober describes his work on the synthesis of chloroplast membrane proteins, which provides information not only about sites of synthesis but also on the control of synthesis in *Chlamydomonas*. Wildman's work on tobacco hybrids has demonstrated joint nuclear and chloroplast synthesis of RUDP carboxylase, a crucial enzyme in photosynthesis. In his paper he shows how these studies have provided information on the molecular evolution of both the nuclear and chloroplast genes that are involved.

The most extensive studies on the genetics of mitochondria have been done with fungi and ciliated protozoa. This work is reviewed by Birky, concentrating on recent studies with yeast. Perlman describes the analysis of the formation and genetic behavior of petite mutants in yeast, involving extensive deletion and duplication of mitochondrial genes, and providing a possible model system for the study of the origin of repeated DNA sequences. Tilney-Bassett gives a thorough review of work on chloroplast genetics in higher plants, with special emphasis on studies of chloroplast mutation and of the mechanism of maternal inheritance, where his own work on the geranium provides the greatest promise of new insights. The deepest genetic analysis of chloroplasts has come from studies on *Chlamydomonas*, pioneered by Sager; her paper concentrates on recent studies of maternal inheritance and leads to some speculations about the evolution of chloroplast genes. Finally, Laughnan describes his recent studies on the genetic basis of cytoplasmic pollen sterility in maize, a phenomenon of great practical importance in modern agriculture. It is widely believed that pollen sterility involves organelles, but Laughnan interprets his most recent data in terms of an episome model.

We are gratified by the enthusiastic response to this first Biosciences Colloquium. More than two hundred persons attended, representing eighteen states and Canada. In our somewhat biased opinion, the colloquium successfully fulfilled its purposes. We wish to thank the Colloquium Series Committee for giving us the opportunity to arrange this event, and the administration of the College of Biological Sciences for their support and encouragement. The staffs of The Ohio State University Division of Continuing Education and the Fawcett Center for Tomorrow provided excellent facilities and assistance for the Colloquium. Our own graduate students gave unstintingly of their time and labor both during and before the Colloquium; we could scarcely have managed without them. We are also grateful to the Ohio State University Press, which has made every effort to enhance the usefulness of this volume by seeing that it appears in print as rapidly as possible. And most important of all, we wish to thank the speakers for accepting our invitation and for their cooperation in promptly submitting manuscripts. It is their contribution that made the Colloquium a success.

OHIO STATE UNIVERSITY BIOSCIENCES COLLOQUIA

Genetics and Biogenesis Of Mitochondria and Chloroplasts

GIUSEPPE ATTARDI, PAOLO COSTANTINO,
JAMES ENGLAND, DENNIS LYNCH, WILLIAM MURPHY,
DEANNA OJALA, JAMES POSAKONY, and BRIAN STORRIE

The Biogenesis of Mitochondria in HeLa Cells: A Molecular and Cellular Study

1

INTRODUCTION

In the last two decades thoughts about the biogenesis of mitochondria have been profoundly influenced by the idea that these organelles have somehow evolved from primitive endosymbiotic bacteria. Originally proposed at the end of the last century on the basis of morphological observations (Altmann, 1890; Benda, 1898), this idea received initial experimental support more than twenty years ago from the discovery, in yeast (Ephrussi et al., 1949) and *Neurospora crassa* (Mitchell and Mitchell, 1952), of the cytoplasmic inheritance of abnormalities of mitochondrial function, which pointed to the existence of cytoplasmic, possibly mitochondrial, genetic determinants. Subsequently, the demonstration that mitochondria contain a unique DNA and a distinctive protein-synthesizing apparatus, with specific ribosomes, tRNAs, amino acyl-tRNA synthetases, initiation and elongation factors (for reviews, see Ashwell and Work, 1970; Borst and Grivell, 1971; Attardi et al., 1973), provided the needed physical basis for the postulated partial reproductive autonomy of mitochondria. The recognition of similarities between bacterial and mitochondrial protein synthesis, such as their sensitivity to antibiotics, mechanism of initiation, and nature of the initiation factors, was generally interpreted as being a strong argument in support of the evolutionary relationship between mitochondria and bacteria, in spite of the evidence suggesting

Division of Biology, California Institute of Technology, Pasadena, California 91125 that most, if not all, of these bacterial-type characteristics probably depend on nuclear DNA-coded components.

It is not surprising that, in this context, growth and division of mitochondria have been traditionally considered to mimic the processes of bacterial reproduction, with the replication of mitochondrial DNA and membrane growth being intimately coordinated and interdependent. In the last few years, however, this picture has proved to be more and more inadequate to describe the process of mitochondriogenesis. In the first place, this picture is not in accord with the increasing number of observations indicating that mitochondrial DNA has no apparent role in controlling the gross formation and division of mitochondria. These processes continue in the absence of a functional mitochondrial DNA, or even in its physical absence, although the organelles thus formed are inactive in oxidative phosphorylation. In the second place, it is now apparent that the mitochondria within each cell are not organelles of fixed size, shape, and number, each dividing into two equal daughter organelles once per cell cycle. Rather, they represent a surprisingly plastic membrane system, whose units undergo, more or less frequently in different cell types, changes in shape and processes of fusion and fission: these changes apparently occur in response to physiological stimuli, and are superimposed upon organelle growth. The cellular complement of mitochondrial membranes, therefore, doubles its mass once per cell cycle following a pattern that lacks the regularity of bacterial growth and division, and that is independent of the replication or even the presence of mitochondrial DNA.

It is, therefore, clear that mitochondriogenesis must be considered as consisting of two processes: the gross formation of the mitochondrial membranes, which appear to be solely controlled by the nuclear genome, and the phenomena that depend, on the contrary, on the expression of the mitochondrial genome, i.e., the differentiation of the mitochondrial *anlagen* into organelles active in oxidative phosphorylation.

Relatively little is known about the mode of gross formation and division of mitochondria, apart from the evidence indicating that, both in *Neurospora crassa* (Luck, 1963, 1965) and in HeLa cells (Storrie and Attardi, 1973*a*), the mitochondrial membranes grow by addition of new components to old structures and that the mitochondrial number increases by division of preexisting organelles. By contrast, information has been accumulating at a rapid pace in recent years on the differentiation of the "inactive" mitochondria into energy-producing organelles. This aspect of mitochondriogenesis has attracted increasing interest among investigators because of the challenging problems it poses in the study of the mechanism and control of gene expression in eukaryotic cells and in the study of membrane structure, function, and biosynthesis.

In the present paper we will summarize the information we have obtained in the last few years concerning the mechanism, products, and control of expression of the mitochondrial genome in a human cell line grown *in vitro,* HeLa cells*; we will further describe some experiments aimed at elucidating the mode and temporal pattern of mitochondrial growth and division during the cell cycle of these cells.

EXPRESSION OF THE MITOCHONDRIAL GENOME

Complete Symmetrical Transcription of Mitochondrial DNA.

Mitochondrial DNA is transcribed in exponentially growing HeLa cells at a very high rate, which can be estimated to approach, at the peak of RNA synthetic activity, i.e., in the G_2 phase of the cell cycle (see below), the rate of transcription of rDNA in the same cells (Attardi and Attardi, 1969).

Figure 1 shows a mitochondrial DNA molecule in the process of being transcribed, spread for electron microscopy by the basic protein film technique (Aloni and Attardi, 1972). The 5 μ open circular DNA molecule is recognizable for its uniform extended appearance, whereas the nascent RNA molecules appear collapsed as bushes. At least 25 bushes of various sizes can be seen attached to the DNA molecule. The occurrence, in the transcription complex shown in figure 1, of growing RNA chains attached at fairly regular intervals to many points along the whole length of DNA is consistent with other experimental evidence, to be discussed below, indicating that mitochondrial DNA in HeLa cells is completely or almost completely transcribed. In particular, direct evidence for a complete transcription of mitochondrial DNA was obtained by saturation hybridization experiments carried out between mitochondrial RNA labedel with [5-^{3}H]uridine to a uniform specific activity and separated DNA strands immobilized on nitrocellulose membranes. As shown in figure 2, the heavy (H) strand is saturated by the labeled RNA at a level corresponding to about

*Unless otherwise specified, the work described here was carried out with the S3 clonal strain of HeLa cells.

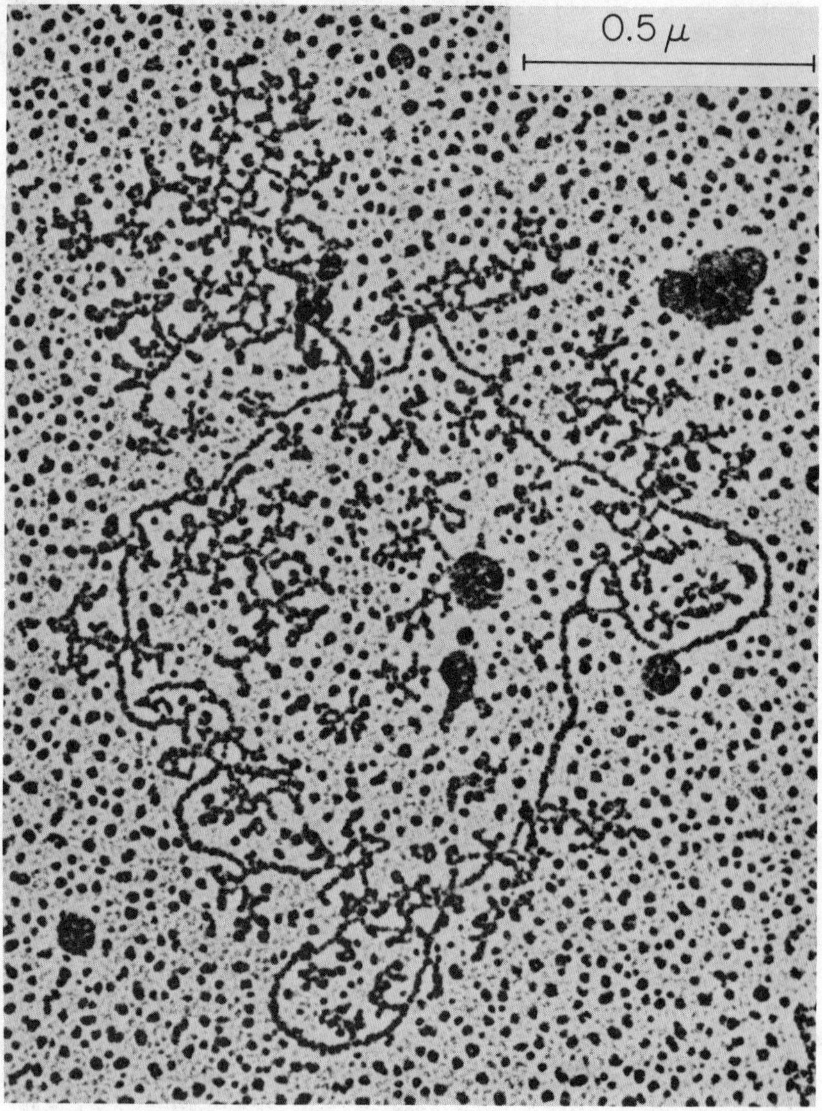

Fig. 1. Transcription complex of HeLa cell mitochondrial DNA. The fast-sedimenting components of an SDS lysate of the mitochondrial fraction, which banded in a CsCl-ethidium bromide density gradient at a density of about 1.78 g/cm^3, were spread for electron microscopy by the basic protein film technique (from Aloni and Attardi, 1972. Grateful acknowledgment is made to the *Journal of Molecular Biology* to reprint this and other figures in this paper.)

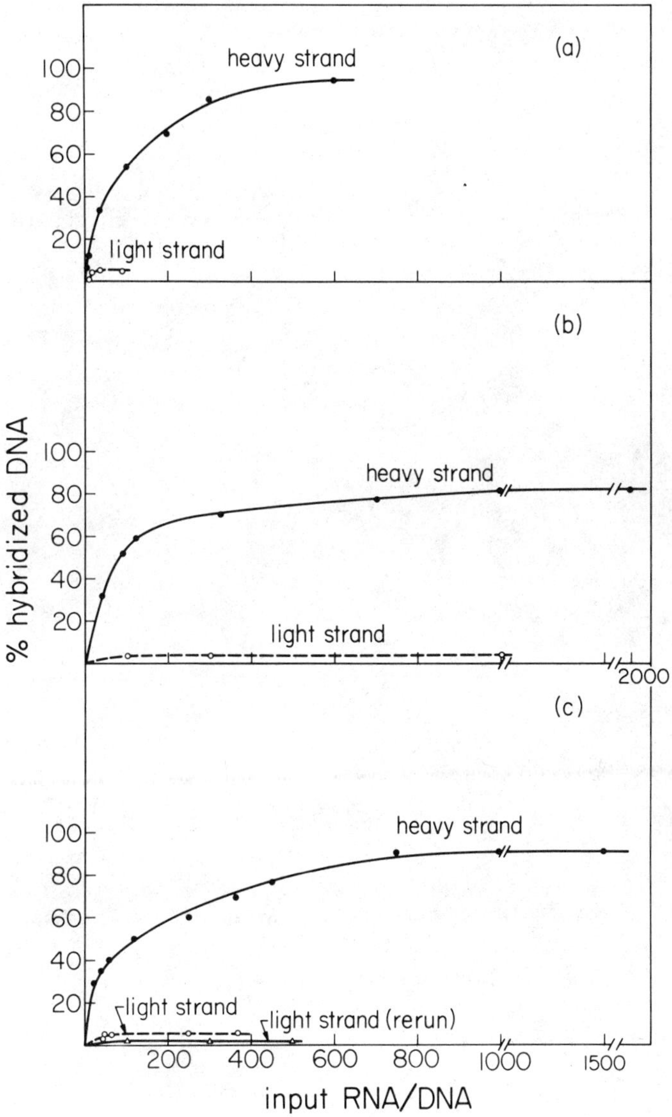

Fig. 2. Hybridization of separated strands of HeLa cell mitochondrial DNA with increasing amounts of RNA from the mitochondrial fraction of HeLa cells uniformly labeled with [5-^{3}H]uridine. The RNA had been fractionated by sucrose gradient centrifugation into components with S values >40 (a), between 22 and 40 (b), and <22 (c) (from Aloni and Attardi, 1971a).

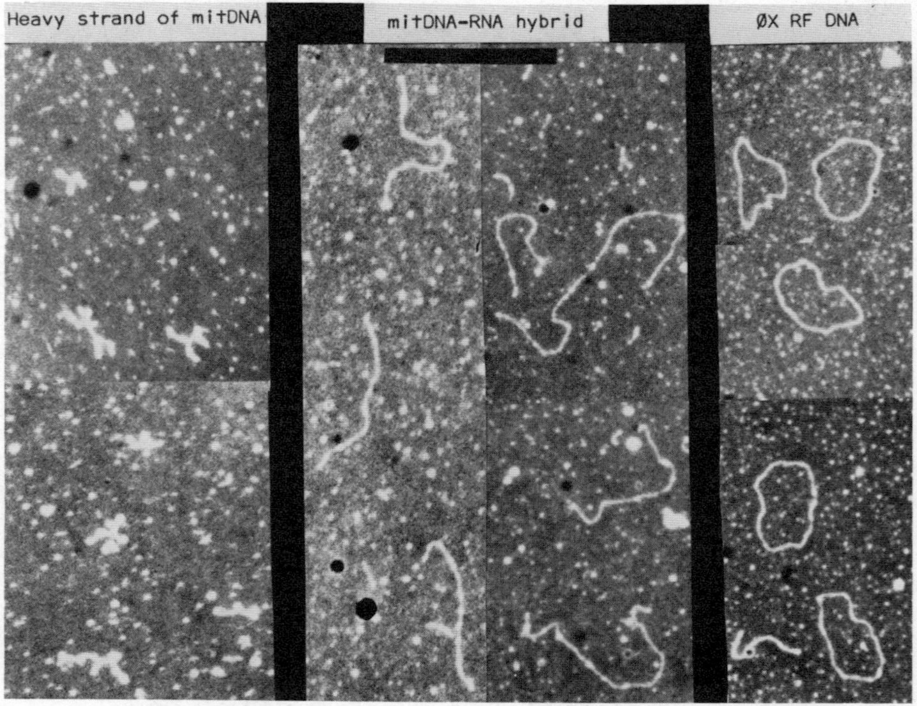

Fig. 3. Electron micrographs of H strand of mitochondrial DNA, mitochondrial RNA-H mitochondrial DNA hybrids (previously banded in a Cs_2SO_4 density gradient) and ϕX RF DNA, spread by the basic protein film technique (from Robberson et al., 1971).

100%, indicating that all or almost all the sequences of the H strand are represented, in complementary form, in mitochondrial RNA (Aloni and Attardi, 1971a). In agreement with this finding, the RNA-DNA hybrids formed with saturating amounts of RNA banded in a Cs_2SO_4 gradient at the density expected for fully base-paired hybrids, and exhibited in the electron microscope a uniform duplex appearance (fig. 3) (Robberson, Aloni, and Attardi, 1971).

Considering that the informational content of animal cell mitochondrial DNA is presumably an indispensable minimum (Borst et al., 1968; Attardi et al, 1970), and that the expression of its genes may be coordinated, it was not surprising to find that this DNA is completely transcribed in HeLa cells. On the contrary, it was an unexpected observation that transcription of this DNA is symmetrical, i.e., that the light (L) strand is also transcribed (Aloni and Attardi, 1971b). In the saturation hybridization experiments discussed above, only a very low level of hybridization was observed between the uniformly [5-^{3}H]uridine-labeled mitochondrial RNA and the L strand (fig. 2). Later experiments showed, however, that this result was due to the low steady state concentration of the L transcripts and to their becoming rapidly unavailable for hybridization with the DNA immobilized on the membrane, due to base-pairing with the excess of H transcripts present in the incubation mixture. In fact, in hybridization experiments carried out in solution in DNA excess and using [5-^{3}H]uridine pulse-labeled mitochondrial RNA, the newly synthesized RNA was found to hybridize to the L strand in substantial proportion. The shorter the labeling time, the greater this proportion was, indicating that an increasing fraction of the labeled RNA was represented by L transcripts; after very short pulses (5 min or less), the H and L transcripts, with due account for their difference in base composition, were labeled to about the same extent, as judged from their relative capacity to hybridize to an excess of H or L strands (Aloni and Attardi, 1971b). These results indicated that the H and L strands of HeLa cell mitochondrial DNA are transcribed at approximately the same rate, with the L transcripts having a much shorter average half-life in mitochondria than the H transcripts and being, in their majority, either rapidly degraded or otherwise removed from the mitochondrial fraction (exported?) after their synthesis.

Recently, it has been possible to partially purify the L transcripts and to carry out saturation hybridization experiments with the L strand (Murphy, Attardi, Tu, and Attardi, unpublished data). Both by analysis of the density in a Cs_2SO_4 gradient of the RNA-DNA hybrids formed

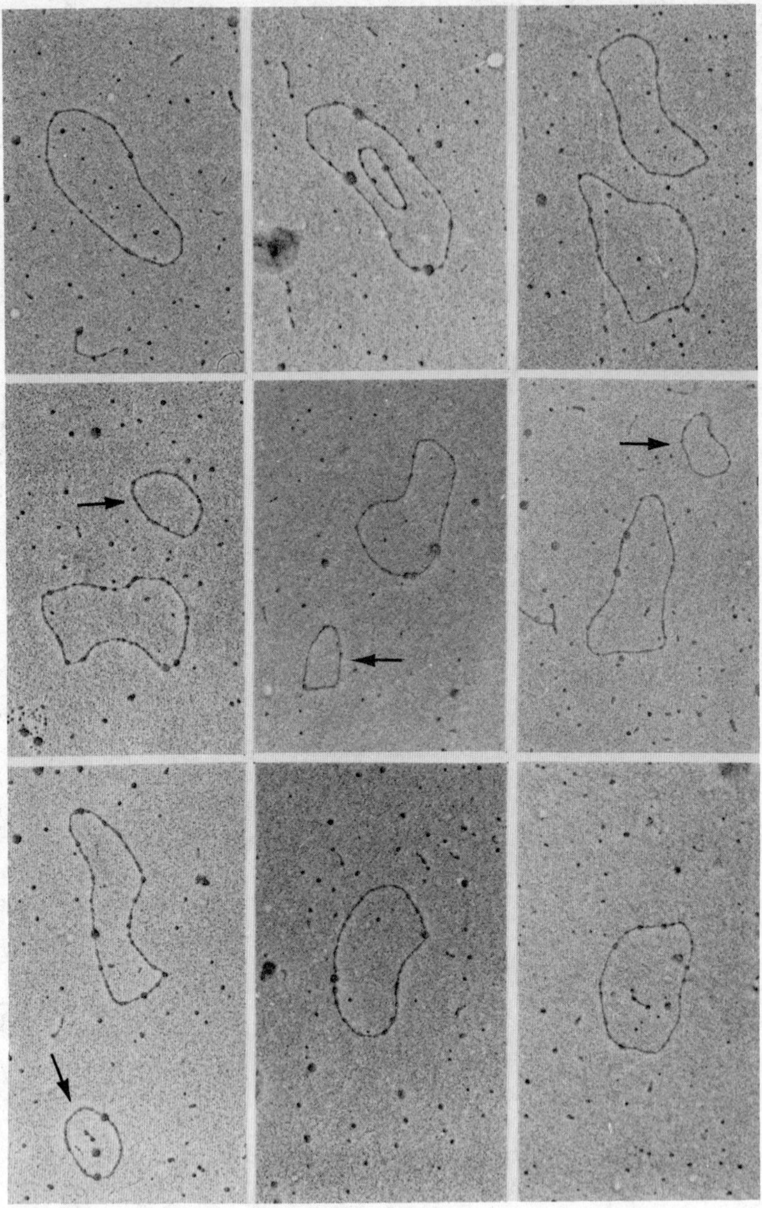

Fig. 4. Electron micrographs of hybrids between L strands of mitochondrial DNA and saturating amounts of partially purified L strand transcripts, spread by the basic protein film technique. ϕX RF DNA is included for size reference (arrow).

with saturating amounts of RNA and by electron microscopy of these hybrids, it has been shown unambiguously that mitochondrial RNA contains sequences complementary to the entire or almost entire length of the L strand. Figure 4 shows some examples of hybrids formed by annealing a preparation consisting of mostly intact, circular or linear, L strands with saturating amounts of L transcripts. The circular shape of the hybrids may reflect either the original structure of the DNA strands or a secondary circularization of linear strands by hybridized RNA. By comparison with RF ϕX174, added as an internal sizing market (1.7 μm), the average length of the circles is approximately 5 μm, as expected for fully base-paired RNA-DNA hybrids involving intact DNA strands.

The above results strongly suggest that the symmetrical transcription of mitochondrial DNA extends to the full length of both strands. The significance of this phenomenon is as yet unknown. After the initial discovery in HeLa cell mitochondrial DNA, the same mode of transcription has been shown to occur in SV40 DNA (Aloni, 1972) and polyoma virus DNA (Aloni and Locker, 1973). Recent experiments aimed at testing the possible occurrence of symmetrical transcription of the nuclear DNA segments coding for heterogeneous nuclear RNA in HeLa cells and immature duck erythrocytes have given negative results (Murphy and Attardi, unpublished data). This suggests that the symmetrical transcription may be related to the closed circular structure of the DNA template.

Primary Gene Products

Whatsoever is the significance of the complete symmetrical transcription of HeLa cell mitochondrial DNA, its existence suggests that post-transcriptional events, in particular cleavage of precursors, may have a considerable importance in the regulation of the formation of the mature RNA products in HeLa cell mitochondria.

In the last few years many discrete RNA components coded for by HeLa cell mitochondrial DNA have been identified. These studies have been greatly facilitated by the possibility of using low doses of actinomycin D to suppress selectively the synthesis of cytoplasmic ribosomal RNA (rRNA), which is found in the ribosomes of the rough endoplasmic reticulum contaminating the mitochondrial fraction (Attardi et al., 1969). Quantitatively, the major stable classes of RNA transcribed from mitochondrial DNA are represented by rRNA and 4 S RNA (fig. 5). The two mitochondrial rRNA species, with a sedimentation coefficient of 16 S and 12 S (Attardi and Attardi, 1971)

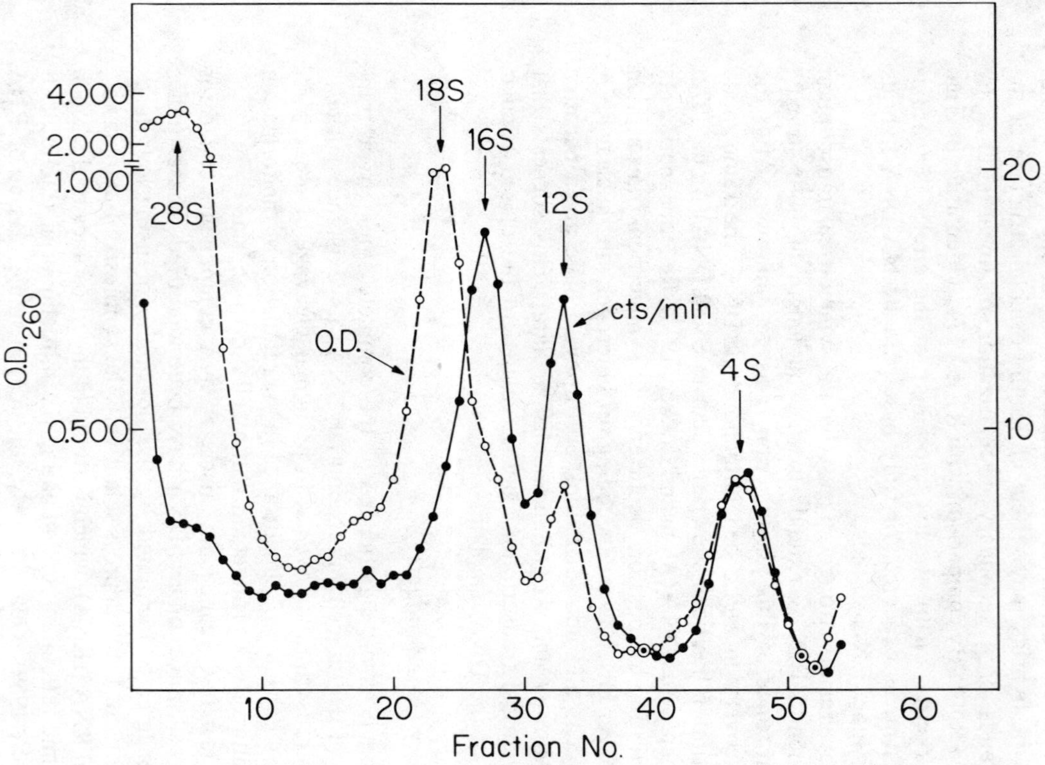

Fig. 5. Sedimentation pattern in sucrose gradient of RNA components sedimenting slower than 28 S rRNA from the mitochondrial fraction of HeLa cells exposed to [5-^{3}H]uridine for 4 hr in the presence of 0.04 μg actinomycin D/ml (to inhibit nuclear rRNA synthesis).

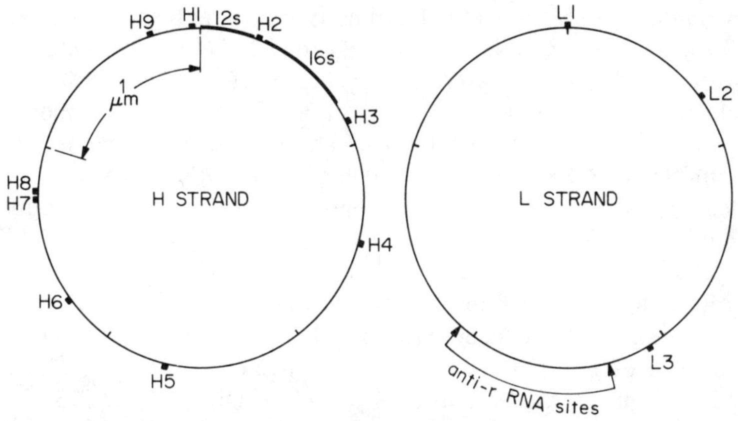

Fig. 6. Circular map of the positions of the complementary sequences for 4 S RNAs on the H and L strands of HeLa mitochondrial DNA and of the 12 S and 16 S rRNA genes on the H strand (modified from Wu et al., 1972). See text for details.

and a molecular weight of 5.4×10^5 and 3.5×10^5 daltons respectively (Robberson, Aloni, Attardi, and Davidson, 1971), are the smallest rRNA species known: they are the structural RNA components, respectively, of the 45 S major subunit and the 35 S minor subunit of the 60 S mitochondrial ribosomes (Attardi and Ojala, 1971; Brega and Vesco, 1971). The 4 S RNA components represent mostly, if not exclusively, mitochondria-specific tRNA species (Buck and Nass, 1969; Galper and Darnell, 1969; Lynch and Attardi, unpublished data).

RNA-DNA hybridization experiments have indicated the presence of one gene for each rRNA component on the H strand of HeLa mitochondrial DNA (Aloni and Attardi, 1971c) and of 12 genes for 4 S RNA: of these, nine appear to be located on the H strand and three on the L strand (Aloni and Attardi, 1971c; Wu et al., 1972). The positions of the two rRNA genes and the 12 4 S RNA genes on the separated strands of HeLa cell mitochondrial DNA have been mapped by electron microscopy of hybrids formed between the separated mitochondrial DNA strands and 12 S RNA, 16 S RNA and ferritin-coupled 4 S RNA (Wu et al., 1972). Figure 6 shows the map constructed from these RNA-DNA hybridization results. The map also contains a preliminary assignment of the position of the anti-rRNA sites on the L strand, determined, in collaboration with M. Wu and N. Davidson, by using as a marker the rRNA-coding sequences purified from the H strand: the orientation of the anti-12 S site and anti-16 S site on the L strand is at present being investigated.

The genes for the rRNA components and 4 S RNA species identified in HeLa cell mitochondrial DNA account for about 25% of the single-strand informational content of a 5 μm-long DNA molecule with no major internal repetitions. The question of whether other informational RNA molecules, in particular mRNA molecules, are coded for by mitochondrial DNA has been the object of dispute in recent years (Dawid, 1970; Swanson, 1971; Reijnders et al., 1972; Mahler and Dawidowicz, 1973). A strong indication of the synthesis of mRNA in animal cell mitochondria, and at the same time a handle for its isolation have been provided by the demonstration of the occurrence of poly(A) sequences of a distinctive size in HeLa cell mitochondrial RNA (Perlman et al., 1973; Attardi et al., 1973; Ojala and Attardi, 1974a), and of a poly(A)-synthesizing enzyme in rat liver mitochondria (Jacob et al., 1972; Jacob and Schindler, 1972).

HeLa cell mitochondrial poly(A), isolated by pancreatic and T1 RNase digestion of the fraction of the organelle RNA that is retained on oligo(dT)-cellulose, migrates in polyacrylamide gel electrophoresis slightly faster than 4 S RNA (fig. 7a), and has been estimated to have a length corresponding to about 60 residues (Hirsch and Penman, 1973). It is therefore considerably shorter than the poly(A) isolated from the mRNA of cytoplasmic polysomes (fig. 7b), which is 150 to 200 residues long (Darnell, Philipson, et al., 1971; Edmonds et al., 1971).

Sedimentation and electrophoretic mobility analysis under denaturing conditions of mitochondrial RNA labeled with [8-^{3}H]adenosine for 2 hr in the presence of 0.1 μg/ml actinomycin D and selected for poly(A) content by oligo(dT)-cellulose chromatography has shown that the majority of the poly(A) stretches are covalently linked to RNA molecules (Ojala and Attardi, 1974a). Furthermore, RNA-DNA hybridization experiments have clearly indicated that the RNA to which mitochondrial poly(A) is linked is coded for by mitochondrial DNA (Ojala and Attardi, 1974a). On the contrary, both the partial ethidium bromide resistance of the synthesis of mitochondrial poly(A) (fig. 7a), and the RNA-DNA hybridization results strongly suggest that this poly(A) is not coded for by mitochondrial DNA, but is added post-transcriptionally, presumably at the 3′-end (Hirsch and Penman, 1973), to mitochondrial DNA-coded RNA molecules (Ojala and Attardi, 1974a). The situation is, therefore, similar to that described for nuclear-cytoplasmic poly(A) (Darnell, Wall, and Tushinski, 1971; Jelinek et al., 1973).

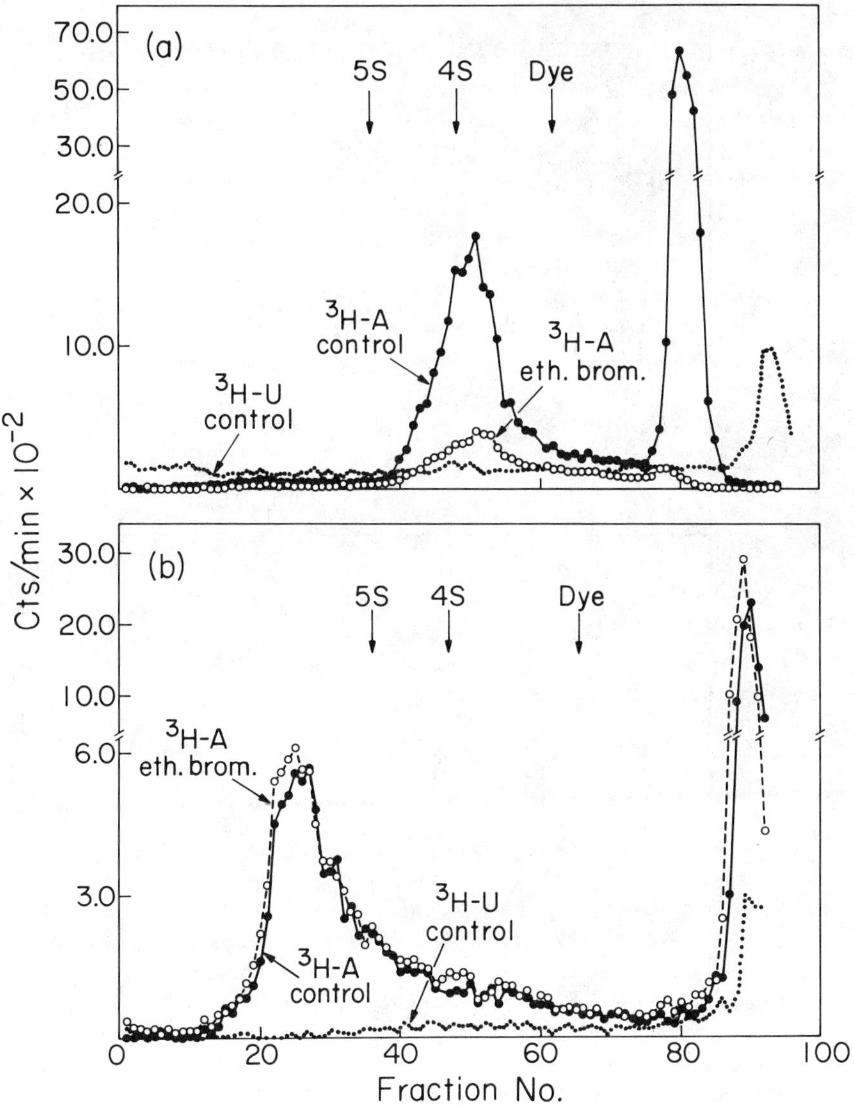

Fig. 7. Polyacrylamide gel electrophoresis pattern of poly(A) present in mitochondrial RNA (a), and cytoplasmic mRNA (b). RNA was extracted from the polysome region of the sucrose gradient sedimentation pattern of a mitochondrial lysate (a), or from free cytoplasmic polysomes (b), of HeLa cells labeled for 2 hr (in the presence of 0.1 μg actinomycin D/ml) with [8-^{3}H]adenosine (in the absence or presence of 1 μg ethidium bromide/ml) or with [5-^{3}H]uridine, then passed through an oligo(dT)-cellulose column; the fraction retained was digested with pancreatic and Tl RNase and then subjected to polyacrylamide gel electrophoresis (from Ojala and Attardi, 1974a).

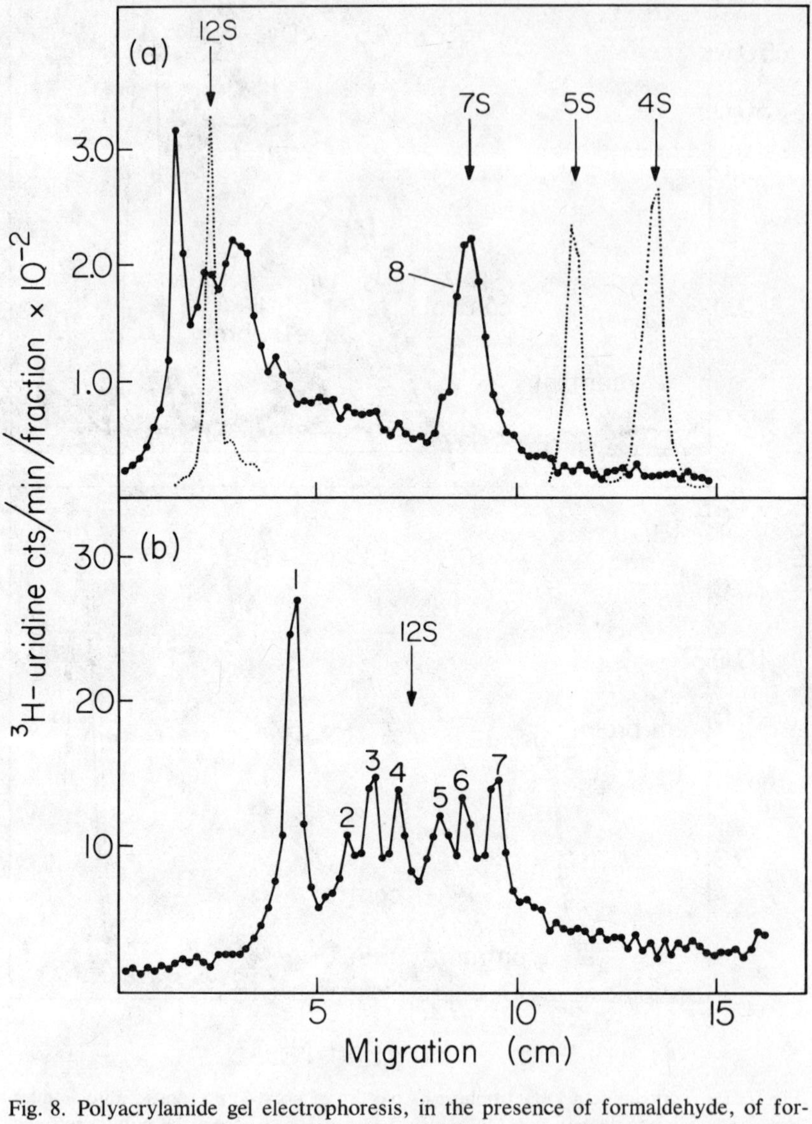

Fig. 8. Polyacrylamide gel electrophoresis, in the presence of formaldehyde, of formaldehyde-treated [8-^{3}H]adenosine-labeled RNA extracted from the mitochondrial polysome region of a sucrose gradient, after centrifugation of a HeLa cell mitochondrial lysate, and selected for poly(A) content by passage through an oligo(dT)-cellulose column. (a) 4 hr run; (b) 16 hr run (from Ohala and Attardi, 1974b). In panel 8a, the dotted lines represent the profiles of mitochondrial 12 S and 4 S RNA and cytoplasmic 5 S RNA markers, which were treated as the experimental sample and run in parallel gels.

In order to resolve discrete components of poly(A)-containing RNA, it has been found necessary to use denaturing conditions for the analysis, due to the great tendency of this RNA to aggregate. Thus, by fractionation through polyacrylamide-formaldehyde gels, the formaldehyde-treated poly(A)-containing RNA isolated from a HeLa cell mitochondrial polysome preparation has been separated into eight distinct components (fig. 8). These components are very reproducible in their occurrence and relative amount if great care is taken to avoid degradation. As shown in table 1, they range in molecular weight between 9×10^4 daltons and 5.3×10^5 daltons, as estimated from their sedimentation velocity under denaturing conditions (Ojala and Attardi, 1974b). The smallest poly(A)-containing RNA component, which has a sedimentation coefficient in the native state of about 7 S, is complementary to the L mitochondrial DNA strand, and the other seven components are complementary to the H strand (table 2). Together with the two mitochondrial rRNA species and with mitochondrial 4 S RNA, the eight poly(A)-containing RNA components, if distinct in sequence, would account for about 70% of the single-strand informational content of HeLa mitochondrial DNA.

On the basis of the known association of poly(A) with eukaryotic mRNA, it seems likely that the discrete poly(A)-containing RNA components described here represent specific mitochondrial DNA-coded mRNA species. In this connection it should be mentioned that the number and size range of these poly(A)-containing RNA compo-

TABLE 1

Molecular Weight of Discrete poly(A)-containing RNA Components Estimated by Different Methods

Component	Molecular weight $\times 10^{-5}$	
	Electrophoresis[a]	Sedimentation[b]
1	4.7	5.3
2	—	3.7
3	3.8	3.0
4	3.6	3.0
5	3.3	2.9
6	3.1	2.8
7	2.9	2.6
8 (7 S RNA)	0.75	0.92
		Σ 24.2

[a]Derived from electrophoretic mobility in polyacrylamide-formaldehyde gels relative to that of mitochondrial RNA markers (from Ojala and Attardi, 1974b).

[b]Derived from sedimentation velocity in sucrose-formaldehyde gradients relative to that of mitochondrial RNA markers (from Ojala and Attardi, 1974b).

TABLE 2

HYBRIDIZATION WITH SEPARATED MITOCHONDRIAL DNA STRANDS OF [5-^{3}H] URIDINE-LABELED POLY(A)-CONTAINING RNA COMPONENTS

Component	PERCENT [^{3}H]RNA HYBRIDIZED	
	"H" strand	"L" strand
1	93	3
2	73	2
3	81	1
4	74	2
5	70	3
6	94	2
7	75	2
8 (7 S RNA)	16	77

The [5-^{3}H]uridine-labeled poly(A)-containing RNA components were purified by two cycles of polyacrylamide gel electrophoresis or by consecutive electrophoresis and sedimentation runs and hybridized in solution with an excess of H or L mitochondrial DNA strands (from Ojala and Attardi, 1974b).

nents correlate fairly well with the number and size of the discrete products of mitochondrial protein synthesis that we have identified in HeLa cells (see below). This correspondence suggests that the poly(A)-containing RNA components may be the templates for the synthesis of the proteins.

Apart from the physiological significance of poly(A), the demonstration of its occurrence in association with mitochondrial DNA-coded RNA has a considerable interest from the evolutionary point of view. In fact, the presence of poly(A) sequences in mRNA seems to be a eukaryotic trait: no poly(A) has been detected so far in bacterial mRNA (Perry et al., 1972).

Whether in addition to mRNA coded for by mitochondrial DNA, HeLa cell mitochondria contain mRNA imported from the nucleus is still an open question. However, it should be mentioned that, at least in yeast, the available evidence argues against a substantial import of nuclear mRNA into mitochondria (Reijnders et al., 1972; Mahler and Dawidowicz, 1973).

Mitochondrial Protein Synthesis

The apparatus.—HeLa cell mitochondria, like those from other sources, possess a "minor" protein-synthesizing apparatus, of which many distinctive components have already been identified. The abovementioned 60 S mitochondria-specific ribosomes (Attardi and Ojala, 1971; Brega and Vesco, 1971) are in their majority associated with mRNA to form polysomes (Ojala and Attardi, 1972). A set of tRNA species and amino acyl-tRNA synthetases, which are, at least in part, mitochondria-specific, have been identified (Lynch and Attardi, unpublished data; Galper and Darnell, 1969). No initiation and elongation factors have so far been isolated from HeLa cell mitochondria; however, the presence of organelle-specific factors has been recognized in mitochondria from yeast (Richter and Lipmann, 1970), *Neurospora crassa* (Grandi and Küntzel, 1970; Sala and Küntzel, 1970), and *Xenopus laevis* (Swanson, 1973).

The investigation of mitochondrial protein synthesis in HeLa cells has been hampered by the fact that it represents, quantitatively, a very minor part of total protein synthesis and that the mitochondrial fraction from these cells is contaminated by rough endoplasmic reticulum and other membrane components (Attardi et al., 1969); furthermore, only a small percentage (7 to 20) of the mitochondrial proteins are synthesized by the organelle-specific protein-synthesizing system (Beattie, 1971; Galper and Darnell, 1971; Coote and Work, 1971). Approaches that have been useful in overcoming these difficulties are the development of mitochondrial protein-synthesizing systems using isolated organelles and the *in vivo* use of selective inhibitors of cytoplasmic protein synthesis.

An *in vitro* mitochondrial protein-synthesizing system utilizing either an exogenous ATP-generating system or oxidative phosphorylation as an energy source has been developed from HeLa cells (Lederman and Attardi, 1970). This system exhibits characteristics of energy requirements and of response to inhibitors of protein synthesis similar to those that have been described for mitochondria from other organisms (Kroon, 1965; Wheeldon and Lehninger, 1966; Linnane, 1968; Küntzel, 1969). Furthermore, the electrophoretic profile of the proteins synthesized *in vitro* is very similar to that of the mitochondrial proteins labeled *in vivo* with a radioactive amino acid in the presence of an inhibitor of cytoplasmic protein synthesis, although the rates of synthesis differ between the two systems (Lederman and Attardi, 1973); this indicates that, qualitatively, the protein-synthesizing activity of the isolated mitochondria reproduces fairly closely the *in vivo* activity. The *in vitro* system has been very useful in our hands for the analysis of the pattern of amino acid utilization for organelle-specific protein synthesis in the absence of cytoplasmic pool effects (see below).

The second approach mentioned above, i.e., the *in vivo* selective labeling of the mitochondrial protein products in the presence of specific inhibitors of cytoplasmic protein synthesis, like cycloheximide

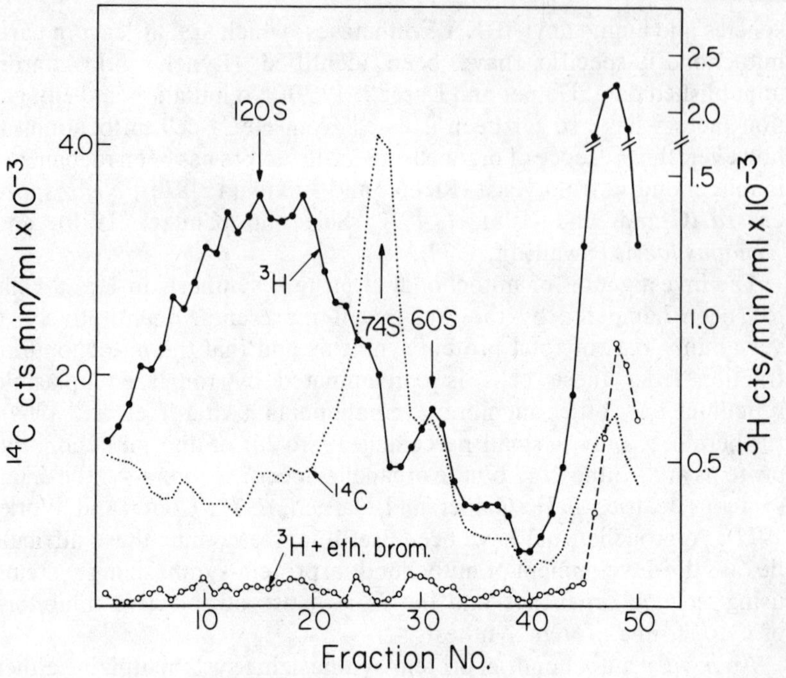

Fig. 9. Sedimentation pattern in sucrose gradient of mitochondrial polysomes from HeLa cells pulse-labeled with L-[4,5-^{3}H]leucine in the presence of 100 μg emetine/ml and in the absence or presence of 1 μg ethidium bromide/ml. The [^{14}C] profile pertains to the structures from cells long-term-labeled with [2-^{14}C]uridine added to the [^{3}H]-labeled cells to provide sedimentation markers (from Ojala and Attardi, 1972).

(Ennis and Lubin, 1964) or emetine (Grollman, 1966), has, however, proven to be of greater utility in the study of mitochondrial protein synthesis in HeLa cells.

Figure 9 shows the sedimentation profile of mitochondrial polysomes obtained by centrifuging in a sucrose gradient a Triton X-100 lysate of the mitochondrial fraction from HeLa cells pulse-labeled with L[4-5-^{3}H]leucine in the presence of emetine at 100 μg/ml. The [^{3}H] radioactivity pattern shows a broad symmetrical band centered around 120 S and extending from about 74 to 200 S, which represents the mitochondrial polysomes, a peak at 60 S, corresponding to the mitochondrial ribosomes, and a smaller peak at 45 S, presumably corresponding to the large mitochondrial ribosomal subunits. Most of the radioactivity in these structures can be chased by puromycin, indicating that it is in nascent polypeptide chains (Ojala and Attardi, 1972). Furthermore, the labeling of these chains is almost completely sensitive to chloramphenicol at 200 μg/ml (Ojala and Attardi, 1972), a specific inhibitor of mitochondrial protein synthesis (Kroon, 1965; Wheeldon and Lehninger, 1966; Linnane, 1968; Lederman and Attardi, 1970), or ethidium bromide (fig. 9), a general inhibitor of mitochondrial macromolecular syntheses (Smith et al., 1971; Zylber et al., 1969; Attardi et al., 1970; Perlman and Penman, 1970a).

The HeLa cell mitochondrial polysomes have been estimated to consist of two to seven 60 S monomers (Ojala and Attardi, 1972). They are, therefore, relatively small as compared with the endoplasmic reticulum-bound and free cytoplasmic polysomes, which cover the range of about 120 to 350 S and contain up to 40 or more ribosome monomers (Rich et al., 1963; A. Bakken, personal communication). HeLa cell mitochondrial polysomes have revealed an unusual resistance to RNase and EDTA, possibly related to the strongly hydrophobic nature of their polypeptide products, which makes the nascent chains particularly sticky, and therefore apt to interact by secondary bonds with the two ribosomal subunits and with other chains on the same polysome (Ojala and Attardi, 1972).

The products.—A considerable amount of evidence (reviewed by Beattie, 1971) indicates that the products of mitochondrial protein synthesis are hydrophobic proteins of the inner mitochondrial membrane. A major source of difficulty in the analysis of these proteins has been their insolubility in aqueous media at neutral pH and their great tendency to aggregate even in the presence of ionic detergents, such as sodium dodecyl sulfate.

In the HeLa cell system, by using appropriate conditions for the preparation of the sample and for electrophoresis, ten discrete components have been reproducibly observed, over a heterogeneous background, in the electrophoretic profiles of the newly synthesized products of mitochondrial protein synthesis labeled *in vivo* with L-[4,5-^{3}H]isoleucine in the presence of emetine (fig. 10) (Costantino and Attardi, 1975). The synthesis of these components is completely sensitive to chloramphenicol at 100 μg/ml. (The chloramphenicol-resistant broad peak near the end of the gel in figure 10, on the basis of the available evidence, does not appear to be due to protein.) It is not known whether each one of the electrophoretic components represents an individual polypeptide species, or whether more than one species is present in some of the peaks. Control experiments, involving a direct sodium dodecyl sulfate lysis of whole pulse-labeled

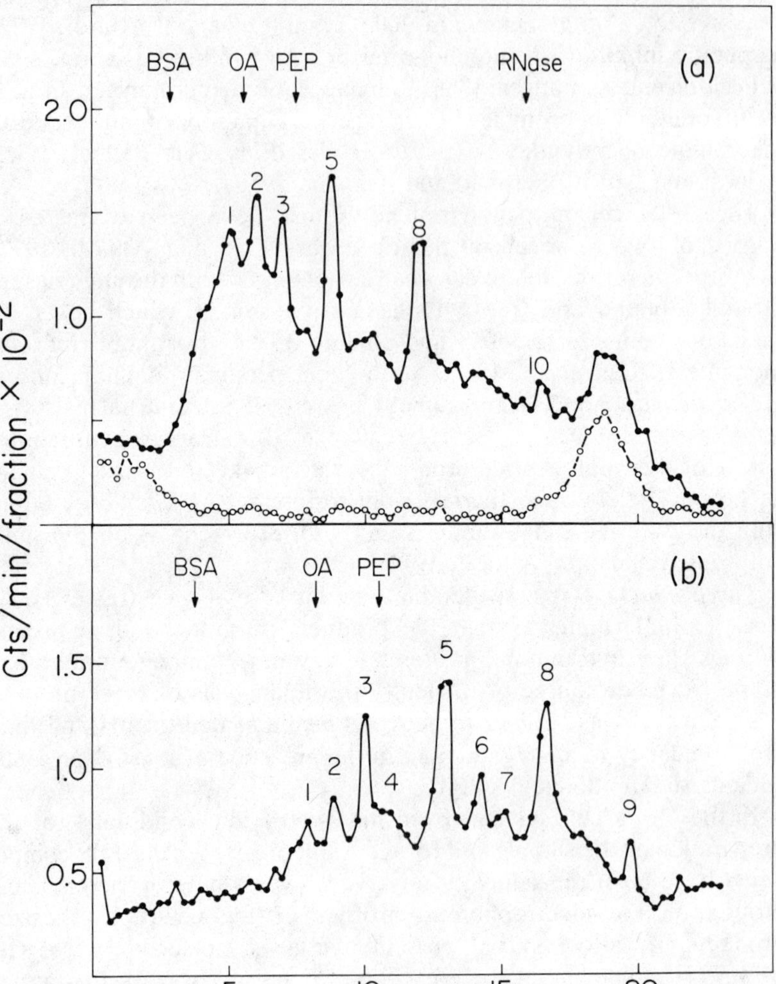

Fig. 10. Electrophoretic profiles of the products of mitochondrial protein synthesis from HeLa cells labeled for 10 min with L-[4,5-^{3}H]isoleucine in the presence of 100 μg of emetine/ml, and in the absence (●) or presence (O) of 100 μg of chloramphenicol/ml (a) 26 hr run; (b) 40 hr run (from Costantino and Attardi, 1975).

TABLE 3

MOLECULAR WEIGHTS OF THE PRODUCTS OF MITOCHONDRIAL PROTEIN SYNTHESIS

Component	Molecular weight
1	42,000
2	39,000
3	35,000
4	31,500
5	27,500
6	24,000
7	22,500
8	19,500
9	15,000
10	11,500

Each value represents the average of several estimates made on the basis of the electrophoretic mobility of the individual components relative to that of standard proteins run in parallel gels.

cells or the use of the protease inhibitor phenylmethylsulfonylfluoride during cell homogenization and fractionation, tend to exclude the possibility that the discrete products of mitochondrial protein synthesis observed in HeLa cells result from enzymic degradation during extraction.

Table 3 shows the molecular weights of the products of mitochondrial protein synthesis identified above, as estimated in many experiments from their electrophoretic mobility, relative to that of standard proteins run in parallel gels. As mentioned above, the gross correspondence in number and size range between these discrete mitochondrially synthesized proteins and the number and size of the poly(A)-containing RNA components transcribed from mitochondrial DNA suggests that these proteins may be specified by mitochondrial genes.

The high degree of hydrophobicity of the products of mitochondrial protein synthesis, which results, at least in part, from the high proportion of hydrophobic amino acids they contain (see below), accounts for their unusual solubility in organic solvents, in particular in chloroform-methanol mixtures. This solubility behavior, which had previously been observed in material from rat liver (Kadenbach, 1971a; Burke and Beattie, 1973) and yeast (Tzagoloff and Akai, 1972; Murray and Linnane, 1972), has been recently confirmed in the HeLa cell system. Here, a selective solubilization of the mitochondrial protein products, representing a 20- to 30-fold purification with respect to the cytoplasmically synthesized proteins, has been obtained by a neutral chloroform-methanol mixture (fig. 11) (Costantino and Attardi, 1975). At least six of the discrete products larger than 20,000 daltons have

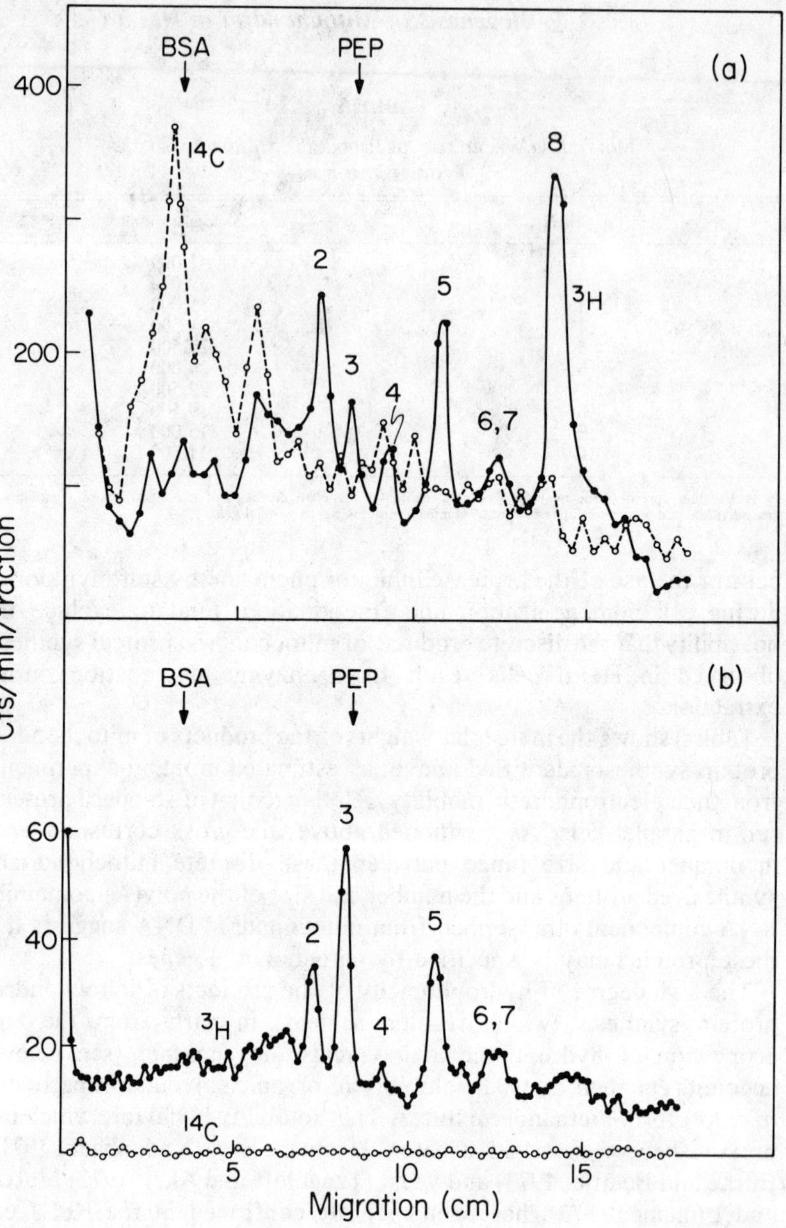

Fig. 11. Electrophoretic profiles of the proteins of the mitochondrial fraction extracted with neutral chloroform-ethanol (b) and of the insoluble residue (a) from HeLa cells long-term-labeled with L-[^{14}C]arginine and L-[^{14}C]lysine and pulse-labeled with L-[4,5-^{3}H]leucine in the presence of 100 μg of emetine/ml (from Costantino and Attardi, 1975).

been found to be extracted to an appreciable extent by neutral chloroform-methanol. These solubility properties will undoubtedly facilitate the isolation in pure form of individual polypeptides synthesized in the organelles.

Although it is not yet possible to make a functional identification of the discrete protein components detected here, it is interesting to notice that both the number and the size range of these components agree reasonably well with the number and size range of the defined molecular species that have been identified in yeast and *Neurospora crassa* mitochondria as products of mitochondrial protein synthesis. These include three subunits of the cytochrome *c* oxidase (Ross et al., 1974; Schatz and Mason, 1974; Sebald et al., 1974; Rubin and Tzagoloff, 1973a,b; Mahler et al., 1974), four subunits of the oligomycin-sensitive ATPase (Tzagoloff and Meagher, 1972), one or two subunits of cytochrome *b* (Weiss and Ziganke, 1974), and one polypeptide necessary for the assembly of cytochrome c_1 (Ross et al., 1974). This correspondence is consistent with the idea that, in animal cells, the products of mitochondrial protein synthesis are components of the same enzymatic complexes of the inner mitochondrial membrane as in lower eukaryotes. Direct evidence is already available concerning the essential role of mitochondrial protein synthesis for the assembly of a functional cytochrome *c* oxidase in animal cells (Kroon and DeVries, 1971; Kadenbach, 1971b), including HeLa cells (Storrie and Attardi, 1972, 1973b; see below). A striking similarity in subunit structure between yeast and beef heart cytochrome *c* oxidase has been recently shown (Rubin and Tzagoloff, 1973a,b).

Mitochondrial Genetic Code and Its Translation

This area, of great interest both from the evolutionary and the functional point of view, has only recently started receiving deserved attention. As discussed above, RNA-DNA hybridization experiments have pointed to the presence in HeLa mitochondrial DNA of 12 genes for 4 S RNA. In view of the evidence suggesting that the population of mitochondrial DNA molecules in each animal cell is substantially homogeneous in sequence (down to the limit of stretches of 100 nucleotides) (Borst, 1969; Clayton et al., 1970), and barring an unlikely microheterogeneity of mitochondrial DNA for tRNA genes, these hybridization data would indicate that the coding capacity for tRNA of HeLa cell mitochondrial DNA is far from complete. Although one cannot exclude the possibility that some tRNA genes may not

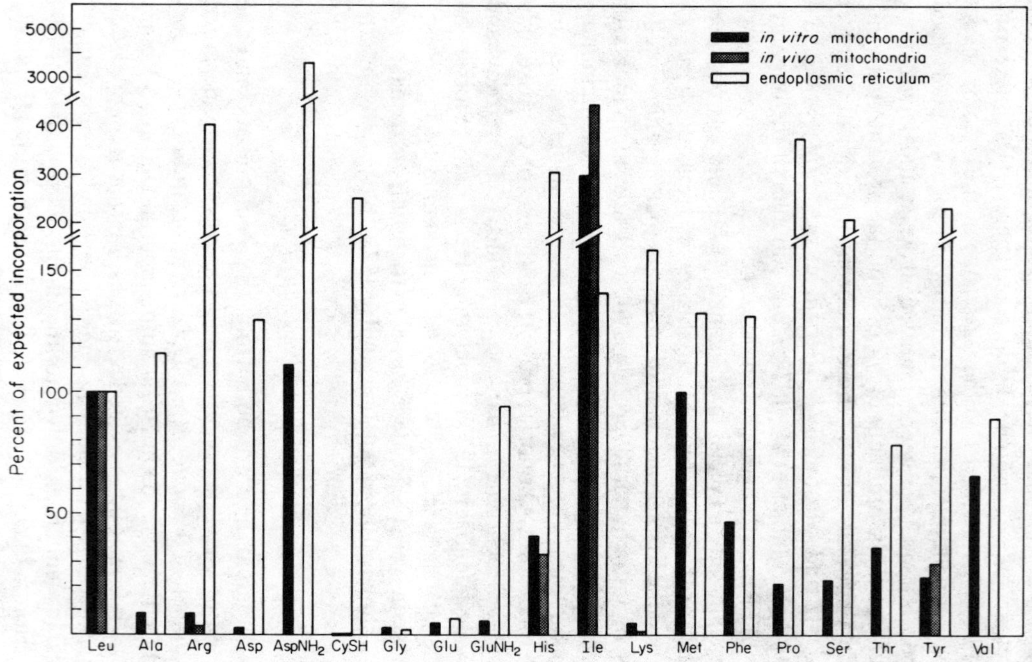

Fig. 12. Percentage of incorporation of various amino acids by HeLa cell mitochondria or endoplasmic reticulum, relative to that expected for the synthesis of an "average" HeLa cell protein. It is assumed that the relative rates of incorporation for the various amino acids into this "average" protein are proportional to their mole percentage in whole HeLa cell protein (Levintow and Darnell, 1960) (from Costantino and Attardi, 1973).

have been detected because of the low concentration of the corresponding tRNA species in mitochondria or because of some unknown restriction in ferritin coupling to certain tRNAs, it is clear that the informational content of HeLa cell mitochondrial DNA for tRNA is well below that corresponding to a full degenerate complement of tRNAs. In a different animal cell system, *Xenopus* oocytes, the reported hybridization plateau of mitochondrial DNA for 4 S RNA corresponds to 15 cistrons (Dawid, 1972).

The above-cited observations have raised the question of whether mitochondrial protein synthesis in HeLa cells and in animal cells in general can be supported only by endogenous tRNA species. In a more general context, it is reasonable to ask whether a genetic system would be viable if only a fraction of the 61 possible amino acid-specifying codons could be read by the translation apparatus which serves that system. Although post-transcriptional modifications of the tRNA synthesized in mitochondria may conceivably increase the range of codon specificities of the primary tRNA transcripts, it is not likely that this mechanism would be sufficient to compensate fully for the deficiency of primary tRNA species. It is clear that an import of selected tRNA species from the cytoplasm could solve the above-discussed dilemma. However, this possibility faces some conceptual difficulties, both as concerns the evolutionary advantage inherent in this situation and as concerns the specific mechanism involved.

With the purpose of obtaining information relevant to these questions, two interrelated experimental approaches have been followed in our laboratory. The first approach has aimed at obtaining information concerning the amino acid composition of mitochondrially synthesized proteins in HeLa cells. For this purpose, the capacity of HeLa cell mitochondria, either isolated or in the intact cell, to incorporate different labeled amino acids into proteins has been investigated (Costantino and Attardi, 1973). As appears in figure 12, eight amino acids (alanine, arginine, aspartic acid, cysteine, glutamic acid, glutamine, glycine, and lysine), which include most of the charged polar ones, showed a very low level, if any at all, of chloramphenicol-sensitive incorporation, relative to that expected for an "average" HeLa cell protein. By contrast, the most hydrophobic amino acids (leucine, isoleucine, valine, phenylalanine, and methionine) were the most actively incorporated by HeLa cell mitochondria. Furthermore, the various labeled amino acids appeared to be utilized by the HeLa cell mitochondrial system in a proportion strikingly different from that observed for protein synthesis *in vitro* by the rough endoplasmic reticulum-bound polysomes. To what extent the apparent constraint that exists in the pattern of amino acid utilization by the HeLa cell mitochondrial protein-synthesizing apparatus reflects pool phenomena is, of course, a crucial question that needs to be answered for an appropriate interpretation of the results. On the other hand, it should be noted that the general pattern of utilization of amino acids by HeLa cell mitochondria is in good agreement with the hydrophobic properties of the protein synthesized in the organelles, arguing against drastic pool effects. It is clear that conclusive evidence on the question of the real utilization for mitochondrial protein synthesis of the amino acids that show a marginal incorporation in pulse labeling experiments will only come from an amino acid analysis of purified proteins synthesized in mitochondria.

The second approach we have employed to determine the transcriptional origin of mitochondrial tRNAs aims at determining the amino acid specificity of the tRNA species coded for by HeLa mitochondrial DNA, by carrying out RNA-DNA hybridization experiments between separated strands of mitochondrial DNA and amino acyl-tRNA complexes labeled in the amino acid moiety. Table 4 lists the amino acids that have been tested so far for their capacity to charge, in the presence of mitochondrial amino acyl-tRNA synthetase preparations, distinct tRNA species hybridizable to mitochondrial DNA. At least twelve tRNA species from HeLa cell mitochondria appear to be capable of hybridizing with mitochondrial DNA; of these, ten are complementary to the H strand and two to the L strand. The observed base sequence homology of leucyl-tRNA and phenylalanyl-tRNA to the H strand and of tyrosyl-tRNA to the L strand is in agreement with findings reported on rat liver material (Nass & Buck, 1970). Two amino acids, histidine and proline, have failed, in repeated tests with different tRNA and amino acyl-tRNA synthetase preparations, to charge to any detectable extent mitochondrial tRNA species hybridizable with mitochondrial DNA.

How closely the twelve tRNA species hybridizable with mitochondrial DNA identified in these experiments correspond to the 4 S RNA species previously shown to base-pair with specific sites on the H or L strand (Aloni and Attardi, 1971c; Wu et al., 1972) cannot as yet be said. In any case, the amino acid specificities so far determined for the mitochondrial DNA-coded tRNA species (table 4) already do not show any obvious correlation with the pattern of amino acid utilization by the HeLa cell mitochondrial protein-synthesizing appara-

TABLE 4

SEQUENCE COMPLEMENTARITY TO HEAVY (H) AND LIGHT (L) MITOCHONDRIAL DNA STRANDS OF AMINO ACYL-tRNAs FROM HELA CELL MITOCHONDRIA

Amino acid specificity	SEQUENCE COMPLEMENTARITY DETECTED		
	H strand	L strand	Neither strand
Arginine	+		
Aspartic acid	+		
Glutamic acid		+	
Glycine	+		
Histidine			+
Isoleucine	+		
Leucine	+		
Lysine	+		
Phenylalanine	+		
Proline			+
Threonine	+		
Tyrosine		+	
Tryptophan	+		
Valine	+		

tus. In fact, on the one hand, there appear to be, among the tRNAs specified by mitochondrial DNA, some species corresponding to amino acids that show only a very low apparent level of incorporation (glycine, glutamic acid, aspartic acid, lysine, and arginine). On the other hand, histidine, which is certainly utilized for mitochondrial protein synthesis, as judged from the electrophoretic profile of the chloramphenicol-sensitive products from cells labeled *in vivo* with [^{3}H]histidine in the presence of emetine, has not so far been found to be able to charge a tRNA species coded for by mitochondrial DNA. If the latter observation is confirmed, it would imply that the tRNA for this amino acid is nuclear-DNA-coded and, therefore, must be imported from the cytoplasm. This possibility is open to direct experimental test. Some evidence suggesting a transport of some tRNA species from the cytoplasm has already been presented for *Tetrahymena pyriformis* mitochondria (Chiu et al., 1974).

Further work along the lines indicated above should be able to answer the question of the origin of the tRNAs utilized for mitochondrial protein synthesis, as well as that of the possible presence in HeLa cell mitochondria of mitochondrial DNA-coded tRNA species which may not be used for translation functions, but rather for regulatory or other functions, as described in bacterial systems (see review by Littauer and Inouye, 1973).

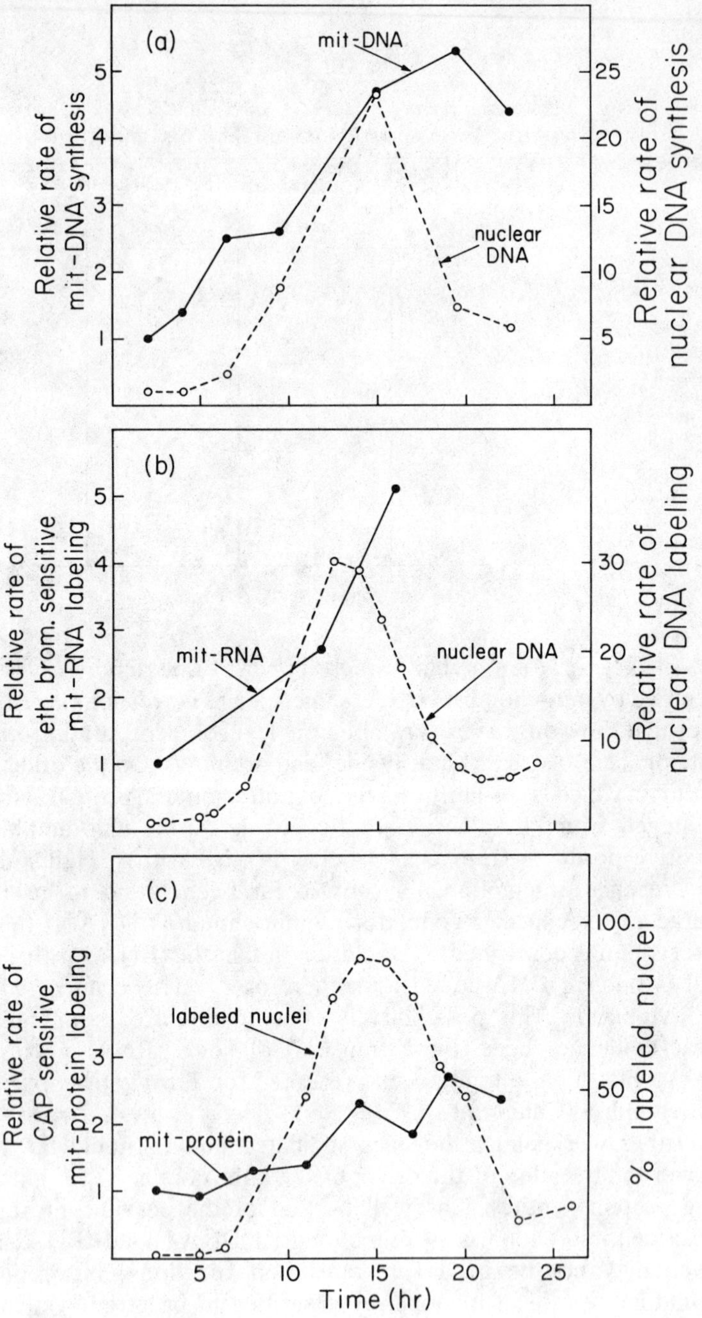
(a)
mit-DNA
nuclear DNA
Relative rate of mit-DNA synthesis
Relative rate of nuclear DNA synthesis
(b)
mit-RNA
nuclear DNA
Relative rate of eth. brom. sensitive mit-RNA labeling
Relative rate of nuclear DNA labeling
(c)
labeled nuclei
mit-protein
Relative rate of CAP sensitive mit-protein labeling
% labeled nuclei
Time (hr)

CONTROL OF EXPRESSION OF THE MITOCHONDRIAL GENOME

Several lines of evidence indicate that the mitochondrial genome is functionally coupled with the nuclear genetic system and that its expression is integrated with the over-all processes of cellular growth and division. In dividing cells, the fact that the average amount of mitochondrial DNA and the average mitochondrial enzymatic activity per cell remain constant from one generation to the next implies some form of control on the part of the extramitochondrial compartment on mitochondrial DNA replication and transcription. Furthermore, the assembly of the mitochondrial protein-synthesizing apparatus and of functionally active mitochondria is under the control of both the nuclear and the mitochondrial genetic systems, pointing to a coordination and interdependence between the two systems.

In order to obtain information on the nature and mechanism of this coupling, two approaches have been followed. First, we have investigated the temporal pattern of mitochondrial DNA, RNA, and protein synthesis during the cell cycle in HeLa cells; second, we have analyzed the role of cytoplasmic protein synthesis in the control of mitochondrial DNA replication and transcription.

Cell Cycle Dependence

Figure 13 summarizes the results of experiments in which the rates of mitochondrial DNA, RNA, and protein synthesis have been measured, at different stages of the cell cycle, in HeLa cells synchronized by the selective detachment technique (Terasima and Tolmach, 1963; Robbins and Marcus, 1964). The rate of both mitochondrial DNA

Fig. 13 (opposite). Relative rates of mitochondrial DNA, RNA, and protein synthesis during the cell cycle in HeLa cells. Synchronous cell populations obtained by selective mitotic detachment were pulse-labeled at different times with [methyl-^{3}H]thymidine or with [5-^{3}H]uridine (in the absence or presence of ethidium bromide) or with L-[4,5-^{3}H]leucine (in the presence of emetine and in absence or presence of chloramphenicol). The labeling data for closed circular mitochondrial DNA were transformed into rates of synthesis by correcting for differences in the specific activity of the intramitochondrial pools of thymidine and its phosphorylated derivatives. No pool correction was applied to the RNA and protein labeling data; however, no significant cell cycle dependent variation was observed in the specific activity of the intramitochondrial pools of UTP and leucine, so that the incorporation data should reflect fairly closely the rates of synthesis. The temporal pattern of nuclear DNA synthesis was obtained by the measurement of incorporation of [2-^{14}C]thymidine into aliquots of the synchronized populations [(a) and (b)] or by the autoradiographic determination of the percentage of labeled nuclei (c).

and mitochondrial RNA synthesis is relatively low in G_1 cell populations, starts accelerating in the S phase, and reaches a maximum in late S and G_2 populations (fig. 13a and b) (Pica-Mattoccia and Attardi, 1971, 1972). No qualitative change was observed in the sedimentation profiles of the ethidium bromide-sensitive, [5-^{3}H]uridine pulse-labeled RNA extracted from cells at different phases of the cell cycle, indicating that the acceleration of synthesis in S and G_2 populations affects uniformly both the ribosomal and 4 S RNA species (Pica-Mattoccia and Attardi, 1971).

The increase in the rate of mitochondrial DNA and RNA synthesis in G_2 relative to G_1 populations is considerably greater than expected from the duplication of the mitochondrial DNA templates that occurs during the cell cycle. It should also be noted that the rate of the two processes, in these experiments, is overestimated in the G_1 populations, due to the presence of contaminating unsynchronized cells in the original selectively detached populations; it is, on the contrary, underestimated in the G_2 populations for the same reason, and also because of the progressive loss of synchrony in the selectively detached mitotic cells, which leads to a substantial contamination of G_2 cells by cells in S phase, mitotic cells, and cells in G_1 phase. From the available evidence it can therefore be concluded that mitochondrial DNA replication and transcription occur predominantly, if not exclusively, in a restricted portion of the cell cycle, covering the S and G_2 phases. At present, there is no direct evidence as concerns the degree of synchrony in DNA or RNA synthetic activity among the mitochondria of each "active" cell, or as concerns the temporal relationship between DNA replication and transcription in the individual organelles. The very similar time pattern of the two processes during the cell cycle suggests that there may be indeed a link between them at the level of the individual mitochondrial DNA molecules.

In contrast to mitochondrial DNA replication and transcription, the rate of mitochondrial protein synthesis in HeLa cell mitochondria increases only about twofold, on a per cell basis, during interphase (fig. 13c), remaining fairly constant, on a per unit mass basis, throughout the cell cycle (England and Attardi, 1974). The results of experiments carried out at 34.5° C, where the G_1 and S phases are better resolved than at 37° C, suggest that the assembly of new portions of functional protein-synthesizing apparatus within the mitochondria does not occur throughout the cell cycle in parallel with the increase in cell mass, but rather mainly in the S and G_2 phases, in correspondence with the synthesis of new mitochondria-specific rRNA and tRNA species.

This process of assembly would add new protein-synthesizing machinery to the preexisting one in such an amount as to match the over-all increase in cell mass, or even in excess of this proportion, if the mitochondrial protein-synthesizing apparatus is somewhat unstable. According to this interpretation, the protein-synthesizing activity measured in the G_1 phase must be that of the preexisting apparatus.

The electrophoretic pattern of the products of mitochondrial protein synthesis remains essentially unchanged in cells in different stages of the cell cycle, suggesting that there is little or no differential control, at the transcription or translation level, of their synthesis which is cell cycle-dependent (England and Attardi, 1974).

Role of Cytoplasmic Protein Synthesis

The observation that mitochondrial DNA and RNA synthesis in HeLa cells are restricted to, or greatly accelerated in, the S phase and, especially, the G_2 phase of the cell cycle indicates that extramitochondrial factors related to the cell cycle, possibly labile nuclear signals, influence the replicative and transcriptive activity of mitochondrial DNA. The existence of labile nuclear signals produced once per cell cycle would ensure a coordination of nuclear chromosome and mitochondrial DNA replication.

With the aim of ascertaining whether this cell cycle-dependent control involves extramitochondrial protein synthesis, the effect of inhibition of cytoplasmic protein synthesis by cycloheximide on mitochondrial DNA and RNA synthesis was investigated.

As shown in figure 14, synthesis of mitochondrial DNA and RNA is depressed gradually during a 4-hour treatment of HeLa cells with cycloheximide, up to 55% and 80% inhibition, respectively. Under the same conditions, total-cell DNA synthesis is inhibited almost completely after only 15 min of cycloheximide treatment, whereas total-cell RNA synthesis is progressively depressed, up to about 60% inhibition. The synthesis of all sedimentation classes of mitochondrial RNA, in particular of the ribosomal and 4 S RNA species, appear to be equally affected by cycloheximide treatment *in vivo*.

Several factors may be responsible for the observed decline in the rate of synthesis of mitochondrial DNA and RNA in the presence of cycloheximide. However, an interesting possibility is that it is due to the progressive disappearance, by turnover or utilization, of some specific protein that needs to be continually produced in the cytoplasm to maintain the normal level of mitochondrial DNA and

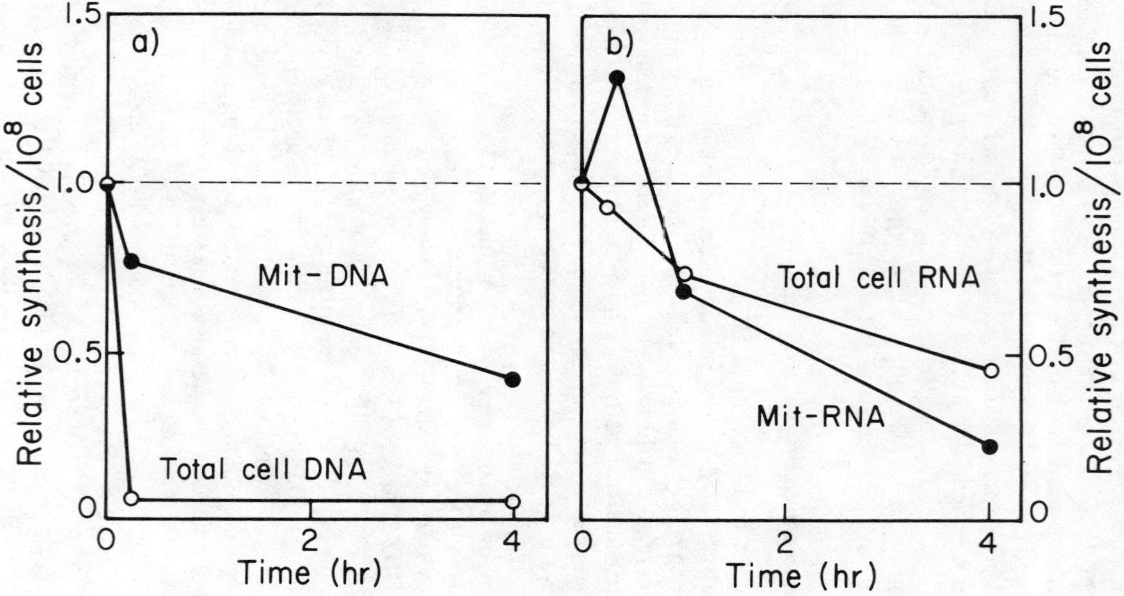

Fig. 14. Effect of cycloheximide pretreatment of HeLa cells on mitochondrial and total cell DNA and RNA synthesis. The rate per cell of nucleic acid synthesis was determined by correcting the [methyl-^{3}H]thymidine and, respectively, [5-^{3}H]uridine labeling data for the changes in specific activity of the intramitochondrial TTP and UTP pools under the same experimental conditions (from Storrie and Attardi, 1972).

RNA synthesis during the "active" period of the cell cycle. It is tempting to speculate that this protein is a signal involved in the cell cycle-dependent regulation of mitochondrial nucleic acid synthesis.

BIOGENESIS OF MITOCHONDRIA

Individuality of Mitochondria

Since the early observations by phase contrast microscopy of living cells in culture, it has become apparent that mitochondria are not organelles of fixed size, shape, and number, but are, on the contrary, endowed with a surprising plasticity. Frequent phenomena of constriction, fragmentation, branching, and fusion were observed and recorded by time-lapse cinematography (Frederic and Chèvremont, 1952; Frederic, 1958). More recently, the genetic and physical evidence of recombination of mitochondrial DNA molecules in yeast (Coen et al., 1970; Carnevali et al., 1969; Shannon et al., 1972) and animal cells (Horak et al., 1974) has provided further support for the idea that mitochondria can undergo fusion and fission processes.

The problem of the individuality of mitochondria has attracted renewed attention in the last few years, as a result of electron microscopic observations, made by serial sectioning techniques, which have indicated the presence of one or very few giant multibranched organelles per cell in yeast and other lower eukaryotic organisms (Calvayrac et al., 1971; Calvayrac and Butow, 1971; Calvayrac et al., 1972; Arnold et al., 1972; Keddie and Barajas, 1969; Hoffmann and Avers, 1973; Osafune, 1973; Grimes et al., 1974a). A similar analysis carried out on rat liver cells has not supported the idea of a single or a few mitochondria per cell, though it has shown that some of the liver mitochondria do have a highly branched, tubular structure (Brandt et al., 1974). On the other hand, the occurrence of very long, thread-like mitochondria, appearing most often to be interconnected, has been reported in diploid human fibroblasts cultivated *in vitro* and examined in the optical microscope after cytological staining (Schneidl, 1974). In many of the above-mentioned studies, the form and number of mitochondria and the type of interrelationships between them—in particular, the extent of fusion of the organelles in the form of multibranched structures—were reported to vary in relationship to the strain (Grimes et al., 1974a), to the phase of the growth cycle (Keddie and Barajas, 1969; Osafune, 1973; Calvayrac et al., 1972), or to the growth or nutritional conditions (Calvayrac

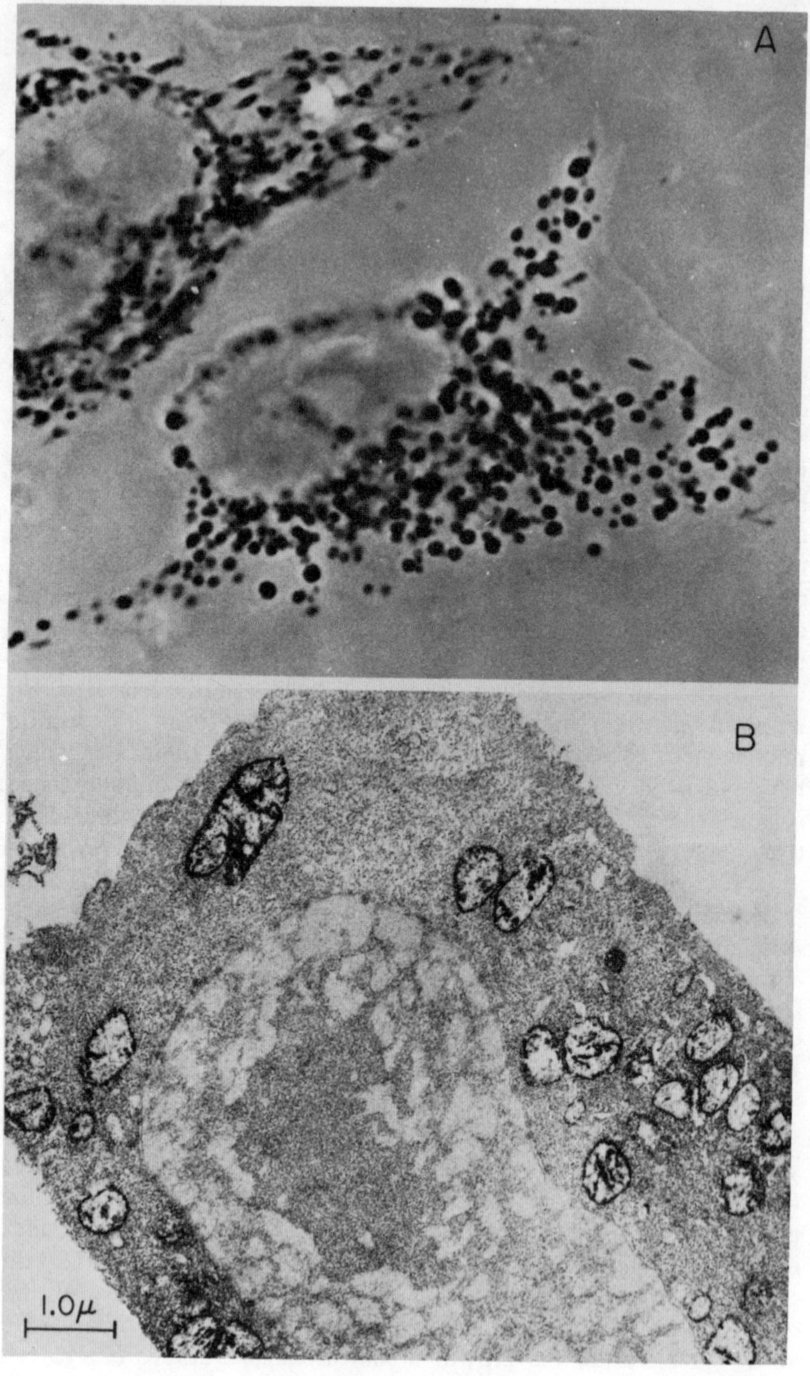
A
B
1.0μ

et al., 1971; Calvayrac and Butow, 1971; Brandt et al., 1974; Grimes et al., 1974b). The above-quoted evidence has given strong support to the idea of the plasticity of mitochondria.

In our laboratory, an electron microscopic examination of about 9,000 mitochondrial profiles in thin sections of exponentially growing HeLa cells (F-315 strain) has revealed a very low frequency (<0.1%) of clearly branched forms. From this, the inference can be drawn that the great majority of the individual mitochondrial profiles examined in these cells derives from the sectioning of distinct organelles. In agreement with this conclusion are the results of an optical microscope examination of HeLa cells (F-315 strain) stained with the diaminobenzidine (DAB) method for detection of cytochrome *c* oxidase in mitochondria (Posakony, England, and Attardi, 1975). A considerable variability among cells in mitochondrial morphology and pattern of DAB staining was observed. The two cells shown in figure 15A are representative of the two most frequent types. In the lower cell, most mitochondria appear as round or oval-shaped bodies, more or less intensely stained. In the peripheral portion of the cell, where the cytoplasm is thinner, the mitochondria appear to be discrete and well separated from each other; in the perinuclear portion of the cytoplasm, the mitochondria are not well resolved, but there is no evidence of a network. In the upper cell, most mitochondria appear as rods lightly stained with DAB, but with an intensely stained region at the center or periphery; here, too, the organelles appear as distinct bodies. In some cells of this type, the rods are longer than shown here. Finally, a few cells contain long filamentous mitochondria with an uneven distribution of the DAB stain, which show some evidence of branching. The reason for the variability in mitochondria morphology among HeLa cells is not known. An examination of DAB-stained HeLa cells in different phases of the cell cycle failed to show any obvious differences in the proportion of various morphological types. It is possible that this variability is related to local differences in conditions of growth on the solid substrate.

Figure 15B shows an electron micrograph of a thin section of a DAB-stained cell. A previous quantitative electron microscopic study had indicated that the number of DAB-positive profiles per cell in

---

Fig. 15. Optical microscope appearance of HeLa cells (F-315 strain) stained by the diaminobenzidine (DAB) method for detection of cytochrome *c* oxidase activity (Seligman et al., 1968) (A), and electron micrograph of a thin section of a DAB-stained cell (B). The solid bar in each picture is 1 μm.

thin sections of HeLa cells is at least as great as the number of morphologically identifiable mitochondria per cell in unstained cells (Storrie and Attardi, 1973a,b). Moreover, the mild fixation conditions and normal osmolarity used for the DAB reaction would tend to exclude any artifactual fragmentation of mitochondria. Therefore, the morphology and interrelationship of mitochondria in the optical microscope pictures of DAB-stained cells should reflect fairly faithfully the *in vivo* situation, supporting the idea of the individuality of the majority of mitochondria in exponentially growing HeLa cells, at least in the strain analyzed and under the growth conditions used here.

The Role of the Mitochondrial Genome in Mitochondriogenesis

It is likely that the expression of the mitochondrial genome occurs mainly, and possibly exclusively, through mitochondrial protein synthesis. Therefore, it should be possible to obtain information about the role of the mitochondrial genome in mitochondriogenesis by examining the effects of a block of mitochondrial protein synthesis on mitochondrial structure, function, and development. This approach has been applied to the HeLa cell system, by taking advantage of the fact that HeLa cells can grow for several generations in the absence of mitochondrial protein synthesis (Storrie and Attardi, 1972, 1973a). As shown in figure 16, growth of these cells continues at a decreasing rate for at least four generations in the presence of chloramphenicol at 40 μg/ml, and for two generations in the presence of the drug at 200 μg/ml, with a slow decrease in cell size at the lower drug concentration. Either concentration of chloramphenicol is sufficient to inhibit more than 90% of the amino acid-incorporating activity that is insensitive to emetine inhibition of cytoplasmic protein synthesis (Perlman and Penman, 1970b; Storrie and Attardi, 1972). The more drastic effects of the higher concentration of chloramphenicol are presumably due to a direct influence of this drug, at 200 μg/ml, on oxidative phosphorylation (Firkin and Linnane, 1968).

As shown in figure 17, mitochondrial DNA and RNA synthesis in HeLa cells continues at a normal rate, on a per cell basis, during 3 days of growth in the presence of chloramphenicol at 40 μg/ml. These results suggest that no protein synthesized in mitochondria is required for the initiation of new rounds of DNA replication, in contrast to what is known in bacteria (see review by Lark, 1969); furthermore, there appears to be little or no direct coupling between translation and transcription, at variance with the situation described

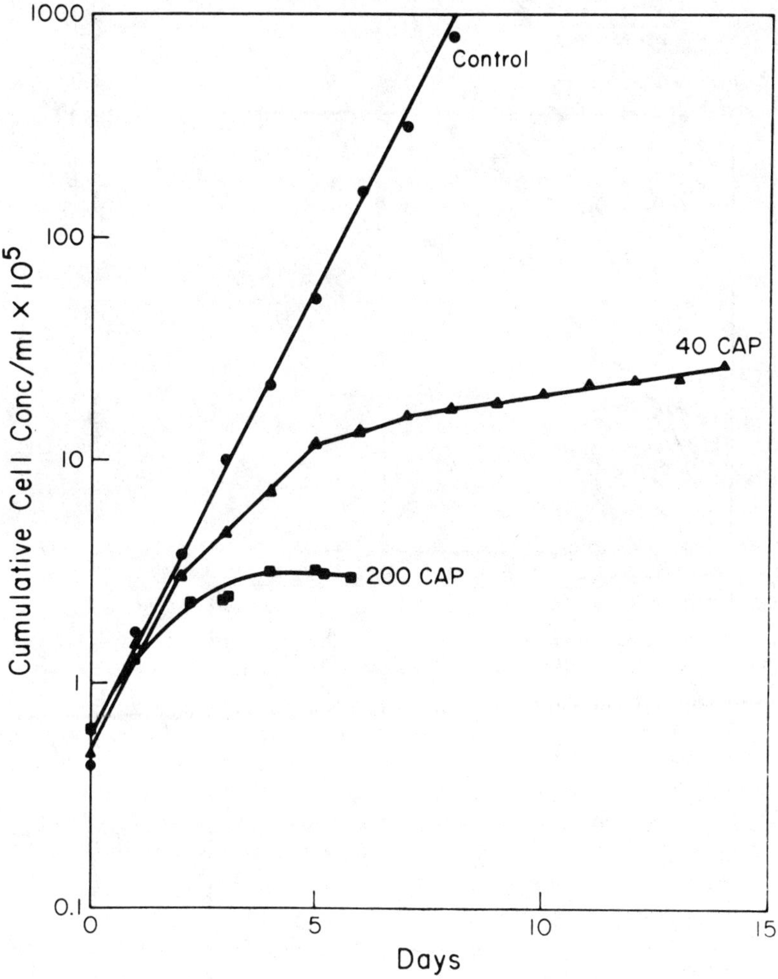

Fig. 16. Growth of HeLa cells in the absence or presence of chloramphenicol at 40 μg/ml or 200 μg/ml. (Modified from Storrie and Attardi, 1972.)

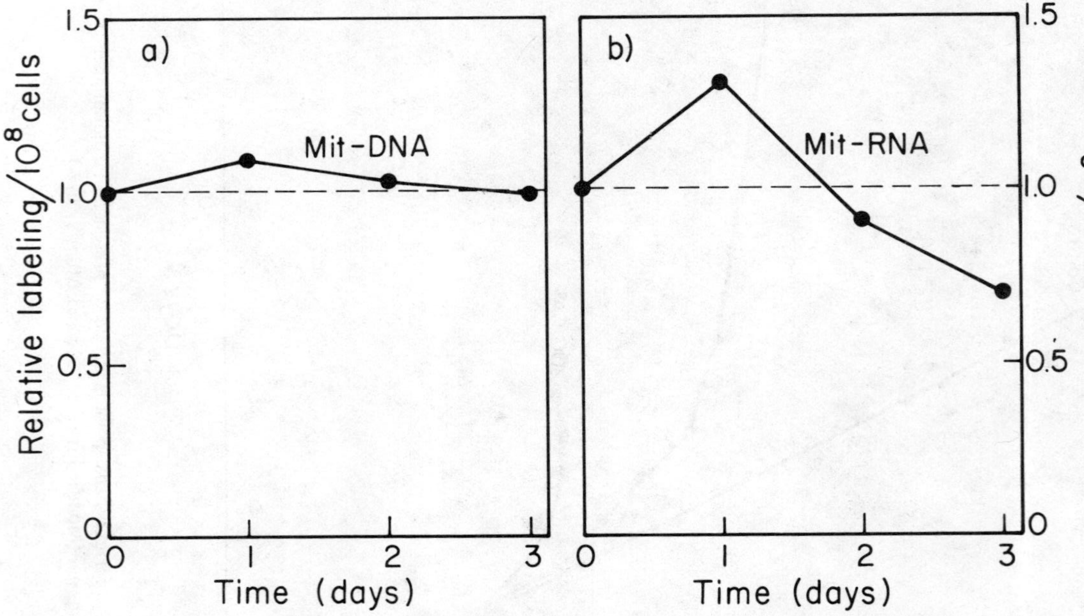

Fig. 17. Effect of chloramphenicol pretreatment of HeLa cells on the synthesis of mitochondrial DNA and mitochondrial RNA. The rates per cell of nucleic acid synthesis were determined by correcting the labeling data for changes in specific activity of the TTP and UTP pools (from Storrie and Attardi, 1972).

for some bacterial operons (Chamberlin, 1970). The results also imply that mitochondrial DNA and RNA polymerases in animal cells are synthesized extramitochondrially, as already shown in yeast (see review by Linnane and Haslam, 1970; Wintersberger and Wintersberger, 1970) and *Neurospora crassa* (Barath and Küntzel, 1972).

Also the mitochondrial protein-synthesizing apparatus, as judged from the *in vivo* incorporation of L-[4,5-^{3}H]leucine into protein of the mitochondrial fraction in the presence of emetine 1 hr after removal of chloramphenicol, continues to be made at a normal or near-to-normal rate, on a per unit cell mass basis, for up to seven days of treatment of the cells with chloramphenicol at 40 μg/ml (about 15 times cell

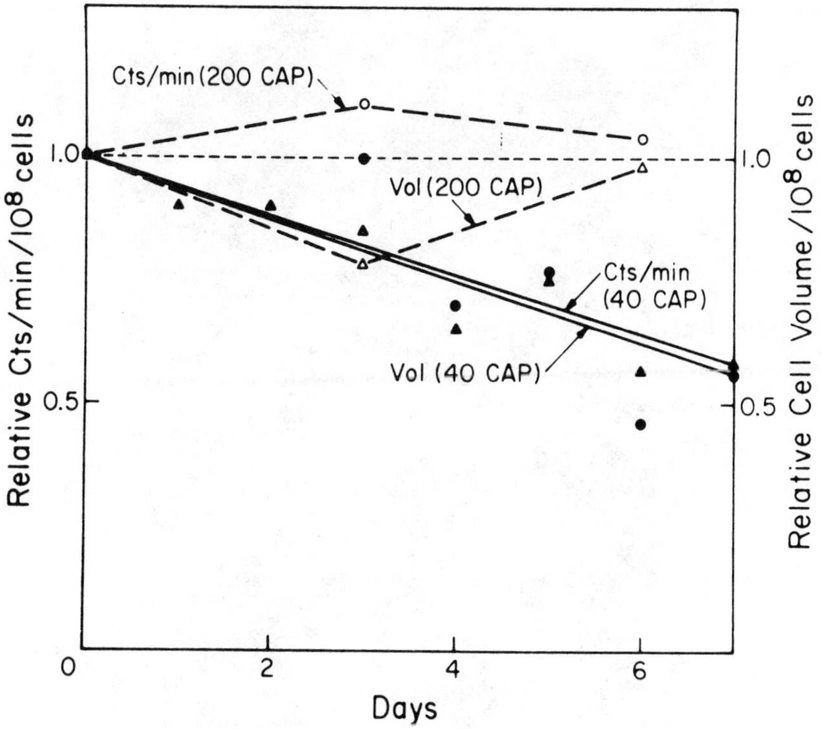

Fig. 18. Effect of chloramphenicol pretreatment of HeLa cells on cell volume and on emetine-insensitive incorporation of L-(4,5-^{3}H)leucine into proteins of the mitochondrial fraction. HeLa cells, after varying periods of growth in the presence of chloramphenicol at 40 or 200 μg/ml, were incubated in chloramphenicol-free medium for 1 hr and then pulse-labeled with (^{3}H)leucine for 30 min in the presence of emetine (from Storrie and Attardi, 1973a. Grateful acknowledgment is made to the *Journal of Cell Biology* to reprint this and other figures in this paper.)

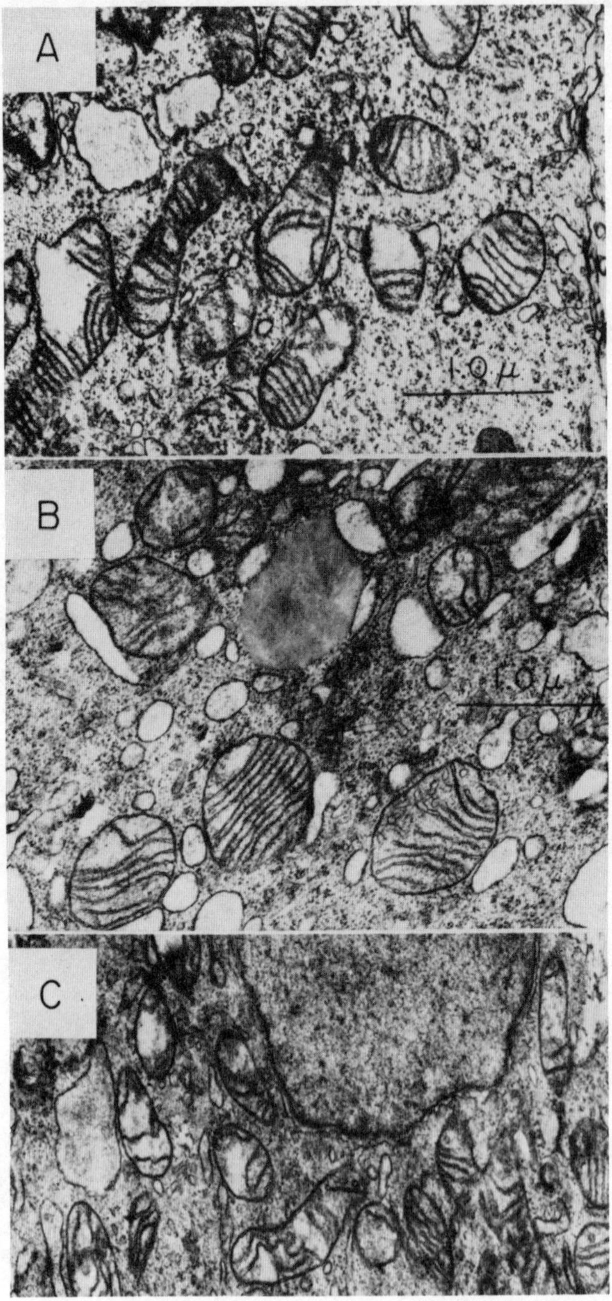
A
1.0 μ
B
1.0 μ
C

mass increase), and up to six days of treatment with the drug at 200 μg/ml (about 5 times cell mass increase) (fig. 18) (Storrie and Attardi, 1973a). These results suggest that the proteins involved in the assembly of the mitochondrial protein-synthesizing machinery, in particular the ribosomal proteins, are made extramitochondrially. The same conclusion has been reached in similar experiments carried out with yeast (Davey et al., 1969).

Figure 19 shows the morphological appearance of mitochondria in HeLa cells grown in the absence (A) or in the presence (B) of chloramphenicol at 200 μg/ml for four days. The mitochondria in the two fields are virtually indistinguishable, and the organelles from both the control and the drug-treated cells appear to be equally rich in cristae.

As shown in table 5, no change in the number of mitochondrial profiles per cell cross-section, in the average mitochondrial profile area and in the amount of cristae per mitochondrion occurs in cells grown in the presence of chloramphenicol at 40 μg/ml for five days or chloramphenicol at 200 μg/ml for four days. These observations imply that mitochondria grow and increase in number normally and that, in particular, the gross assembly of the inner mitochondrial membrane continues to take place in the absence of mitochondrial protein synthesis. However, in 4 to 5% of the mitochondrial profiles, the cristae appear to be arranged in unusual, circular, looped, or whorled configuration (Storrie and Attardi, 1973a); it is possible that this altered organization of the cristae is a secondary phenomenon, not directly related to the lack of products of mitochondrial protein synthesis. Qualitatively similar observations have been reported by King et al., (1972).

Although the mitochondria made during growth in the presence of chloramphenicol appear to be normal in number and, for the most part, also in morphology, they are functionally defective. This was shown by an analysis of the behavior of cytochrome *c* oxidase activity during growth of HeLa cells in the presence of chloramphenicol. As appears in figure 20, there is, under such conditions, a progressive decrease in cytochrome *c* oxidase activity per cell, which closely approximates that expected from dilution of the preexisting enzyme

---

Fig. 19 (opposite). Morphological appearance of mitochondria in HeLa cells grown in the absence (A), or for 4 days in the presence of chloramphenicol at 200 μg/ml (B), or for 2 days in the presence of ethidium bromide at 1 μg/ml (C). The solid bar in each picture is 1 μm (from Storrie and Attardi, 1973a, and unpublished data).

TABLE 5

EFFECT OF CHLORAMPHENICOL (CAP) AND ETHIDIUM BROMIDE (ETH. BROM.) TREATMENT OF HELA CELLS ON SIZE AND NUMBER PER CELL CROSS-SECTION OF MITOCHONDRIAL PROFILES AND ON TOTAL CRISTAE LENGTH PER UNIT OF MITOCHONDRIAL PROFILE AREA OR OF OUTER MEMBRANE LENGTH

Treatment	No. mitoch. profiles/cell section	Average mitoch. profile area (μm^2)	TOTAL CRISTAE LENGTH 0.25 μm^2 mitoch. profile area (μm)	Total cristae length per unit of outer membrane contour length (μm)
Control	17.8 ± 1.4	0.23 ± 0.014	3.63 ± 0.31	1.43 ± 0.13
40 µg CAP/ml (5 days)	19.9 ± 1.3	0.25 ± 0.018	3.80 ± 0.16	1.79 ± 0.22
200 µg CAP/ml (4 days)	18.5 ± 1.5	0.23 ± 0.021	4.16 ± 0.32	1.69 ± 0.21
1.0 µg Eth. Brom./ml (2 days)	18.7 ± 1.2	0.27 ± 0.021	3.54 ± 0.16	1.93 ± 0.11

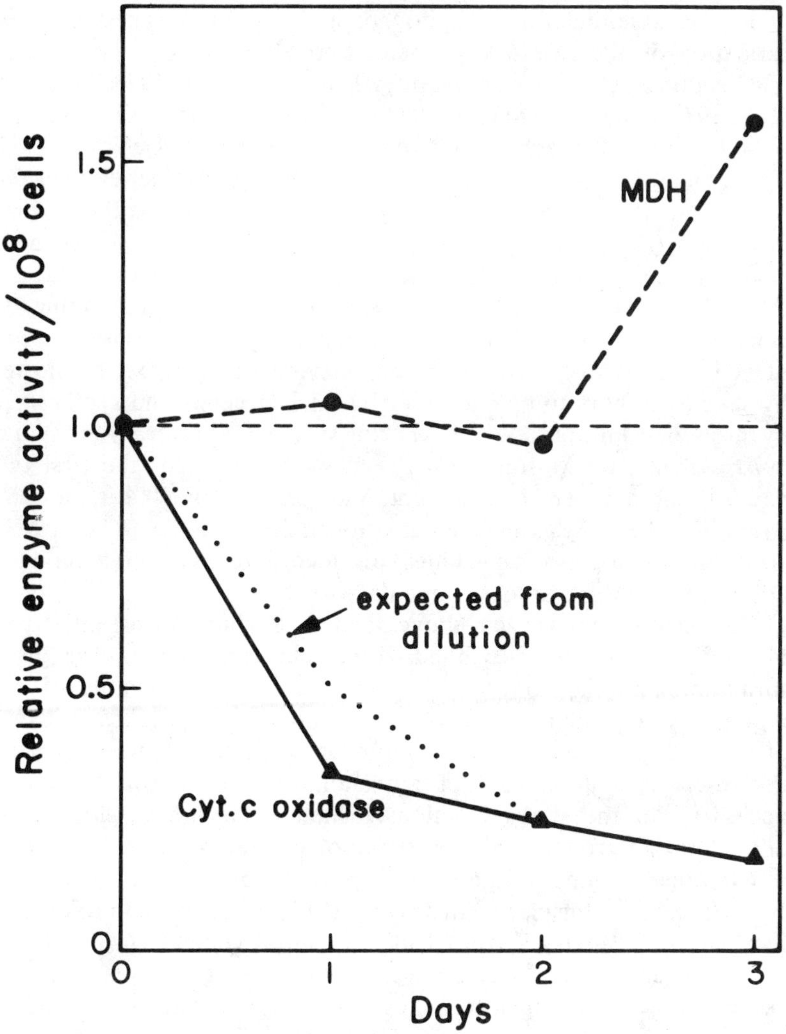

Fig. 20. Effect of chloramphenicol pretreatment of HeLa cells on the activity of the mitochondrial malate dehydrogenase and cytochrome *c* oxidase (modified from Storrie and Attardi, 1972).

activity. This result, which is in agreement with previous observations by Firkin and Linnane (1968), is also consistent with what is known about the essential role of mitochondrial protein synthesis for the formation of the respiratory chain, both in lower eukaryotic cells and in animal cells (see reviews by Ashwell and Work, 1970; Tzagoloff et al., 1973; Schatz and Mason, 1974), and in particular for the assembly of a functional cytochrome *c* oxidase (Kroon and DeVries, 1971; Kadenbach, 1971b; Rubin and Tzagoloff, 1973a,b; Mahler et al., 1974; Ross et al., 1974; Sebald et al., 1974). It seems likely that the slowing down and eventual cessation of cell growth in the presence of chloramphenicol occurs when the preexisting respiratory capacity of the cell has been diluted to the point of becoming rate-limiting. In contrast to the cytochrome *c* oxidase activity, the activity per cell of malate dehydrogenase, a matrix enzyme that has been reported by several authors to be coded for by nuclear genes and synthesized by the cytoplasmic protein-synthesizing system (see Ashwell and Work, 1970; Schatz and Mason, 1974), remains constant for the first two days of chloramphenicol treatment, and increases by 60% in the third day; this indicates that malate dehydrogenase continues to be synthesized by the cytoplasmic machinery and incorporated into mitochondria at a normal rate for at least three days.

The results summarized above tend to exclude an essential role of the products of mitochondrial protein synthesis in the gross formation and division of mitochondria, in the assembly of the mitochondrial protein-synthesizing apparatus, and in the maintenance of normal replication of mitochondrial DNA. Of course, the possibility that there is a large pool of mitochondrially synthesized proteins necessary for the above functions, which would not be sufficiently diluted during growth in the presence of chloramphenicol to become rate limiting, cannot be rigorously excluded; however, this possibility seems unlikely, considering that there is at least a tenfold or a fourfold increase in cell mass and total mitochondrial volume during growth in the presence of the drug at 40 μg/ml or 200 μg/ml, respectively.

Further evidence against an essential role of the mitochondrial genome in mitochondriogenesis comes from inhibition experiments with ethidium bromide, a general inhibitor of mitochondrial macromolecular syntheses. HeLa cells, in the presence of the drug at 1 μg/ml, go through a little more than one generation of growth in two days without any change in size, and then stop growing. As shown in table 5, the number of mitochondrial profiles per cell cross-section, the average mitochondrial profile area, and the amount of cristae per mitochondrion appear to be unaltered after two-day treatment with the drug at 1 μg/ml. However, the cristae show a somewhat disorganized arrangement, with a wavy appearance (fig. 19c) and a frequent circular, looped, whorled, or fenestrated configuration.

In HeLa cells, ethidium bromide, at a concentration of 1 μg/ml, inhibits rapidly more than 95 percent of mitochondrial DNA (Smith et al., 1971; Leibowitz, 1971) and RNA synthesis (Zylber et al., 1969; Attardi et al., 1970), besides blocking almost completely mitochondrial protein synthesis (Perlman and Penman, 1970a; Ojala and Attardi, 1972). Therefore, the normal formation of mitochondria for one generation in ethidium bromide-treated cells suggests that mitochondrial growth and division in these cells, besides not requiring mitochondrial protein synthesis, is also independent of concomitant mitochondrial DNA or RNA synthesis.

The conclusion derived from the inhibition studies described above—i.e., in HeLa cells the mitochondrial genome does not play any essential role for the gross formation and division of mitochondria, for the assembly of the framework of the inner mitochondrial membrane, and for its own replication—extends to mammalian cells the information already available for yeast on this subject. In fact, in both petite mutants of this organism containing a defective DNA (Linnane and Haslam, 1970; Coen et al., 1970) or possibly lacking completely DNA (Nagley and Linnane, 1970), and wild-type strains grown on a fermentable substrate in the presence of chloramphenicol (Davey et al., 1969), growth and division of mitochondria and replication of mitochondrial DNA, whenever present, are clearly not affected.

Mode of Formation of Mitochondria

There is evidence, in *Neurospora crassa,* that the mitochondrial membranes grow by addition of new components to old structures and that the mitochondrial number increases by division of preexisting organelles (Luck, 1963, 1965). That the same situation holds for animal cell mitochondria is suggested by experiments in which the distribution of the preexisting cytochrome *c* oxidase activity during cell growth in the absence of mitochondrial protein synthesis was analyzed by electron microscopy of thin sections of DAB-stained cells (Storrie and Attardi, 1973b).

A constant average number of cytochrome *c* oxidase positive (DAB-stained) organelle profiles per cell section and a uniform decrease in DAB staining of the mitochondrial profile population (fig. 21) were

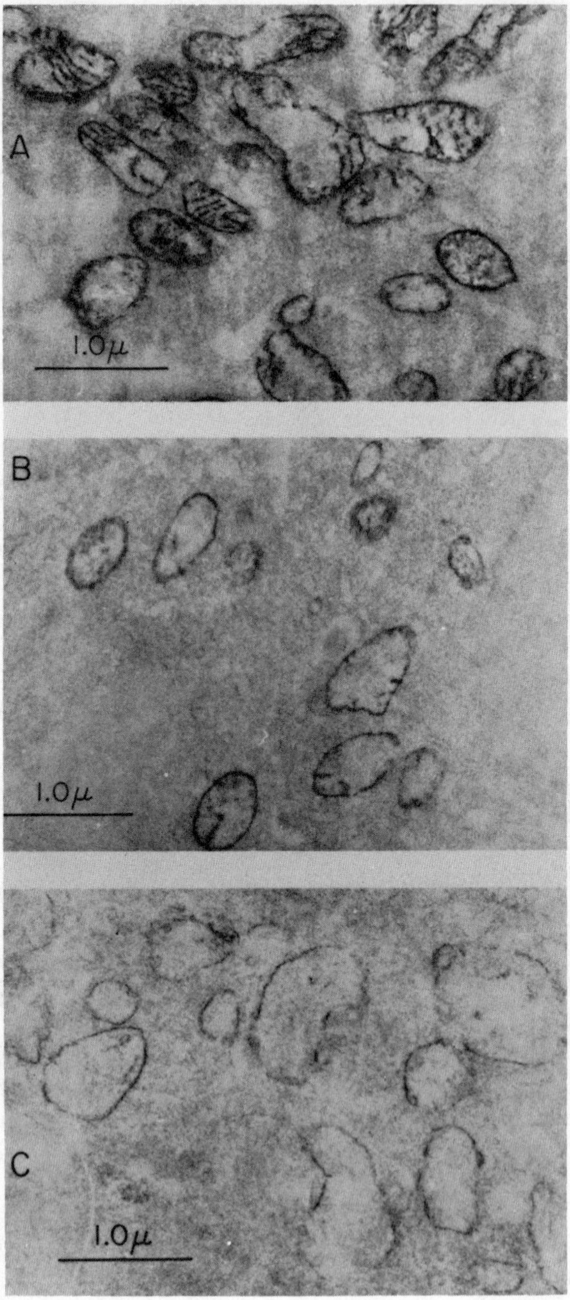
A
1.0μ
B
1.0μ
C
1.0μ

found after as many as four generations of growth of HeLa cells in the presence of chloramphenicol at 40 μg/ml. These findings and the previous observation (Storrie and Attardi, 1973a) that no change in the average mitochondrial profile area occurs under the same conditions are compatible with the idea that HeLa cell mitochondria, like those of *Neurospora*, arise by growth and division of preexisting organelles. The fairly uniform decrease in the intensity of the histochemical staining of the cytochrome *c* oxidase along the inner mitochondrial membrane in drug-treated cells, as compared to control mitochondria, may result from a random insertion of newly synthesized cytoplasmic proteins, individually or in complexes, into the inner mitochondrial membrane, and a consequent dilution of the preexisting cytochrome *c* oxidase activity. However, a localized growth of the inner membrane with an *in vivo* redistribution of the preexisting cytochrome *c* oxidase complex (either by migration in the plane of the membrane or by transport into the matrix and back to a different site in the membrane) could equally well produce a fairly uniform concentration of enzyme activity along the membrane. Unfortunately, no data are as yet available concerning the diffusion rates of proteins or protein-lipid complexes in the inner mitochondrial membrane to allow a judgment about the plausibility of the above-discussed diffusion model. A post-fixation diffusion of the DAB reaction product, which would lead to the same uniform distribution of the stain, although less likely, cannot likewise be excluded.

Temporal Pattern of Mitochondrial Growth and Division during the Cell Cycle

We have presented above the evidence indicating that replication and transcription of mitochondrial DNA in HeLa cells are restricted to, or greatly accelerated in, the S and G_2 phases of the cell cycle. To gain some understanding of the factors that control growth and division of mitochondria in these cells, it was considered of great interest to examine whether these processes also have a fixed temporal relationship with the events of the cell cycle. This question was investigated by a morphometric analysis of electron micrographs of synchronized cells (Posakony, England and Attardi, unpublished data). For this purpose, synchronized populations (consisting of 85 to 90%

Fig. 21 (opposite). DAB-stained mitochondria in HeLa cells grown in the absence (A) or presence of 40 μg chloramphenicol/ml for 3 days (B), or 5 days (C). (From Storrie and Attardi, 1973b.)

mitotic cells) of HeLa cells (F-315 strain) were obtained by the selective detachment technique. At various times during the cell cycle (fig. 22), samples were withdrawn, pulse-labeled with [methyl-^{3}H]thymidine, and then prepared for electron microscopy. A technique involving correlation of electron microscopy of thin sections with autoradiography of adjacent thick sections allowed the recognition, and therefore the exclusion, from the "G_1" and "G_2" populations, of contaminating S phase cells (synthesizing nuclear DNA), and from the "S" population, of contaminating non-DNA-synthesizing cells.

Measurements were made on the cellular, nuclear, and mitochondrial profiles in electron micrographs of thin sections using a Hewlett-Packard Digitizer-calculator apparatus. As shown in table 6, the fractional cross-section area of mitochondria (total mitochondrial profile area/cytoplasm area), which is equivalent to the fractional mitochondrial volume and is shape independent (Weibel, 1969), remains constant during the cell cycle, implying that the total mitochondrial mass increases in proportion to the cell mass. Similarly, the total number of mitochondrial profiles per unit of cytoplasmic area and the average area of mitochondrial profiles do not vary significantly in different cellular stages. Since no appreciable change in mitochondrial shape occurs during the cell cycle, as indicated by the close similarity in the distributions for different time points of mitochondrial profile area or of outer membrane contour length or of major to minor axis ratio, and by the optical microscope appearance of DAB-stained cells at various times during the cycle, it is reasonable to interpret the above-mentioned two-dimensional data as indicative of a constancy in the number of organelles per unit of cytoplasmic volume and in the average mitochondrial volume during the cell cycle. These results imply that, in the whole cell population, mitochondrial membranes grow and divide throughout the cycle. Furthermore, since no inverse relationship was found, in individual cells at any time point in the cell cycle, between average mitochondrial profile area and number of mitochondrial profiles per unit of cytoplasmic cross-section area (as expected if, in each cell, division of mitochondria occurred at a critical time, which could be anywhere in the cycle and different from cell to cell), it must be concluded that, also in the individual cells, growth and division of the organelles occur throughout the cycle.

The above observations do not say whether the synthesis in the cytoplasm of the components necessary for the gross assembly of mitochondrial membranes also occurs throughout the cell cycle, although this seems to be likely. Above, mention was made of

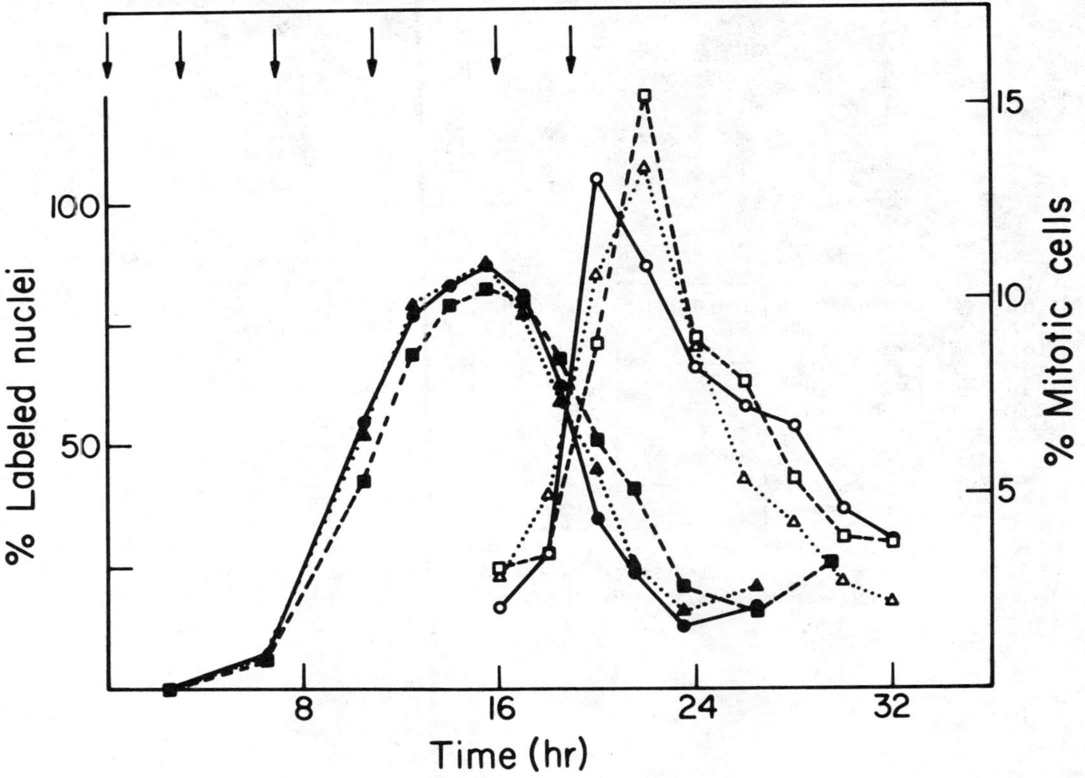

Fig. 22. Percentage of labeled nuclei (filled symbols) and of mitotic cells (open symbols) during the cell cycle in three different synchronous HeLa cell populations (F-315 strain) obtained by selective detachment, which were used in the morphometric analysis described in the text.

TABLE 6

BEHAVIOR DURING THE CELL CYCLE OF SIZE, NUMBER PER UNIT OF CYTOPLASMIC AREA, AND OUTER MEMBRANE CONTOUR LENGTH OF MITOCHONDRIAL PROFILES

Time after mitosis (hr)	Cell area (μm^2)	CYT. AREA	Cyt. area (μm^2)	TOTAL MITOCH. PROFILE AREA	NO. MITOCH. PROFILES	Average mitoch. profile area (μm^2)	Average outer membrane contour length (μm)	Average axis ratio
		Cell area (%)		Cyt. area (%)	μm^2 Cyt. area			
0	197.5 ± 9.7	67.3 ± 2.1[a]	131.3 ± 7.7	10.9 ± 0.5	0.57 ± 0.03	0.193 ± 0.006	1.64 ± 0.02	1.77 ± 0.04
3	117.6 ± 5.0	68.4 ± 2.1	79.1 ± 3.3	10.6 ± 0.5	0.52 ± 0.02	0.202 ± 0.008	1.70 ± 0.04	1.90 ± 0.06
7	124.4 ± 7.6	64.2 ± 2.0	78.4 ± 4.9	9.4 ± 0.4	0.46 ± 0.02	0.205 ± 0.008	1.69 ± 0.03	1.86 ± 0.06
11	128.9 ± 5.7	66.8 ± 2.4	86.2 ± 4.9	10.8 ± 0.5	0.53 ± 0.02	0.204 ± 0.006	1.69 ± 0.03	1.84 ± 0.05
16	144.4 ± 6.1	65.9 ± 1.7	95.1 ± 4.6	10.9 ± 0.7	0.54 ± 0.03	0.203 ± 0.007	1.69 ± 0.03	1.94 ± 0.08
19	170.6 ± 7.1	71.3 ± 2.2	118.4 ± 3.5	10.6 ± 0.5	0.51 ± 0.03	0.210 ± 0.008	1.73 ± 0.03	1.91 ± 0.08

[a] Estimated as average of values for other time points.

experiments indicating that the various components of the inner mitochondrial membrane that are products of mitochondrial protein synthesis are synthesized throughout the cell cycle at a fairly constant rate per unit of cell mass and at the same relative rate; in this case, no evidence was obtained as to whether these proteins are continuously utilized for the assembly of an active inner mitochondrial membrane. However, the continuous formation during the cell cycle of the framework of mitochondrial membranes suggests that the differentiation of these newly formed membranes for the function of oxidative phosphorylation does indeed occur throughout the cycle.

The continuous formation and division of mitochondria during the cell cycle should be contrasted with the temporal behavior of mitochondrial DNA replication. Clearly, the situation is very different from that observed in bacteria, where DNA replication shows a great regularity in relation to cell growth and division, suggesting a precise coordination and interdependence of membrane growth and chromosome replication (Jacob et al., 1963).

As concerns the mode of division of mitochondria, no information was obtained in the present work that would support unambiguously either the elongation-constriction or the transverse partition mechanism (Tandler et al., 1969; Tandler and Hoppel, 1973).

CONCLUSIONS AND SPECULATIONS

Growth and Differentiation of Mitochondria: Two Phases of Mitochondriogenesis

The picture of mitochondriogenesis that has emerged from the work carried out on the HeLa cell system and from the parallel work carried out in lower eukaryotes, especially yeast and *Neurospora,* is one of a process in which the formation of the inner mitochondrial membrane is independent of, and may possibly precede, under physiological conditions, its differentiation for the function of oxidative phosphorylation. The mitochondrial membranes and the spaces they enclose constitute a potential, if not actual, continuum, which, in exponentially dividing cells, grows in proportion to the increase in cell mass. This closed system provides the structural framework and the chemical environment within which the differentiation process takes place leading to organelles capable of oxidative phosphorylation.

Although the formation of this mitochondrial framework appears to be under exclusive nuclear control, and to continue even in the absence of a functional mitochondrial genome, the differentiation of the primordial organelles into mitochondria active in oxidative phosphorylation is under the control of both the nuclear and the mitochondrial genome. Justification for distinguishing two phases in the biogenesis of mitochondria comes not only from the difference in genetic control but also from a wealth of experimental evidence indicating that the two processes, which are normally intimately coordinated, can, under certain conditions, be dissociated. In fact, a dissociation occurs physiologically during anaerobic growth and, less dramatically, during glucose repression in yeast (see review by Linnane and Haslam, 1970). Likewise, in some animal cells, there is evidence of incomplete development of the respiratory units of the inner mitochondrial membrane at early stages of embryonic development (Lang, 1965; Lennie and Birt, 1967; Haddock et al., 1969), or during growth in low oxygen tension (Pious, 1970). The different temporal pattern during the cell cycle of the gross formation of mitochondria and of mitochondrial DNA replication and transcription, which has been demonstrated in HeLa cells, also points to a different control of the processes of growth and differentiation in mitochondriogenesis. Finally, a dissociation of mitochondrial growth and differentiation can be produced in yeast, as in mammalian cells, by the use of inhibitors of mitochondrial protein synthesis (Clark-Walker and Linnane, 1967; Mahler and Perlman, 1971; Storrie and Attardi, 1973a).

The existence of a separate control for the gross formation of mitochondria and their differentiation into organelles active in oxidative phosphorylation may provide the cell with a greater flexibility of adaptation and with the ability to respond selectively and rapidly to increased respiratory demands; this may occur by addition of new respiratory units to the available sites in the inner mitochondrial membrane, if these normally are not saturated. Here may indeed lie the evolutionary advantage of maintaining through more than one billion years of evolution, in a separate compartment, a protein-synthesizing system distinct from the cytoplasmic machinery, specialized for the synthesis of a few proteins having a key role in oxidative phosphorylation.

The Mitochondrial Genome: A Eukaryotic Cell Multigene Family

Within the above frame of reference, the individual mitochondrial DNA molecules within each cell do not appear like minichromosomes endowed with the capacity of autonomous replication and expression that control the reproduction of mitochondria. Rather, they have to be considered, regardless of their evolutionary origin, as members of a special multigene family of the eukaryotic cells, which preside over a differentiation process of the inner mitochondrial membrane. Replication and expression of this mitochondrial multigene family is subjected to extramitochondrial control and integrated with the over-all processes of cellular growth. The location of this multigene family outside the nucleus (a characteristic that distinguishes it from the other known multigene families) can be satisfactorily accounted for by assuming it to be advantageous to the cell, from the point of view of regulation, to have these genes and their products close to the site of their activity. The same argument would give an explanation for the fact that the members of this family are unlinked to each other, in view of the generally discontinuous character of mitochondria. The existence, in the mitochondrial compartment, of a distinctive translation system serving the mitochondrial genome and in part coded by the genome itself may reflect the need for the cell to maintain within the mitochondria an adequate potential of protein synthesizing capacity unavailable for general protein systhesis and specialized, in terms of regulation or of genetic code, for the translation of mitochondrial messages; the requirement that some highly hydrophobic polypeptides be synthesized inside mitochondria for their proper integration into the inner membrane (Borst, 1969) may also play a role in this respect, although it would not by itself account for the unique composition of the mitochondrial protein-synthesizing machinery.

Various lines of evidence support the idea that the mitochondrial genome may share with the nuclear genome several properties related to mechanism of expression and mode of control. Thus, the dependence of mitochondrial DNA replication and transcription on continuous cytoplasmic protein synthesis (as opposed to their independence of mitochondrial protein synthesis) has its parallel in a similar behavior of nuclear DNA and RNA synthesis. The presence of poly(A), though of distinctive length, in mitochondrial DNA-coded RNA is a typical eukaryotic trait, suggesting a similarity of control mechanisms for the nuclear and mitochondrial genomes. The simultaneous stimulation in some mammalian cell lines of nuclear and mitochondrial DNA synthesis by SV40 (Levine, 1971) or polyoma virus infection (Vesco and Basilico, 1971) and the absence of stimulation of synthesis of both DNAs by the same viral infections in other cell lines (Levine, 1971), as well as the enhancement of both nuclear and mitochondrial

DNA synthesis by fresh serum in confluent monolayers, are other striking examples of parallelism of response, and presumably of control, of the two portions of the animal cell genome. The recent demonstration of histones associated with the closed circular DNAs from SV40 (Frearson and Crawford, 1972; Hall et al., 1973; Lake et al., 1973) and polyoma virus (Green et al., 1971), which have a very similar supercoiled structure to that of mitochondrial DNA, suggests the possibility of a similar occurrence of histones in association with mitochondrial DNA.

A further reason for believing that the mitochondrial and nuclear genomes have some common traits of organization and control is that they are tightly coupled both evolutionarily and functionally. Evolutionary coupling is indicated by the fact that all or almost all proteins, which recognize mitochondrial nucleic acid sequences, are coded for by nuclear genes. Functional coupling is implicit in the participation of both nuclear DNA-coded and mitochondrial DNA-coded components in the assembly of functional mitochondria. The evidence for the existence in HeLa cells of a cell cycle dependent control, at the transcription level, of the expression of the mitochondrial genome has been mentioned above. This transcriptional control is essential for increasing the over-all mitochondrial protein-synthesizing capacity of the cell in proportion to its mass. In addition to this cell cycle dependent control, there may be regulation of expression of the mitochondrial genome in response to physiological stimuli. The modulation of the mitochondrial function that has been described in response to thyroid hormones (Tata, 1966), oxygen tension (Pious, 1970), or changes in respiratory demands (Hamberger et al., 1969) may indeed involve a translational or transcriptional control of the expression of the mitochondrial genes. Inasmuch as the assembly of the functional units of oxidative phosphorylation requires both nuclear DNA-coded and mitochondrial DNA-coded components, it is very likely that, both in the cell cycle dependent control and in any possible physiological modulation of the respiratory function, the expression of the mitochondrial genome is coordinated with that of the relevant nuclear genes. How this coupling between nuclear and mitochondrial genome operates, whether it is actuated by positive or negative control, whether the primary target of regulatory signals are the nuclear or the mitochondrial genes, are central questions for an understanding of the control of assembly of the mitochondrial oxidative phosphorylation system. And herein may lie the answer to some of the basic laws of expression and control of the eukaryotic genome.

ACKNOWLEDGMENTS

The work described in this paper has been supported by grant GM-11726 from the National Institutes of Health. The valuable assistance of Mr. P. Koen with the electron microscopy and of Mrs. Arger Drew, Miss Gloria Engel, and Miss Rossella Ummarino is gratefully acknowledged. The purified L mitochondrial DNA strand preparation utilized in the hybridization saturation experiments with L transcripts was a gift of Dr. J. Flory, and the ϕX174 RF a gift of Mr. Welcome Bender.

REFERENCES

Aloni, Y. 1972. Extensive symmetrical transcription of simian virus 40 DNA in virus-yielding cells. Proc. Nat. Acad. Sci. USA. 69:2404–9.

Aloni, Y., and G. Attardi. 1971a. Expression of the mitochondrial genome in HeLa cells. II. Evidence for complete transcription of mitochondrial DNA. J. Mol. Biol. 55:251–67.

Aloni, Y., and G. Attardi. 1971b. Expression of the mitochondrial genome in HeLa cells. VII. Symmetric *in vivo* transcription of mitochondrial DNA in HeLa cells. Proc. Nat. Acad. Sci. USA. 68:1957–61.

Aloni, Y., and G. Attardi. 1971c. Expression of the mitochondrial genome in HeLa cells. IV. Titration of mitochondrial genes for 16 S, 12 S and 4 S RNA. J. Mol. Biol. 55:271–76.

Aloni, Y., and G. Attardi. 1972. Expression of the mitochondrial genome in HeLa cells. XI. Isolation and characterization of transcription complexes in mitochondrial DNA. J. Mol. Biol. 70:363–73.

Aloni, Y., and H. Locker. 1973. Symmetrical *in vivo* transcription of polyoma DNA and the separation of self-complementary viral and cell RNA. Virology 54:495–505.

Altmann, R. 1890. Die Elementarorganismen und ihre Beziehungen zu den Zellen. Veit and Co., Leipzig.

Arnold, C. G., O. Schimmer, F. Schötz, and H. Bathelt. 1972. Die Mitochondrien von *Chlamydomonas reinhardii*. Arch. Mikrobiol. 81:50–67.

Ashwell, M., and T. S. Work. 1970. The biogenesis of mitochondria. Ann. Rev. Biochem. 39:251–90.

Attardi, B., B. Cravioto, and G. Attardi. 1969. Membrane-bound ribosomes in HeLa cells. I. Their proportion to total cell ribosomes and their association with messenger RNA. J. Mol. Biol. 44:47–70.

Attardi, B., and G. Attardi. 1971. Expression of the mitochondrial genome in HeLa cells. I. Properties of the discrete RNA components from the mitochondrial fraction. J. Mol. Biol. 55:231–49.

Attardi, G., and B. Attardi. 1969. The informational role of mitochondrial DNA, pp. 245–83. *In* E. W. Hanly (ed.), Symposium on problems in biology: RNA in development. University of Utah Press, Salt Lake City.

Attardi, G., Y. Aloni, B. Attardi, D. Ojala, L. Pica-Mattoccia, D. Robberson, and

B. Storrie. 1970. Transcription of mitochondrial DNA in HeLa cells. Cold Spring Harbor Symp. Quant. Biol. 35:599-619.

Attardi, G., and D. Ojala. 1971. Mitochondrial ribosomes in HeLa cells. Nature New Biol. 229:133-36.

Attardi, G., P. Costantino, J. England, M. Lederman, D. Ojala, and B. Storrie. 1973. Mitochondrial protein synthesis in HeLa cells. Acta Endocrinologica, Suppl. no. 180:263-93.

Barath, Z., and H. Küntzel. 1972. Induction of mitochondrial RNA polymerase in *Neurospora crassa*. Nature New Biol. 240:195-97.

Beattie, D. S. 1971. The synthesis of mitochondrial proteins. Subcell. Biochem. 1:1-23.

Benda, C. 1898. Über die Spermatogenese der Vertebraten und höherer Evertebraten. II. Theil: Die Histiogenese der Spermien. Arch. Anat. Physiol. Pp. 393-98.

Borst, P. 1969. Size, structure and information content of mitochondrial DNA, pp. 260-66. *In* N. Boardman, A. Linnane, and R. Smillie (eds.) Autonomy and biogenesis of mitochondria and chloroplasts. North-Holland, Amsterdam.

Borst, P., and L. A. Grivell. 1971. Mitochondrial ribosomes. FEBS Letters 13:73-88.

Borst, P., E. F. J. Van Bruggen, and G. J. C. M. Ruttenberg. 1968. Size and structure of mitochondrial DNA, pp. 51-70. *In* E. C. Slater, J. M. Tager, S. Papa, E. Quagliariello, eds. Biochemical aspects of the biogenesis of mitochondria. Adriatic Editrice, Bari.

Brandt, J. T., A. P. Martin, F. V. Lucas, and M. L. Vorbeck. 1974. The structure of rat liver mitochondria: A reevaluation. Biochem. Biophys. Res. Commun. 59:1097-1103.

Brega, A., and C. Vesco. 1971. Ribonucleoprotein particles involved in HeLa mitochondrial protein synthesis. Nature New Biol. 229:136-39.

Buck, C. A., and M. M. K. Nass. 1969. Studies on mitochondrial tRNA from animal cells. I. A comparison of mitochondrial and cytoplasmic tRNA and aminoacyl-tRNA synthetases. J. Mol. Biol. 41:67-82.

Burke, J. P., and D. S. Beattie. 1973. The synthesis of proteolipid protein by isolated rat liver mitochondria. Biochem. Biophys. Res. Commun. 51:349-56.

Calvayrac, R., and R. A. Butow. 1971. Action de l'antimycine A sur la respiration et la structure des mitochondries d'*Euglena gracilis* Z. Arch. Mikrobiol. 80:62-69.

Calvayrac, R., R. A. Butow, and M. Lefort-Tran. 1972. Cyclic replication of DNA and changes in mitochondrial morphology during the cell cycle of *Euglena gracilis* (Z). Exp. Cell Res. 71:422-32.

Calvayrac, R., F. van Leute, and R. A. Butow. 1971. *Euglena gracilis:* Formation of giant mitochondria. Science 173:252-54.

Carnevali, F., G. Morpurgo, and G. Tecce. 1969. Cytoplasmic DNA from petite colonies of *Saccharomyces cerevisiae:* A hypothesis on the nature of the mutation. Science 163:1331-33.

Chamberlin, M. 1970. Transcription 1970: A summary. Cold Spring Harbor Symp. Quant. Biol. 35:851-73.

Chiu, N., A. O. S. Chiu, and Y. Suyama. 1974. Three isoaccepting forms of leucyl transfer RNA in mitochondria. J. Mol. Biol. 82:441-57.

Clark-Walker, G. D., and A. W. Linnane. 1967. The biogenesis of mitochondria in

Saccharomyces cerevisiae: A comparison between cytoplasmic respiratory-deficient mutant yeast and chloramphenicol-inhibited wild type cells. J. Cell Biol. 34:1-14.

Clayton, D. A., R. W. Davis, and J. Vinograd. 1970. Homology and structural relationships between the dimeric and monomeric circular forms of mitochondrial DNA from human leukemic leucocytes. J. Mol. Biol. 47:137-53.

Coen, D., J. Deutsch, J. Netter, E. Petrochilo, and P. P. Slonimski. 1970. Mitochondrial genetics. Symp. Soc. Exp. Biol. 24:449-96.

Coote, J. L., and T. S. Work. 1971. Proteins coded by mitochondrial DNA of mammalian cells. Europ. J. Biochem. 23:564-74.

Costantino, P., and G. Attardi. 1973. Atypical pattern of utilization of amino acids for mitochondrial protein synthesis in HeLa cells. Proc. Nat. Acad. Sci. USA. 70:1490-94.

Costantino, P., and G. Attardi. 1975. Identification of discrete electrophoretic components among the products of mitochondrial protein synthesis in HeLa cells. J. Mol. Biol. (in press).

Darnell, J. E., L. Philipson, R. Wall, and M. Adesnik. 1971. Polyadenylic acid sequences: Role in conversion of nuclear RNA into messenger RNA. Science 174:507-10.

Darnell, J. E., R. Wall, and R. J. Tushinski. 1971. An adenylic acid-rich sequence in messenger RNA of HeLa cells and its possible relationship to reiterated sites in DNA. Proc. Nat. Acad. Sci. USA. 68:1321-25.

Davey, P. J., R. Yu, and A. W. Linnane. 1969. The intracellular site of formation of the mitochondrial protein synthetic system. Biochem. Biophys. Res. Commun. 36:30-34.

Dawid, I. B. 1970. The nature of mitochondrial RNA in oocytes of *Xenopus laevis* and its relation to mitochondrial DNA. Symp. Soc. Exp. Biol. 24:227-46.

Dawid, I. B. 1972. Mitochondrial RNA in *Xenopus laevis*. I. The expression of the mitochondrial genome. J. Mol. Biol. 63:201-16.

Edmonds, M., M. H. Vaughan, and H. Nakazato. 1971. Polyadenylic acid sequences in the heterogeneous nuclear RNA and rapidly labeled polyribosomal RNA of HeLa cells: Possible evidence for a precursor relationship. Proc. Nat. Acad. Sci. USA. 68:1336-40.

England, J. M., and G. Attardi. 1974. Expression of the mitochondrial genome in HeLa cells. XXI. Mitochondrial protein synthesis during the cell cycle. J. Mol. Biol. 85:433-44.

Ennis, H. L., and M. Lubin. 1964. Cycloheximide: Aspects of inhibition of protein synthesis in mammalian cells. Science 146:1474-76.

Ephrussi, B., H. Hottinguer, and J. Tavlitzki. 1949. Action de l'acriflavine sur les levures. II. Étude génétique du mutant "petite colonie." Ann. Inst. Pasteur 76:419-50.

Firkin, F. C., and A. W. Linnane. 1968. Differential effects of chloramphenicol on the growth and respiration of mammalian cells. Biochem. Biophys. Res. Commun. 32:398-402.

Frearson, P. M., and L. V. Crawford. 1972. Polyoma virus basic proteins. J. Gen. Virol. 14:141-55.

Frederic, J. 1958. Recherches cytologiques sur le chondriome normal où soumis à l'expérimentation dans des cellules vivantes cultivées *in vitro*. Arch. Biol. 69:167-349.

Frederic, J., and M. Chèvremont. 1952. Recherches sur les chondriosomes de cellules vivantes par la microscopie et la microcinématographie en contraste de phase. Arch. Biol. 63:109-31.

Galper, J. B., and J. E. Darnell. 1969. The presence of N-formyl-methionyl-tRNA in HeLa cell mitochondria. Biochem. Biophys. Res. Commun. 34:205-14.

Galper, J. B., and J. E. Darnell. 1971. Mitochondrial protein synthesis in HeLa cells. J. Mol. Biol. 57:363-67.

Grandi, M., and H. Küntzel. 1970. Mitochondrial peptide chain elongation factors from *Neurospora crassa*. FEBS Letters 10:25-28.

Green, M. H., H. I. Miller, and S. Hendler. 1971. Isolation of a polyoma-nucleoprotein complex from infected mouse-cell cultures. Proc. Nat. Acad. Sci. USA. 68:1032-36.

Grimes, G. W., H. R. Mahler, and P. S. Perlman. 1974a. Nuclear gene dosage effects on mitochondrial mass and DNA. J. Cell Biol. 61:565-74.

Grimes, G. W., H. R. Mahler, and P. S. Perlman. 1974b. Mitochondrial morphology. Science 185:630-31.

Grollman, A. P. 1966. Structural basis for inhibition of protein synthesis by emetine and cycloheximide based on an analogy between ipecac alkaloids and glutarimide antibiotics. Proc. Nat. Acad. Sci. USA. 56:1867-74.

Haddock, J., S. Jakovicic, D. Hsia, and G. S. Getz. 1969. Mitochondrial development in liver of fetal and newborn rats. Fed. Proc. 28:855. (abst.).

Hall, M. R., W. Meinke, and D. A. Goldstein. 1973. Nucleoprotein complexes containing replicating simian virus 40 DNA: Comparison with polyoma nucleoprotein complexes. J. Virol. 12:901-8.

Hamberger, A., N. Gregson, and A. L. Lehninger. 1969. The effect of acute exercise on amino acid incorporation into mitochondria of rabbit tissues. Biochim. Biophys. Acta 186:373-83.

Hirsch, M., and S. Penman. 1973. Mitochondrial polyadenylic acid-containing RNA: Localization and characterization. J. Mol. Biol. 80:379-91.

Hoffman, H.-P., and C. J. Avers. 1973. Mitochondrion of yeast: Ultrastructural evidence for one giant, branched organelle per cell. Science 181:749-51.

Horak, I., H. G. Coon, and I. B. Dawid. 1974. Interspecific recombination of mitochondrial DNA molecules in hybrid somatic cells. Proc. Nat. Acad. Sci. USA. 71:1828-32.

Jacob, F., S. Brenner, and F. Cuzin. 1963. On the regulation of DNA replication in bacteria. Cold Spring Harbor Symp. Quant. Biol. 28:329-48.

Jacob, J. T., and D. G. Schindler. 1972. Polyriboadenylate polymerase solubilized from rat liver mitochondria. Biochem. Biophys. Res. Commun. 48:126-234.

Jacob, J. T., D. G. Schindler, and H. P. Morris. 1972. Mitochondrial polyriboadenylate polymerase: Relative lack of activity in hepatomas. Science 178:639-40.

Jelinek, W., M. Adesnik, M. Salditt, D. Sheiness, R. Wall, G. Molloy, L. Philipson, and J. E. Darnell. 1973. Further evidence on the nuclear origin and transfer to the cytoplasm of polyadenylic acid sequences in mammalian cell RNA. J. Mol. Biol. 75:515-32.

Kadenbach, B. 1971a. Isolation and characterization of a peptide synthesized in mitochondria. Biochem. Biophys. Res. Commun. 44:724-30.

Kadenbach, B. 1971b. Biosynthesis of mitochondrial cytochromes, pp. 360–71. *In* N. Boardman, A. Linnane, and R. Smillie (eds.), Autonomy and biogenesis of mitochondria and chloroplasts. North Holland, Amsterdam.

Keddie, F., and Barajas, L. 1969. Three-dimensional reconstruction of *Pityrosporum* yeast cells based on serial section electron microscopy. J. Ultrastruc. Res. 29:260–75.

King, M. E., G. C. Goodman, and D. W. King. 1972. Respiratory enzymes and mitochondrial morphology of HeLa and L cells treated with chloramphenicol and ethidium bromide. J. Cell Biol. 53:127–42.

Kroon, A. M. 1965. Protein synthesis in mitochondria. III. On the effects of inhibitors on the incorporation of amino acids into protein by intact mitochondria and digitonin fractions. Biochim. Biophys. Acta 108:275–84.

Kroon, A. M., and H. DeVries. 1971. Mitochondriogenesis in animal cells: Studies with different inhibitors, pp. 318–27. *In* N. Boardman, A. Linnane, and R. Smillie (eds.), Autonomy and biogenesis of mitochondria and chloroplasts. North Holland, Amsterdam.

Küntzel, H. 1969. Proteins of mitochondrial and cytoplasmic ribosomes from *Neurospora crassa*. Nature 222:142–46.

Lake, R. S., S. Barban, and N. P. Salzman. 1973. Resolution and identification of the core deoxynucleoproteins of the simian virus 40. Biochem. Biophys. Res. Commun. 54:640–47.

Lang, G. A. 1965. Respiratory enzymes in the heart and liver of the natal and postnatal rat. Biochem. J. 95:365–71.

Lark, K. G. 1969. Initiation and control of DNA synthesis. Ann. Rev. Biochem. 38:569–604.

Lederman, M., and G. Attardi. 1970. *In vitro* protein synthesis in a mitochondrial fraction from HeLa cells: Sensitivity to antibiotics and ethidium bromide. Biochem. Biophys. Res. Commun. 40:1492–1500.

Lederman, M., and G. Attardi. 1973. Expression of the mitochondrial genome in HeLa cells. XVI. Electrophoretic properties of the products of *in vivo* and *in vitro* mitochondrial protein synthesis. J. Mol. Biol. 78:275–83.

Leibowitz, R. D. 1971. The effect of ethidium bromide on mitochondrial DNA synthesis and mitochondrial DNA structure in HeLa cells. J. Cell Biol. 51:116–22.

Lennie, R. W., and L. M. Birt. 1967. Aspects of the development of flight-muscle sarcosomes in the sheep blowfly, *Lucilia cuprina*, in relation to changes in the distribution of protein and some respiratory enzymes during metamorphosis. Biochem. J. 102:338–50.

Levine, A. J. 1971. Induction of mitochondrial DNA synthesis in monkey cells infected by simian virus 40 and (or) treated with calf serum. Proc. Nat. Acad. Sci. USA. 68:717-20.

Levintow, L. and J. E. Darnell. 1960. A simplified procedure for purification of large amounts of polio virus: Characterization and amino acid analysis of type 1 polio virus. J. Biol. Chem. 235:70–73.

Linnane, A. W. 1968. The nature of mitochondrial RNA and some characteristics of the protein-synthesizing system of mitochondria isolated from antibiotic-sensitive and resistant yeasts, pp. 333-53. *In* E. C. Slater, J. M. Tager, S. Papa, and E.

Quagliariello (eds.), Biochemical aspects of the biogenesis of mitochondria. Adriatica Editrice, Bari.

Linnane, A. W., and J. M. Haslam. 1970. The biogenesis of yeast mitochondria. Current Topics in Cellular Regulation 2:101-72.

Littauer, U. Z., and H. Inouye. 1973. Regulation of tRNA. Ann. Rev. Biochem. 42:439-70.

Luck, D. J. L. 1963. Genesis of mitochondria in *Neurospora crassa*. Proc. Nat. Acad. Sci. USA. 49:233-40.

Luck, D. J. L. 1965. Formation of a mitochondria in *Neurospora*. A study based on mitochondrial density changes. J. Cell Biol. 24:461

Mahler, H. R., and P. S. Perlman. 1971. Mitochondriogenesis analyzed by blocks on mitochondrial translation and transcription. Biochemistry 10:2979-90.

Mahler, H. R., and K. Dawidowicz. 1973. Autonomy of mitochondria in *Saccharomyces cerevisiae* in their production of messenger RNA. Proc. Nat. Acad. Sci. USA. 70:111-14.

Mahler, H. R., F. Feldman, S. H. Phan, P. Hamill, and K. Dawidowicz. 1974. Initiation, identification and integration of mitochondrial proteins. pp. 423-41. *In* A. M. Kroon, and C. Saccone (eds.), Biogenesis of mitochondria and chloroplasts. Academic Press, London.

Mitchell, M. B., and H. K. Mitchell. 1952. A case of "maternal" inheritance in *Neurospora crassa*. Proc. Nat. Acad. Sci. USA. 38:442-49.

Murray, D. R., and A. W. Linnane. 1972. Synthesis of proteolipid protein by yeast mitochondria. Biochem. Biophys. Res. Commun. 49:855-62.

Nagley, P., and A. W. Linnane. 1970. Mitochondrial DNA deficient petite mutants of yeast. Biochem. Biophys. Res. Commun. 39:989-96.

Nass, M. M. K., and C. A. Buck. 1970. Studies on mitochondrial tRNA from animal cells. II. Hybridization of aminoacyl-tRNA from rat liver mitochondria with heavy and light complementary strands of mitochondrial DNA. J. Mol. Biol. 54:187-98.

Ojala, D., and G. Attardi. 1972. Expression of the mitochondrial genome in HeLa cells. X. Properties of mitochondrial polysomes. J. Mol. Biol. 65:273-89.

Ojala, D., and G. Attardi. 1974a. Expression of the mitochondrial genome in HeLa cells. XIX. Occurrence in mitochondria of polyadenylic acid sequences, "free" and covalently linked to mitochondrial DNA-coded RNA. J. Mol. Biol. 82:151-76.

Ojala, D., and G. Attardi. 1974b. Identification and partial characterization of multiple discrete polyadenylic acid-containing RNA components coded for by HeLa cell mitochondrial DNA. J. Mol. Biol. 88:205-19.

Osafune, T. 1973. 3-dimensional structures of giant mitochondria, dictyosomes, and "concentric lamellar bodies" formed during the cell cycle of *Euglena gracilis* (Z) in synchronous cùlture. J. Electron Microscopy 22:51-61.

Perlman, S., and S. Penman. 1970a. Protein-synthesizing structures associated with mitochondria. Nature 227:133-37.

Perlman, S., and S. Penman. 1970b. Mitochondrial protein synthesis: Resistance to emetine and response to RNA synthesis inhibitors. Biochem. Biophys. Res. Commun. 40:941-48

Perlman, S., H. T. Abelson, and S. Penman. 1973. Mitochondrial protein synthesis:

RNA with the properties of eukaryotic messenger RNA. Proc. Nat. Acad. Sci. USA. 70:350–53.

Perry, R. P., D. E. Kelley, and J. La Torre. 1972. Lack of polyadenylate acid sequences in the messenger RNA of *E. coli.* Biochem. Biophys. Res. Commun. 48:1592–1600.

Pica-Mattoccia, L., and G. Attardi. 1971. Expression of the mitochondrial genome in HeLa cells. V. Transcription of mitochondrial DNA in relationship to the cell cycle. J. Mol. Biol. 57:615–21.

Pica-Mattoccia, L., and G. Attardi. 1972. Expression of the mitochondrial genome in HeLa cells. IX. Replication of mitochondrial DNA in relationship to the cell cycle in HeLa cells. J. Mol. Biol. 64:465–84.

Pious, D. A. 1970. Induction of cytochromes in human cells by oxygen. Proc. Nat. Acad. Sci. USA. 65:1001–8.

Posakony, J. W., J. M. England, and G. Attardi. 1975. Morphological heterogeneity of HeLa cell mitochondria visualized by a modified diaminobenzidine staining technique. J. Cell Science (in press).

Reijnders, L., C. M. Kleisen, L. A. Grivell, and P. Borst. 1972. Hybridization studies with yeast mitochondrial RNA's. Biochim. Biophys. Acta 272:396–407.

Rich, A., S. Penman, Y. Becker, J. Darnell, and C. Hall. 1963. Polyribosomes: Size in normal and polio-infected HeLa cells. Science 142:1658–63.

Richter, D., and F. Lipmann. 1970. Separation of mitochondrial and cytoplasmic peptide chain elongation factors from yeast. Biochemistry 9:5065–70.

Robberson, D., Y. Aloni, and G. Attardi. 1971. Electron microscopic visualization of mitochondrial RNA-DNA hybrids. J. Mol. Biol. 55:267–70.

Robberson, D., Y. Aloni, G. Attardi, and N. Davidson. 1971. Size determination of mitochondrial ribosomal RNA by electron microscopy. J. Mol. Biol. 60:473–84.

Robbins, E., and P. I. Marcus. 1964. Mitotically synchronized mammalian cells: A simple method for obtaining large populations. Science 144:1152–53.

Ross, E., E. Ebner, R. D. Poyton, T. R. Mason, B. Ono, and G. Schatz. 1974. The biosynthesis of mitochondrial cytochromes, pp. 477–83. *In* A. M. Kroon and C. Saccone (eds.), Biogenesis of mitochondria and chloroplasts. Academic Press, London.

Rubin, M. S., and A. Tzagoloff. 1973a. Assembly of the mitochondrial membrane system. IX. Purification, characterization, and subunit structure of yeast and beef cytochrome oxidase. J. Biol. Chem. 248:4269–74.

Rubin, M. S., and A. Tzagoloff. 1973b. Assembly of the mitochondrial membrane system. X. Mitochondrial synthesis of three of the subunit proteins of yeast cytochrome oxidase. J. Biol. Chem. 248:4275–79.

Sala, F., and H. Küntzel. 1970. Peptide chain initiation in homologous and heterologous systems from mitochondria and bacteria. Europ. J. Biochem. 15:280–86.

Schatz, G., and T. L. Mason. 1974. The biosynthesis of mitochondrial proteins. Ann. Rev. Biochem. 43:51–87.

Schneidl, W. 1974. Veranderungen des Mitochondrienbestandes während des Zellzyklus. Cytobiologie 8:403–11.

Sebald, W., Th. Hofstötter, D. Hacker, and Th. Bücher. 1969. Incorporation of amino acids into mitochondrial protein of the flight muscle of *Locusta migratoria in vitro* and *in vivo* in the presence of cycloheximide. FEBS Letters 2:177–80.

Sebald, W., W. Machleidt, and J. Otto. 1974. Cooperation of mitochondrial and cytoplasmic protein synthesis in the formation of cytochrome *c* oxidase, pp. 453-63. *In* A. M. Kroon, and C. Saccone (eds.), Biogenesis of mitochondria and chloroplasts. Academic Press, London.

Seligman, A. M., M. J. Karnovsky, H. L. Wasserkrug, and J. S. Hanker. 1968. Nondroplet ultrastructural demonstration of cytochrome oxidase activity with a polymerizing osmiophilic reagent, diaminobenzidine. J. Cell Biol. 38:1-14.

Shannon, C., A. Rao, S. Douglass, and R. S. Criddle. 1972. Recombination in yeast mitochondrial DNA. J. Supramolec. Struc. 1:145-52.

Smith, C. A., J. M. Jordan, and J. Vinograd. 1971. *In vivo* effects of intercalating drugs on the superhelix density of mitochondrial DNA isolated from human and mouse cells in culture. J. Mol. Biol. 59:255-72.

Storrie, B., and G. Attardi. 1972. Expression of the mitochondrial genome in HeLa cells. XIII. Effect of selective inhibition of cytoplasmic or mitochondrial protein synthesis on mitochondrial nucleic acid synthesis. J. Mol. Biol. 71:177-99.

Storrie, B., and G. Attardi. 1973a. Expression of the mitochondrial genome in HeLa cells. XV. Effect of inhibition of mitochondrial protein synthesis on mitochondrial formation. J. Cell Biol. 56:819-31.

Storrie, B., and G. Attardi. 1973b. Expression of the mitochondrial genome in HeLa cells. XVIII. Heterogeneity of isolated HeLa cell mitochondria as assayed for their enzymatic and *in vivo* biosynthetic activities. J. Cell Biol. 56:833-38.

Swanson, R. F. 1971. Incorporation of high molecular weight polynucleotides by isolated mitochondria. Nature 231:31-34.

Swanson, R. F. 1973. Specificity of mitochondrial and cytoplasmic ribosomes and elongation factors from *Xenopus laevis*. Biochemistry 12:2142-46.

Tandler, B., R. A. Erlandson, S. L. Smith, and E. L. Wynder. 1969. Riboflavin and mouse hepatic cell structure and function. II. Division of mitochondria during recovery from simple deficiency. J. Cell Biol. 41:477-93.

Tandler, B., and C. L. Hoppel. 1973. Division of giant mitochondria during recovery from cuprizone intoxication. J. Cell Biol. 56:266-72.

Tata, J. R. 1966. The regulation of mitochondrial structure and function by thyroid hormones under physiological conditions, pp. 489-507. *In* J. M. Tager, S. Papa, E. Quagliariello, and E. C. Slater (eds.), Symp. on regulation of metabolic processes in mitochondria (BBA Library, vol. 7). Elsevier, Amsterdam.

Terasima, T., and L. J. Tolmach. 1963. Growth and nucleic acid synthesis in synchronously dividing populations of HeLa cells. Exp. Cell Res. 30:344-62.

Tzagoloff, A., and A. Akai. 1972. Assembly of the mitochondrial membrane system. VIII. Properties of the products of mitochondrial protein synthesis in yeast. J. Biol. Chem. 247:6517-27.

Tzagoloff, A., and P. Meagher. 1972. Assembly of the mitochondrial membrane system. VI. Mitochondrial synthesis of subunit proteins of the rutamycin-sensitive adenosine triphosphatase. J. Biol. Chem. 247:594-603.

Tzagoloff, A., M. S. Rubin, and M. F. Sierra. 1973. Biosynthesis of mitochondrial enzymes. Biochim. Biophys. Acta 301:71-104.

Vesco, C., and C. Basilico. 1971. Induction of mitochondrial DNA synthesis by polyoma virus. Nature 229:336-38.

Weibel, E. 1969. Stereological principles for morphometry in electron microscopic cytology. Int. Rev. Cytol. 26:235–302.

Weiss, H., and B. Ziganke. 1974. Biogenesis of cytochrome b in *Neurospora crassa*, pp. 491–500. *In* A. M. Kroon, and C. Saccone (eds.), Biogenesis of mitochondria and chloroplasts. Academic Press, London.

Wheeldon, L. W., and A. L. Lehninger. 1966. Energy-linked synthesis and decay of membrane proteins in isolated rat liver mitochondria. Biochemistry 5:3533–45.

Wintersberger, U., and E. Wintersberger. 1970. Studies on deoxyribonucleic acid polymerase from yeast. 2. Partial purification and characterization of mitochondrial DNA polymerase from wild-type and respiration-deficient yeast cells. Europ. J. Biochem. 13:20–27.

Wu, M., N. Davidson, G. Attardi, and Y. Aloni. 1972. Expression of the mitochondrial genome in HeLa cells. XIV. The relative positions of the 4S genes and of the ribosomal RNA genes in mitochondrial DNA. J. Mol. Biol. 71:81–93.

Zylber, E., C. Vesco, and S. Penman. 1969. Selective inhibition of the synthesis of mitochondria-associated RNA by ethidium bromide. J. Mol. Biol. 44:195–204.

HENRY R. MAHLER, ROBERTO N. BASTOS,
URS FLURY, CHI CHUNG LIN, AND
SEM H. PHAN

Mitochondrial Biogenesis in Fungi

2

INTRODUCTION

It is appropriate that a paper of this volume be devoted to a discussion of mitochondrial biogenesis in fungi. For it was with the fungi *Saccharomyces cerevisiae* and *Neurospora crassa* that non-Mendelian, cytoplasmic inheritance of respiratory competence—a mitochondrial trait—was first demonstrated (Ephrussi et al., 1949; Mitchell and Mitchell, 1952; for details see Sager, 1972); it was studies on the same two fungi that provided the first unambiguous, biochemical demonstration of discrete and distinct species of mitochondrial (mt)DNA and RNA as well as of the latter as mitochondrial gene products since they were specified by the former (Luck and Reich, 1964; Schatz et al., 1964; Tewari et al., 1965; Wintersberger, 1973; Rifkin et al., 1967; Barnett and Brown, 1967; Küntzel and Noll, 1967; Fukuhara, 1968; for reviews see Sager, 1972; Mahler, 1973a); it was the discovery of antibiotic resistance mutations in *S. cerevisiae* by Wilkie, Linnane, Slonimski, and their collaborators (Wilkie et al., 1967; Thomas and Wilkie, 1968a,b; Linnane et al., 1968; Coen et al., 1970) that not only furnished independent proof that the cytoplasmic genophore was subsumed in mtDNA but opened up the whole field of mitochondrial genetics; and finally, it was with the same two fungi that the mitochondrial contribution to its own biogenesis first became clearly outlined and defined in molecular terms (reviewed in Mahler,

Mitochondrial Biogenesis Group, Chemical Laboratories, Indiana University, Bloomington, Indiana 47401

1973a; Tzagoloff et al., 1973; Schatz and Mason, 1974). Since then these discoveries, which, except for the first set, only go back some ten years, have been extended both in depth and in breadth. As a result they have led to a fairly impressive documentation of the existence and the molecular characteristics of a separate and unique system for the maintenance, duplication, and expression of genetic information within the organelle in all eucaryotic cells; a system that, though essential for the construction and replication of this organelle, cannot do so without the active collaboration of, and continued interaction with, the classical nucleo-cytosolic one. Some of the most impressive recent work has been done by Dr. Attardi with animal cells and will be covered by him. We shall restrict ourselves to studies and findings on ascomycetes and principally on *S. cerevisiae*, the organism we have been using in our own investigation. Since a number of recent symposia and reviews have dealt with many of the general findings and inferences to be discussed (Miller, 1970; Schatz, 1970; Linnane and Haslam, 1970; Boardman et al., 1971; Borst, 1972; Getz, 1972; Kroon, Agsterribe and deVries, 1972; Linnane et al., 1972; Mahler, 1973a; Kroon and Saccone, 1974; Schatz and Mason, 1974; Tzagoloff, 1974) our plan is first to summarize the current state of the art, utilizing principally these general sources, without going into excessive experimental detail. Then we shall turn our attention to several important problems, as yet unsolved, and to some of the means employed by us in a search for their solution.

THE MITOCHONDRIAL SYSTEM

Nature of the System

In ascomycetes, as in all other cells examined, the mitochondrial system consists of a set of discrete and unique nucleic acids and proteins; some of them are required for the synthesis of these nucleic acids, many others as components of an equally distinct system of polypeptide synthesis, devoted to the formation of a highly specific set of constituents of the organelle. We shall briefly describe first the properties of mitochondrial (mt)DNAs and RNAs, then turn to their participation in the mitochondrial system of protein synthesis, and devote most of our attention to the nature of the products of the latter and to their implication as possible mitochondrial gene products.

Mitochondrial DNA

Wild-type DNA.—The mtDNA of yeast is easily distinguishable and separable from the nuclear (n)DNA of the same cell *not* because it is covalently circular—only a minority of molecules exhibit this property—but because of its unique size and base composition (table 1). These properties form the basis of its isolation and purification, using techniques such as rate or bouyant density sedimentation or chromatography on hydroxyapatite or polylysine-kieselguhr. With *S. cerevisiae,* to which the data apply, all ordinary methods, even when considerable care is exercised to avoid shear or nuclease action, lead principally to the isolation of molecules with a $M_r \simeq 25 \times 10^6$. However, molecules twice that size, some of them circular, can be identified in crude lysates by means of electron microscopy; and since renaturation kinetics indicate a complexity corresponding to the same size, it is generally believed that (1) the true molecular weight of mtDNA lies in the range of 50×10^6 and (2) the molecule exhibits little, if any, long-range sequence reiteration. For *Neurospora* the results are very similar except that the molecule is somewhat smaller and circular, at least in one particular strain (contour length =

TABLE 1

MITOCHONDRIAL DNA IN S. CEREVISIAE

Property	Strain: ρ+	Strain: ρ−
Mass		
Per molecule (daltons × 10^6)	50	<25 → ≪2 (DNA^0)[a]
Per haploid nucleus (molecules)	44 ± 2	~40 or ≪2 (DNA^0)
Per mitochondrion[b] (molecules)	4 ± 1	
Base composition (% A + T)	80	76 → ≥ 96
Topography		
Length (μm)	26	<14[d]
Circularity	rare	common
Oligomers	absent	common
Kinetic complexity (×10^6)	50	<20 → 0.02
Repetitive sequences	rare, small	common, any size
Degradation[c]	easy	difficult

[a] Also referred to as ρ^0.
[b] Fully derepressed; in other physiological states this value may equal the total complement.
[c] Spontaneous, during isolation, or in response to ethidium.
[d] Larger linear molecules are sometimes observed; these are oligomers.

20 μm) and richer in G+C (35%, but with almost the same decrement relative to nDNA, of 20%, as in yeast). Other entries shown in the table deal with the apparent ease of degradation of the molecules (either during isolation already alluded to, or in the course of mutagenesis by ethidium bromide [EtdBr], which will be described later and discussed extensively by Dr. Perlman) and their average number per haploid cell and its mitochondria. These latter data were determined by Grimes et al., (1974) in the course of a study, designed to establish this parameter and its relation to the mass and number of organelles, in a series of isogenic haploid and diploid cells growing exponentially on a respiratory carbon source. We found all of the to be directly proportional to chromosomal gene dosage, in other words, a constancy of all of them when expressed per unit nuclear genome. The number of mitochondria per cell is strongly dependent on growth rate and carbon source and in these haploid strains varies from as few as one or two per cell under conditions unfavorable for mitochondrial development up to 25 or so under conditions favorable for development. In contrast the number of mtDNA molecules per cell remains constant at the value indicated regardless of these variations.

The DNA of ρ^- *mutants.*—The second column compares all these properties of wild type mtDNA with those exhibited by the family of cytoplasmic respiration-deficient, petite, ρ^- strains, the classical extrachromosomal mutation in this organism (Ephrussi, 1950; Preer, 1971, Sager, 1972; Perlman, this volume). This event is now known to be due in the first instance to relatively massive deletions of base sequences (Slonimski et al., 1968; Goldring et al., 1971; Nagley and Linnane, 1972; Borst, 1972; Sanders, Flavell, et al., 1973a; Mahler, 1973a; Faye et al., 1973), varying from <50% retention of wild-type sequences down to a few hundred nucleotides. The mass of the molecules as measured either directly on electron micrographs or by kinetic complexity varies from about 10 down to about 0.2 μm ($\sim 4 \times 10^5$ daltons). Some strains contain very little and probably no mtDNA at all (Nagley and Linnane, 1970; Goldring et al., 1970). In consequence the mtDNA of different stable mutant cell lines varies in a virtually continuous fashion in base composition and informational content. Base composition ranges from values equaling or exceeding the wild type in G+C content (Mounolou et al., 1968; Faye et al., 1973) to extremely aberrant forms that contain practically only A+T (Mehrotra and Mahler, 1968; Bernardi et al., 1968; Hollenberg et al., 1972a,b). The informational content—as measured by the retention of mitochondrial genetic markers (Bolotin et al., 1971; Nagley and Linnane, 1972; Mahler, 1973a; Deutsch et al., 1974; Perlman, this volume); by hybridization with isolated mitochondrial rRNAs or tRNAs, with *in vitro* transcripts of wild-type mtDNA, or with that species directly (Cohen et al., 1972; Casey et al., 1972; Fukuhara et al., 1974; Gordon and Rabinowitz, 1973; Gordon et al., 1974; Lazowska et al., 1974)—exhibits a similar variability, from considerable retention of *meaningful* sequences to none at all. The latter type of "senseless"—sometimes called ρ^0—petite can thus be of two varieties, one containing no mtDNA at all (DNA^0) and others of DNA devoid of meaning. Clearly, the latter will depend on the definition of what constitutes "meaning," e.g., an antibiotic resistance marker, a tRNA (Carnevali et al., 1973), or some even shorter sequence. Except in these two extreme cases, however, there is no immediately apparent correlation between base sequence retention and another characteristic genetic property of petite strains, that of vegetative dominance (suppressiveness) upon being crossed to wild-type (Nagley and Linnane, 1970, 1972; Moustacchi, 1972; Michaelis et al., 1971). Finally, since the ρ^- strains that do contain mtDNA all appear to do so in normal amounts, it is immediately apparent that the sequences retained in these strains must be present in multiple copies. In addition to a possible *inter*molecular repetition of separate molecules, there is also good evidence for *intra*molecular, multiply reiterated, sequences; these are frequently arranged in tandem repeats, and occasionally inverted in direction (Sanders et al., 1973; Faye et al., 1973; Rabinowitz et al., 1974; Michel et al., 1974; Locker et al., 1974).

Other mitochondrial mutations.—Other, less traumatic, changes in mtDNA produce the family of mutations conferring resistance to a variety of antibiotic inhibitors of mitochondrial function. At least four loci are concerned with sensitivity to inhibitors of mitochondrial protein synthesis (chloramphenicol, erythromycin, spiramycin, and paromomycin); one, the ω locus, controls polarity of transmission of the first three of these genes; and four additional ones affect the sensitivity of the mitochondrial ATPase to *its* characteristic inhibitors (oligomycin, venturicidin, triethyltin, and others). No details of the changes involved either in the responsible base sequence of mtDNA (whether transitions, transversions, frame shifts, and so on, or even whether missense or nonsense) or in the resultant gene product are as yet known. These mutants and the topic of mitochondrial genetics in general are discussed in greater detail by Drs. Birky and Griffiths in this volume.

Replication, repair, and mutagenesis of mtDNA.—Few details also are available concerning the individual enzymes and reactions responsible for these three processes, all of which are known to be performed by mitochondria on their DNA (see Borst, 1972). What *is* known is that a number of mutants, many of them isolated for phenotypes concerned with *nuclear* functions and all but one of them exhibiting the classical Mendelian segregation pattern, appear also to affect some (though not always the *corresponding*) mitochondrial process. For instance, cells—the chromosomes of which are either recombination deficient or radiation sensitive—may possess mitochondria that are unaffected with regard to these two functions, but instead are resistant to mutagenesis by EtdBr. The existence of such mutants will no doubt prove of great value in defining the precise reaction affected in *both* systems. Some results will be described later on in this presentation and discussed further by Dr. Perlman (this volume).

Mitochondrial RNA

Ribosomal RNA.—The known properties of mitochondrial rRNAs of selected fungi are summarized in table 2 (for references see Borst and Grivell, 1971; Mahler, 1973a; Raff and Mahler, 1974). Several interesting features emerge from such a compilation. First, such mitochondrial rRNAs constitute a discrete family, distinct not only from the cytoribosomal RNAs in the same cell, but also from bacterial rRNA on the basis of size, base composition, methylation (1.4 per 100 nucleotides in *Neurospora*), and (not shown here) conformation. Second, there is evidently a close correlation between base composition

TABLE 2

MITOCHONDRIAL rRNA OF ASCOMYCETES

Parameter	Yeast	*Neurospora*	*Aspergillus*
Base comp. (% G+C)			
mt DNA [vs. nDNA]	18–20 [36]	39–42 [51]	30
mt rRNA [vs. cyto rRNA] . . .	26 [47]	35–38 [50]	30–32 [52]
methylation		Yes	
Size			
Length (μm)	0.92 + 0.46		0.91 + 0.47
Mass (daltons × 10^6)	1.30 + 0.70	1.30 + 0.70	1.30 + 0.70
Precursor			
Mass (daltons × 10^6)		2.40	

of these mtRNAs and the mtDNA of the same cell. Since these rRNAs constitute mitochondrial gene products encoded in unique sequences (specifically) in that molecule, this observation indicates that such sequences must occupy a significant fraction in it. Third, since the rRNAs of animal mitochondria are quite different from those of fungi in size, base composition (and probably methylation) (as are their respective DNAs with regard to the first two parameters) it is evident that all these mitochondrial entities have undergone quite profound changes in the course of evolution (Raff and Mahler, 1972, 1974). Fourth, none of the mitochondria examined contain the third, small ("5 S," 120 nucleotide-long) ribosomal RNA found in bacteria and readily detectable in the cytoribosomes of all eucaryotic, including fungal, cells. Fifth, the first transcriptional product, constituting the common precursor of the two large rRNAs also appears to be unique: in *Neurospora* Kuriyama and Luck (1974) have shown it to consist of a molecule of $M_r = 2.4 \times 10^6$ (32 S), in contrast to 2.5×10^6 (35 S) for the nuclear precursor of yeast cytosolic rRNA and to 2.1×10^6 (30 S) for the precursor in *E. coli* (Nikolaev et al., 1974).

Mitochondrial tRNAs.—Discrete and distinct species of mitochondrial transfer RNAs have been identified in a number of fungi. In *Saccharomyces* the RNA-DNA hybridization plateau (0.9% of the genome) corresponds to about 20 such molecules (Reijnders and Borst, 1972). In both yeast and *Neurospora* mitochondria at least a dozen or so of them have been unambiguously characterized, either by amino acid acceptance and ligase specificity, or by hybridization of the corresponding aminoacylated species with mtDNA (Epler, 1969; Casey et al., 1972; Rabinowitz et al., 1974). The species so characterized include the chain initiating molecule fMettRNA_F^{Met} (Epler et al., 1970; Halbreich and Rabinowitz, 1971), the formylation of which is brought about by a transformylase restricted entirely to mitochondria (Halbreich and Rabinowitz, 1971; Dawidowicz, 1972).

Mitochondrial mRNA.—Beyond its existence and functions in mitochondrial protein synthesis little is as yet known with respect to mitochondrial mRNA in fungi (see Borst, 1972; Reijnders et al., 1972; Küntzel et al., 1973b). We are still in the dark concerning the details of its molecular organization, such as length of senseful and other (e.g., poly rA—Cooper and Avers [1974]) tracts, and even the nature of the template—in other words whether *any* of the mRNA translated on mitochondrial polysomes (see below) originates in the nucleus. That *most* of it is of intramitochondrial origin is indicated by four lines of evidence:

TABLE 3

EVIDENCE FOR TRANSLATIONAL CONTROL IN TS $^{-}$ 187 AFTER SHIFT-UP

Incorporation of (% of control)	Cytosol			Mitochondria		
	23°	36°		23°	36°	
		0–30 min	30–60 min		0–30 min	30–60 min
Histidine (protein)	(100)	−20	−20	(100)	100	−90
Formate (protein)	(100)	<15	<5	(100)	100	−45
Formate (fMet-puro)				(100)	50	100

Incorporation is expressed as percentage of the control (23°) rate prior to shift-up; the protocol: 60 min at 23°; followed by shift to 36° with two sets of measurements during the first and second 30-minute periods, as indicated. Negative values represent the extent of decay of counts previously incorporated.

1. Its provision to mitochondrial polysomes and its resultant function is resistant to treatments that result in the complete disruption of the analogous processes on the polysomes of the cytosol and vice versa. Some data, bearing out this inference, obtained in our laboratory (Dawidowicz and Mahler, 1973) are summarized in table 3. Here we have used a temperature-sensitive mutant, originally characterized by Hartwell and his collaborators (Hutchison et al., 1969; Hartwell, 1970) to provide a reversible block—at the non-permissive temperature of 36°—in the supply of all *nuclear* transcripts, resulting in the decay of cytosolic polysomes with a $t_{1/2}$ of 23 min, and EtdBr to block transcription (and translation) in *mitochondria* (South and Mahler, 1968; Fukuhara and Kujawa, 1970; Mahler and Perlman, 1971; Kroon et al., 1972). As expressions of functional mRNA we have looked for rapid and linear incorporation of labeled uracil into the RNA of the extramitochondrial and intramitochondrial compartments; the stability there of polysomes carrying nascent polypeptide chains; the ability of the system to support the incorporation of labeled leucine into its proteins and that of formate into nascent chains, or its transfer to puromycin—two reactions quite specific for mitochondria (Mahler et al., 1972; Feldman and Mahler, 1974). These events have been studied under conditions where we have interrupted the supply of the transcripts of (i) nuclear; (ii) both nuclear and mitochondrial; and (iii) mitochondrial origin. We see that under conditions when the flux of nuclear mRNA is disrupted and restored there is response only in the cytosol and not in the mitochondria, whereas the converse holds for the mRNA formed in mitochondria. Both treatments are required in combination to impede the supply of *total* cellular message; but the two are quite specific as to locale of utilization, and neither is exacerbated by the simultaneous presence of the other. The most parsimonious model to account for these observations is one that postulates that, at least under these conditions, all mRNA translated in mitochondria has been transcribed from mtDNA.

2. Particles that resemble C-type RNA viruses in their morphology have been identified in several species of fungi and isolated and characterized by Küntzel et al. (1973a,b) from *abn*(ormal)-1, a cytoplasmic (presumably mitochondrial) mutant with impaired respiration of *Neurospora*. The particles contain single-stranded RNA, complementary to 4.5% of the sequences of either *abn*-1 or wild-type mtDNA and with no homology to the cytosolic or mitochondrial rRNA of either strain. This stable RNA could function as a messenger for *in vitro* protein synthesis, specifying a protein of M_r = 11,000 resem-

bling one of the two proteins of the viral capsid, and known to be synthesized by mitochondria *in vivo*. It thus seems that this RNA constitutes a stable form of mt mRNA, overproduced in the mitochondria of *abn*-1, as a consequence of this lesion.

3. Scragg (1974) has provided some evidence that an RNA transcribed *in vitro* by a purified mitochondrial RNA polymerase from *Saccharomyces* can be used as a mRNA in a heterologous (*E. coli*) translational system to produce polypeptides capable of immunological cross reactions with membrane proteins isolated from yeast mitochondria.

4. The most unambiguous evidence for mitochondrial specification of polypeptide messages would be provided by the demonstration that a mutation in mtDNA produces a lesion in a polypeptide known to have been translated on mitoribosomes (see below). Some preliminary evidence by Criddle's, Somlo's and Griffith's groups (Shannon et al., 1973; Somlo et al., 1974; Griffiths, this volume) indicates that mitochondria, known to carry alleles for oligomycin resistance, contain ATPases with an alteration in their membrane factor, an entity that—in the wild type (Tzagoloff et al., 1973)—utilizes polypeptides synthesized on mitoribosomes (see below). The isolation and characterization of a different mutant class with a lesion bearing on this problem will be described below.

Mitochondrial transcription and polymerase.—We also know little about the details of mitochondrial transcription and its regulation, and in particular the important problem of the size and number of transcriptional units is completely unsolved. However, the DNA-dependent RNA polymerase implicated in this process has been isolated both from yeast (Eccleshall and Criddle, 1974; Scragg, 1974; Wintersberger, 1973) and *Neurospora* (Küntzel and Schäfer, 1971; Schäfer and Küntzel, 1972). The protein from the latter source and the reaction catalyzed by it have been investigated in detail. It is a protein containing a single polypeptide chain with M_r = 64,000, insensitive to α-amanitin, but sensitive to 6 μg/ml of rifampicin. It is thus clearly distinct from the polymerases of both procaryotes and eucaryotic nuclei.

Mitochondrial Translational System

Mitochondrial ribosomes and polysomes.—All protein-synthesizing systems investigated so far, regardless of origin, contain the ribonucleoprotein particles known as ribosomes as the central core of their machinery. Mitochondria are no exception. As shown in table 4 (data

TABLE 4

MITORIBOSOMES AND CYTORIBOSOMES OF ASCOMYCETES

Parameter	Mitoribosomes	Cytoribosomes
Dimensions (Angstroms)	265 × 210	258 × 220
Sedimentation Coeff. (s)		
Ribosome	72 – 78	80
Large subunit	50 – 58	60
Small subunit	35 – 40	40
Density (g/cm^3)		
Ribosome	1.46 – 1.48	1.52
Composition (% RNA)		
Ribosome	46	55
M_r ($\times 10^{-6}$)		
Ribosome	4.16	4.49
Large subunit	2.47	
Small subunit	1.69	
Number of proteins in		
Large subunit	30	
Small subunit	23	
Presence of 5S RNA	no	yes

from Borst and Grivell, 1971; Raff and Mahler, 1974), mitoribosomes of fungal mitochondria, as is characteristic of ribosomes in general, consist of two nonidentical subunits. Each of these contains its own rRNA (already discussed) and a set of associated proteins—one of which, localized in the large subunit, is responsible for the catalysis of the chain-elongating, peptide bond-forming, step in polypeptide synthesis. The various properties summarized in the table set apart the ribosomes of mitochondria not only from those of the cytosol of the same cell but from their procaryotic counterparts as well. In both these systems ribosomes actively engaged in protein synthesis become linearly arrayed on mRNA to form polysomes. Does this attribute also apply to mitoribosomes? The consensus appears to be in the affirmative, although the interpretation of at least some of the experiments reported is obscured by the known tendency of the particles to aggregation, independent of mRNA, caused by association of the nascent polypeptide chains by virtue of their highly hydrophobic nature (Ledermann and Attardi, 1973; Kroon et al., 1974). However, we have described polysomes that do respond by their dissociation either to the removal or disruption of their mRNA, or to the transfer of their nascent chains to puromycin (Dawidowicz and Mahler, 1973; Mahler and Dawidowicz, 1973).

Other components.—In addition to the ribosomes—their RNAs and proteins—the mRNAs, and tRNAs already discussed, mitochondrial protein synthesis also depends critically on four additional groups of enzymes, or at least proteins, designed to fulfill a characteristic function. These consist of a minimum of twenty amino acid-tRNA ligases or synthetases plus the transformylase already described, two or three initiation factors, three elongation and three termination factors. Although not studied in all its details, the pattern established so far indicates that all these mitochondrial proteins, including those intrinsic to their ribosomes, are distinct from their counterparts in the cell sap (reviewed in Raff and Mahler, 1974; Mahler and Raff, 1974) and are probably imported into the mitochondria.

Inhibitors.—It has become axiomatic in recent years to accept and interpret all inhibition data in terms of a postulate due originally to Linnane and his collaborators (Clark-Walker and Linnane, 1966; Huang et al., 1966) that all mitochondrial ribosomes conform to a bacterial "70 S" pattern in sharp contrast to the characteristically eucaryotic "80 S" cytoribosomes of the same cell (table 5). Provided appropriate safeguards are maintained in both their execution and interpretation, experiments using such site-specific inhibitors can be and have been of the greatest utility in probing the mechanism and products of mitochondrial protein synthesis, especially in intact cells. For ascomycetes the following conclusions appear warranted on the basis of evidence currently available (Mahler, 1973a,c; Mahler and Raff, 1974).

1. Those inhibitors of cytoribosomes that bind to their larger (60 S) subunits, such as the glutarimides (e.g., cycloheximide, CHX), emetine and anisomycin, appear specific for this and completely inactive for the mitochondrial mode of polypeptide synthesis. In particular, any labeled (or otherwise identified) product found in mitochondria under conditions of significant reduction of synthesis of cytosolic and other *mitochondrial* products can be considered to have originated in the particles.

2. Conversely, the groups of bacterial inhibitors that bind to their large (50 S) subunits such as chloramphenicol (CAP), mikamycin (vernamycin), carbomycin, and spiramycin appear to be effective and specific inhibitors of mitochondrial protein synthesis both *in vivo* and *in vitro*. Other macrolide antibiotics, such as lincomycin and erythromycin, known to be also highly effective in the bacterial system are potent inhibitors for mitochondrial protein synthesis in yeast but not in *Neurospora*.

TABLE 5

INHIBITORS OF PROTEIN SYNTHESIS

Acting on ribosomal systems of the 80S type	Acting on ribosomal systems of the 70S and 80S types	Acting on ribosomal systems of the 70S type	Acting on mitochondrial ribosomes
Anisomycin[a]	Actinobolin	Aminoglycosides	Aminoglycosides(?)[b]
Diphtheria toxin	Aurintricarboxylic acid	Chloramphenicol	*Chloramphenicol*
Emetine	Blasticidin S	Lincomycin	Lincomycin[c]
Endomycin	Bottromycin A_2	Macrolides	Macrolides
Glutarimide group:	Edeine	Carbomycin	Carbomycin
Actiphenol	Fusidic acid	Erythromycin	Erythromycin[c]
Cycloheximide	Gougerotin	Spiramycin	Spiramycin
Streptimidone	Nucleocidin	Mikamycin	Mikamycin
	Pactamycin	(Vernamycin)	(Vernamycin)
Streptovitacin A	Poly-dextran-sulphate		
Pederin	Puromycin	Siomycin	Siomycin
Phenomycin	Sparsomycin	Thiostrepton	*Thiostrepton*
Tenuazonic acid	Tetracycline group	(Bryamycin)	(Bryamycin)
Tylophora alkaloids:[c]	Chlortetracycline		
Cryptopleurine	Deoxycycline		
Tylocrebrine	Oxytetracycline		
Tylophorine	Tetracycline		

[a] Most generally useful inhibitors are in italics.

[b] Incompletely studied; in yeast neomycin and paromomycin appear to be relatively effective and discriminating mitochondrial inhibitors *in vitro* but are ineffective with animal mitochondria; other common members of this group (e.g. streptomycin, kanamycin) are ineffective even with unicellular eucaryotes.

[c] Ineffective with mitochondria from animals.

Furthermore, although their inhibitory action on bacterial ribosomes—and its absence in strains resistant to these antibiotics—appear to be interpretable in a relatively straightforward manner in terms of their binding to specific proteins of the large subunits, the situation in yeast, at least, appears to be much more complicated. The phenotype patterns resulting from different mutations in the various alleles controlling the resistance to these antibiotics are complex, overlapping, and different from those reported for bacteria. Although it has recently become possible to ascribe three of them—the products of the so-called R_{I}, R_{II}, and R_{III} loci—to alterations in the mitochondrial ribosome (Grivell et al., 1973), it is its rRNA rather than its protein that is the entity affected for mutants at R_{III} and probably at R_{II}; R_{I} mutants are not associated with either rRNA species (Faye et al., 1974).

3. Fusidic acid, which is believed to act by stabilizing the, usually transient, intermediate complex between the large ribosomal subunit, elongation factor G and a guanosine nucleoside diphosphate or triphosphate, is an effective inhibitor in the bacterial and cytoribosomal systems, including those of *Neurospora* and yeast. It is, however, ineffective at least under certain conditions with the mitochondrial system of both these fungi (Grandi et al., 1971; Dawidowicz, 1972). Its utility may also be decreased by virtue of its reported interference with the mitochondrial ATPase (Kroon and Arendtzen, 1972). All these ambiguities are eliminated by the use of thiostrepton or siomycin, two inhibitors that appear effective in both bacteria and mitochondria by complexing their large ribosomal subunits with both elongation factors G and T (for references see Mahler, 1973a, and Haselkorn and Rothman-Denes, 1973).

Products of Mitochondrial Protein Synthesis

Methods used for their identification.—A number of techniques have been employed for this purpose (reviewed in Sager, 1972; Mahler, 1973a; Tzagoloff et al., 1973; Schatz and Mason, 1974). The one most generally applicable, and used with both yeast and *Neurospora*, is to look for the formation of products insensitive to CHX and sensitive to CAP (or euflavine or EtdBr). The yeast system lends itself in addition to two other approaches.

The first is the use of ρ^- and ρ^0 mutants. All these mutants regardless of the severity of the alteration in their mtDNA are deficient in mitochondrial protein synthesis and even polypeptide chain initiation (Mahler et al., 1972; Mahler, Bastos, et al., 1974; Mahler, Feldman, et al., 1974) and exhibit the same phenotype with respect to the proteins retained by the organelle (Perlman and Mahler, 1970). Therefore, none of these proteins can have been synthesized (or in the case of ρ^0 mutants even specified) by the mitochondrial system. Since mitochondria *are* found in these mutants, it is clear that the mitochondrial system is *not* required for the biogenesis of the organelle proper.

The second utilizes formate as a tag for mitochondrially synthesized polypeptides. We have been able to show that not only do such polypeptides carry an N-terminal formyl group on their nacent chains as expected, but—due to the absence of a deformylase or other enzymes capable of cleaving peptide bonds at or close to the N-terminus—retain this label even upon transfer of the completed chain from its site of synthesis to its site of utilization (Mahler et al., 1972; Feldman and Mahler, 1974).

Nature of the products.—The results of the application of these techniques to the problem of the nature of the intramitochondrial products formed by cells, or spheroplasts, *in vivo* are summarized in table 6, which addresses itself to the over-all quantitative and functional aspects of the mitochondrial contribution.

Such studies on mitochondria have now been extended to a direct examination of those proteins for which a mitochondrial contribution had been inferred from this earlier work: these include cytochrome oxidase, which has been studied both in yeast and *Neurospora* (Schatz and Mason, 1974; Ross et al., 1974; Tzagoloff et al., 1974; Werner, 1974; Sebald et al., 1973); a soluble, but oligomycin-sensitive ATPase from yeast (Tzagoloff et al., 1973; Tzagoloff et al., 1974); and cytochrome *b* from *Neurospora* (Weiss and Ziganke, 1974a,b). The results obtained with regard to the subunit composition of these entities and their mode of biosynthesis is summarized in table 7. It should

TABLE 6

PRODUCTS OF MITOCHONDRIAL PROTEIN SYNTHESIS IN ASCOMYCETES

1. Location	Inner membrane only
2. Extent	10 ± 2.5% of total or ~20% of inner membrane
3. Function	Cytochrome *c* oxidase; NADH ($CoQH_2$): cytochrome *c* reductase; oligomycin-sensitive ATPase
4. Entities	Cytochrome aa_3; cytochrome b_{566} (b_T); membrane factors (CF_o)
5. Polypeptides	8–10; M_r ($\times 10^{-3}$): ~ 40(3), ~ 30(3–4), ~ 20, ~ 10

TABLE 7

SUBUNIT COMPOSITION OF ISOLATED MITOCHONDRIAL COMPONENTS
(Molecular weights × 10^{-3})

Cytochrome Oxidase			Oligomycin-sens. ATPase	Cytochrome *b*
Yeast		*Neurospora*	Yeast	*Neurospora*
(a)	(b)	(c)	(a)	(d)
			58.8 (1)	
			54.0 (2)	
40	*40*	*41* (1)	38.5 (3)	
27.3	*33*	*28.5* (2)	31.0 (4)	
25.0	*22*	*21* (3)	*29.0* (5)	*30*
13.8	14.5	16 (4)	*22.0* (6)	
13	12.7	14 (5a)	18.5 (7)	
10.2	12.7	11.5 (5b)	*12*, 12 (8a and b)	
9.5	4.6	10 (6)	*7.5* (9)	

Subunits translated by mitochonrdia are indicated in italics; subunit number in parentheses.
[a] Tzagoloff, Rubin, and Sierra (1973)
[b] Poyton and Schatz (1975)
[c] Sebald, Machleit, and Otto (1973)
[d] Weiss and Ziganke (1974a,b)

be added that recent studies by Weiss (personal communication) suggest the biogenesis of his preparation of cytochrome *b* ($M_r \simeq$ 60,000, with subunits of $M_r \simeq$ 30,000) to be susceptible to inhibition by both CAP and CHX. Therefore, the protein either consists of a mixture of two entities, only one of which is of mitochondrial origin, or it contains two non-identical subunits, one each of cytoplasmic and mitochondrial origin. It thus becomes evident from all these data that the obligatory cooperation between the two systems of gene expression, required for the construction of the organelle, also finds its counterpart at higher levels of resolution including the inner membrane, its constituent enzyme complexes, and the individual proteins into which they can be resolved.

Three questions arise immediately: (1) How many more mitochondrial proteins synthesized by the organelle remain to be identified? (2) What is the function of each? (3) What is the mode of regulation and integration of the two sets of polypeptides in the construction of these proteins by themselves and as integral parts of the inner membrane? Here we will discuss only the first, and defer the other two to the next part of this presentation.

The problem can be approached in both qualitative and quantitative terms. First, a comparison of the data summarized in table 6 with those of table 7 suggests that, providing the various electrophoretically identifiable components are indeed present in roughly equimolar amounts, most, if not all, of them can be accounted for by discrete polypeptide subunits of known proteins. The same conclusion is reached from a quantitative comparison of the total mitochondrial contribution to the inner membrane with its identified components (see table 8). The inference appears inescapable that very few, if any, *majority* proteins, i.e., those present in amounts stoichiometric to the ones already identified and described, remain to be found. This is not to deny the possible existence of other mitochondrial translational products, particularly ones concerned with reactions involving mtDNA or RNA. However, such proteins must be either few in number or present in much lower amounts—which would raise interesting problems concerning regulation either of the transcription of the mRNA responsible for their specification, or of its translation.

As a corollary, then, one would expect that most of the polypeptides involved in the various functions of mitochondrial nucleic acids, including their replication and their participation in protein synthesis can*not* be products of the mitochondrial translational system. Fortu-

TABLE 8

BALANCE SHEET FOR MITOCHONDRIAL TRANSLATIONAL PRODUCTS

Complex	Description	Molecular Weight X 10^{-5}: Total	Molecular Weight X 10^{-5}: Subunits synthesized in mitochondria
I	NADH: CoQ reductase	7.0	
II	Succinate: CoQ reductase	2.0	
III	$CoQH_2$: cyt *c* reductase	2.3	2(?) × 0.3
IV	Cyt *c* oxidase	2 × 1.33	2 × 0.4
			2 × 0.3
			2 × 0.2
V	ATPase complex	3.6	0.3
			2 × 0.2
			4(?) × 0.075
	Total	17.6	3.4 (19%)
	Fraction of inner membrane	50%	9.5%

Based on Mahler (1973a); Schatz and Mason (1974).

Assumption: M_r of cyt *c* oxidase in membrane equals that of solubilized enzyme (2.5 × 10^5), and therefore contains two sets of subunits (M_r = 2.7 × 10^5).

nately there is direct and independent proof of this inference based on the methods generally employed with other mitochondrial proteins. This applies at least to the proteins involved in the replication and transcription of mtDNA, the construction of mitoribosomes, the aminoacid-tRNA ligases, the transformylase and the elongation factors (reviewed in Mahler, 1973a; Raff and Mahler, 1972; 1974).

Identity of polypeptides specified by mtDNA with those translated on mitoribosomes.—We have already briefly addressed the problem of possible import of nuclear mRNA to be translated by mitoribosomes. In addition there is no evidence for the accumulation or deficiency of any additional polypeptides, or other products, in mitochondria (or elsewhere) under conditions when mitochondrial transcription rather than translation is blocked. Although not yet completely excluded, it would therefore appear that import of nuclear mRNA is probably not significant. There is no evidence either, for the, *a priori* even less likely, export of mitochondrial mRNA for translation on cytoribosomes, and we shall not even consider it.

The problem of the possible export of a polypeptide synthesized on mitochondrial polysomes and exported to either the nucleus or the cytosol there to fulfill a specific—most likely regulatory—function needs also to be considered. Although not properly a part of the question defined by the subheading, it is an interesting one, particularly since such a protein would escape detection by any method that restricted itself entirely to an examination of mitochondrial components. In fact, there is some indirect evidence for the existence of such entities in both yeast and *Neurospora,* fulfilling a negative regulatory function ("repressor") in each instance: in yeast the production of cytochrome *c* is increased preferentially in a number of regimes (Sherman and Slonimski, 1964; Lamb and Rojanapo, 1973; Perlman and Mahler, 1974) that interfere with the transcription or translation of mitochondrial entities (in petites, and in the presence of EtdBr, euflavine, or CAP in wild-type); in *Neurospora* it is the mitochondrial DNA-dependent RNA polymerase and protein factors involved in protein synthesis that become affected in this fashion (Barath and Küntzel, 1972).

INTERACTIONS BETWEEN THE TWO SYSTEMS OF GENE EXPRESSION DURING MITOCHONDRIOGENESIS

From the foregoing it is evident that mitochondriogenesis, and even maintenance of supramacromolecular integrity of the organelle, are far from autonomous. Instead, these processes require the simultaneous and cooperative participation of both systems of gene expression present in the eucaryotic cell, that of the mitochondrion as well as the more classical one located in the nucleus and the cytosol. The collaborative nature of this enterprise is evident at all levels of organization from the whole organelle down to the individual protein molecule. Recent work in a number of laboratories has therefore become directed more and more to a new aspect of the problem: the nature and mechanism of these collaborative interactions, and the matter and manner that regulate them. The distant and implicit goal in all this is the eventual reconstruction of the biogenetic sequence *in vitro*. The next sections will describe some of the problems and our own attempts at their resolution.

Interactions in Cytochrome Oxidase

Rationale.—Conceptually, at least, the simplest and most direct approach to studying the nature of the contributions of, and the interactions between, the two systems of gene expression is provided by an examination of various possible functions of their products in a single purified protein, singly, and in association with other components of the membrane. This approach is being pursued in several laboratories, particularly those of Schatz (Schatz et al., 1974) and Tzagoloff (Tzagoloff et al., 1974). In our laboratory we have concentrated our attention on cytochrome oxidase with the aim of resolving a highly purified and active preparation (containing Tween 20 in place of the bulk phospholipid of the mitochondrial enzymes), previously described by us (Shakespeare and Mahler, 1971) into smaller entities retaining enzymatic activity or other defined functional attributes (such as prosthetic groups, etc.) (table 9). When such a preparation (B of table 9) is dissociated and examined on polyacrylamide with SDS we see only six polypeptides in most gel systems (fig. 1) with extrapolated (Ferguson, 1964) masses of 44.0, 37.0, 26.5, 14.2, 11.8, and 10.0 Kdaltons. The seventh polypeptide (mass $\simeq$ 13 Kdaltons) is revealed only in 16% gels. The starting material thus contains seven subunits, and resembles the enzymes described by others (table 7).

S. H. Phan has succeeded in resolving this enzyme by means of chromatography on leucine-agarose columns, which leads to the quantitative removal of the two largest subunits with a substantial retention of heme *a* and Cu, as well as of enzymatic activity (Prepara-

TABLE 9

PROPERTIES OF CYTOCHROME OXIDASE PREPARATIONS

	A	B	C	C	D
Parameter					
Enzymatic activity					
Specific (units[a]/mg)	239 ± 49	431 ± 62	794 ± 30		≤100 (250)[c]
Recovery (%)	(100)	90	73		
Heme a					
Specific (nmoles/mg)	5.2 ± 0.3	7.1 ± 0.2	11.1 ± 0.6	12.5	20.4
Recovery (%)	(100)	89	60	(100)	86
Copper					
Specific (nmoles/mg)	9.1 ± 0.7	11.9 ± 0.4	13.4 ± 1.0	14.2	23.2
Recovery (%)	(100)	81.1	44.6	(100)	85
Subunits					
Number (mass)	7	7	5(≤25[b])		4(≤15[b])

[a] k/min

[b] K daltons

[c] recycled in 0.1% SDS

Preparation A is a partially purified enzyme which was then subjected to further purification by chromatography on Sephadex G200 S either directly (Preparation B) or after prior dissociation on a column of L-leucine-sepharose (Preparation C). Such preparations can then be dissociated further by chromatography on G200 S at pH 8.0 in the presence of ≤ 0.9% SDS to yield Preparation D.

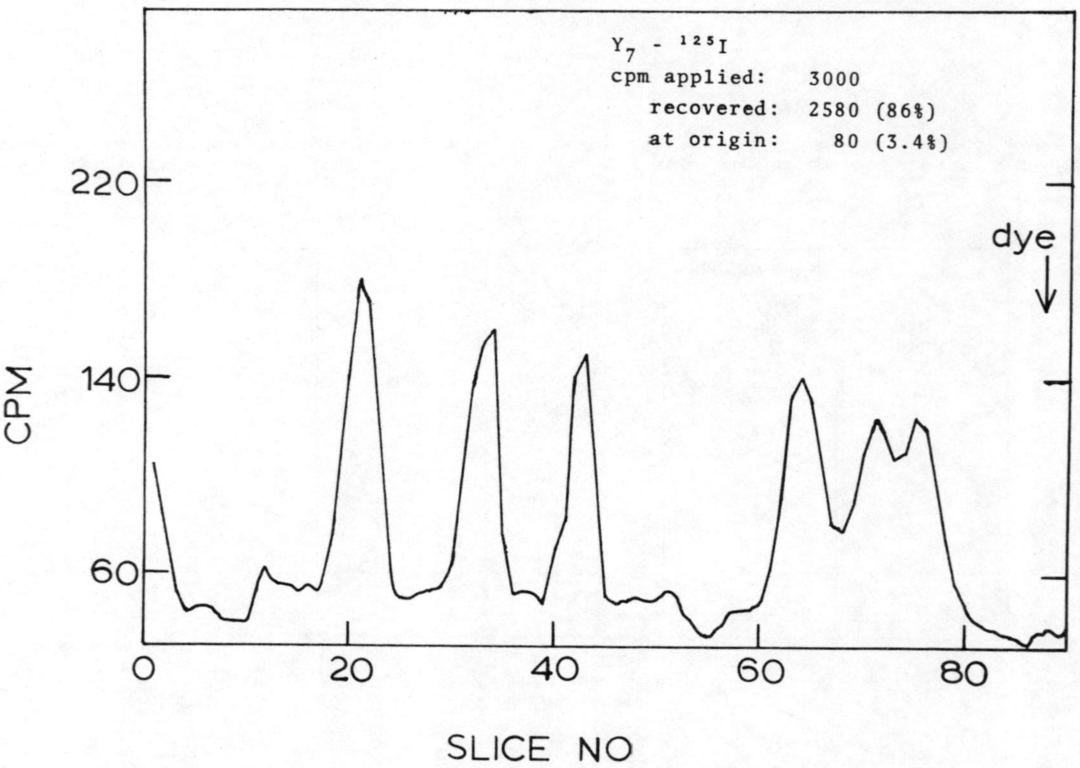

Fig. 1. Subunit composition of a Y_7 preparation of cytochrome *c* oxidase

TABLE 10

RESOLUTION OF PURE CYTOCHROME OXIDASE (TYPE B)

Stage	Protein (mg)	Enzymatic Activity		Heme *a*	
		Specific	Total ($\times 10^{-3}$)	Specific	Total
1. Type B (Y_7)	25.0	789	19.7	8.7	218
2. After Leu-agarose	16.8	882	14.8	12.0	200
3. After Sephadex G200 S, 0.1% SDS, 0.05 *M* tris pH 8.0	13.6	1125	15.3	14.0	190
4. Type C (Y_5)	13.6	972	13.2	14.1	192

For definitions of units and types of preparations, see table 9.

tions C of tables 9 and 10). This removal of the large subunits can be accomplished with enzymes isolated either from yeast or beef heart, and we can start either with a partially (table 9) or highly purified (table 10) enzyme. Its success has been monitored not only by the staining patterns in a number of different gel systems, but also with a radioactive enzyme obtained by iodination *in vitro* with ^{125}I (fig. 2). Further resolution can then be achieved by chromatography on Sephadex G200 S at pH 8.0 in the presence of 0.9–1.0% SDS or by repeated recycling at lower SDS concentrations. Under these conditions subunit 3 is detached and can be separated from the remainder of the molecule. The properties of the resultant preparation are also shown in table 9. These results permit us to conclude that the large (>25 Kdaltons), mitochondrially synthesized, polypeptides are *not* essential for enzymatic activity and that most probably the attachment sites for both prosthetic groups are not furnished by *any* of the polypeptides sharing this mode of biogenesis. There is some confirmatory evidence for this rather startling inference: (1) Mitochondria of ρ^- mutants, which are now known to lack these three polypeptides but to contain all but one (the smallest of mass = 10 K, in table 7) of the cytoplasmically synthesized ones (Ebner, Mennucci, and Schatz, 1973; Ebner, Mason, and Schatz, 1973), also contain normal amounts of copper (Mackler et al., 1965). (2) An antiserum against subunits 1 and 2 kindly provided to us by Dr. Poyton inhibits the seven subunit enzyme to the extent of 82%, while inhibition of the five subunit enzyme under identical conditions is only 25%. The reason for the occurrence of *any* inhibition—and the accompanying immunological cross-reaction—is not yet known but is under investigation.

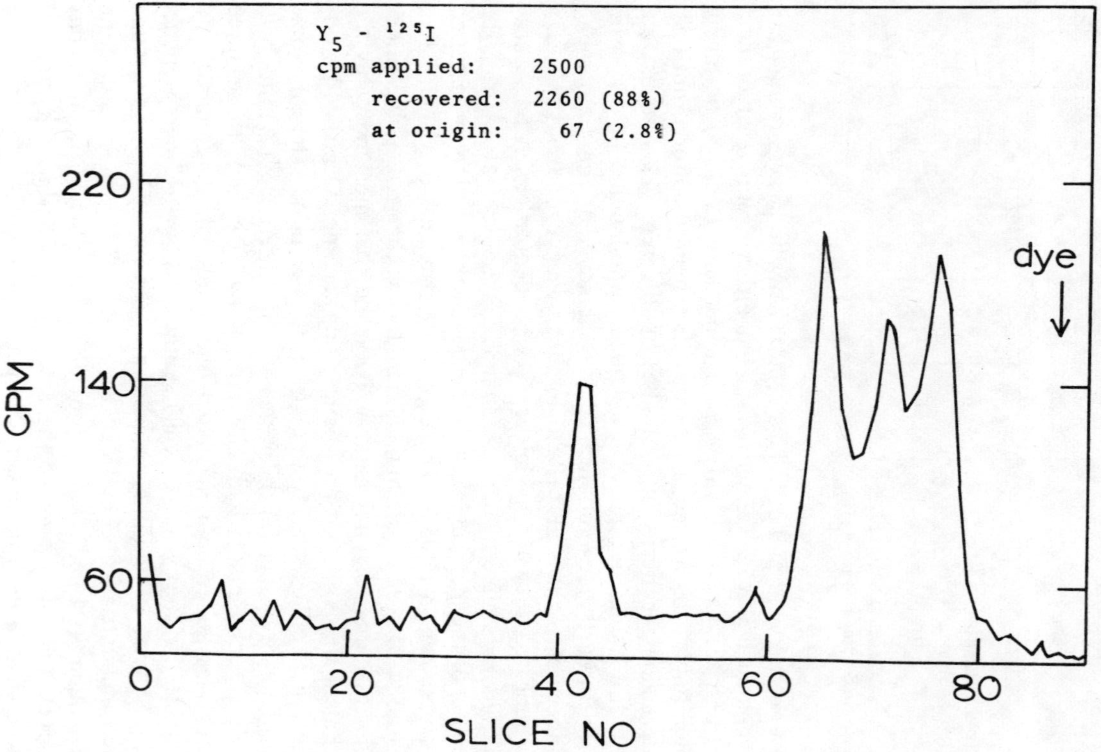

Fig. 2. Subunit composition of a Y_5 preparation of cytochrome *c* oxidase

A New Class of Mitochondrial Mutants

Rationale.—As already briefly discussed, a major handicap in the study of mitochondrial gene expression and its regulation has been a lack of mutants in which an alteration in mtDNA produces a lesion in a mitochondrial—and locally synthesized—polypeptide. We therefore devised a screening paradigm for respiration-deficient (RD) point mutants arising in consequence of pairing errors due to the incorporation of the base analogue 2-aminopurine specificially into mtDNA (Shannon et al., 1973; Mahler and Assimos, unpublished). The procedure used raised the incidence of RD mutants by a factor of 3; of some 400 obtained in this manner we found 12 with a measurable but low reversion rate, which suggested that they were not ρ^- mutants and qualified them for further study. Of them, four in turn exhibited a phenotype sufficiently different from ρ^- to warrant additional exploration, and we have studied one of them, strain 73/1, in some detail.

Genotype and phenotype.—Characteristics of the genotype and phenotype of the new mutants, as exemplified by strain 73/1, are summarized in table 11. The various criteria used for identification of mitochondrial mutations and their application have been described by Sager (1972), Slonimski and his collaborators (Coen et al., 1970; Bolotin et al., 1971), Preer (1971), and Linnane et al. (Linnane and Haslam, 1970; Linnane et al., 1972). Table 11 also provides a comparison with the characteristics of suppressive ρ^- mutants, which the new mutants resemble in their genotype, as well as of certain, respiration-deficient (*pet*) mutants of nuclear specification (Sherman and Slonimski, 1964; Ebner et al., 1973a,b; Goffeau et al., 1973) but analogous phenotype. The results shown form the basis of our conclusions that we are indeed dealing with a novel class of mitochondrial mutants, with a pleiotropic phenotype. An additional feature worth mentioning is the fact that, although the mitochondrial ATPase has been rendered oligomycin-resistant, it is present in normal amounts in mitochondria isolated at 0°. These and other (Bastos and Mahler, 1974b) observations suggest that at least some of the polypeptides of the membrane attachment sites (also called membrane factors, CF_0 or base piece) are still present, but perhaps in altered form.

Search for altered polypeptides.—The table also indicates that—again unlike all ρ^- mutants examined—mitochondria of mutant 73/1 have retained the capacity for protein synthesis *in vivo:* they are

TABLE 11

COMPARISON OF NEW RESPIRATION-DEFICIENT (RD) MUTANTS WITH *PET* AND SUPPRESSIVE PETITE MUTANTS

	MUTANT		
	73/1	Suppressive ρ^-	*pet*
Genotype			
Complementation with ρ^0 [a]	No	No	Yes
Meiotic segregation (tetrad analysis)	0:4	0:4	2:2
Mitotic segregation of RS diploids	Yes	Yes	No
Mitotic segregation of RS diploids after conversion to ρ^0 [b]	No	No	No
Phenotype			
Functional respiratory chain	No	No	No
Cytochrome oxidase	Absent	Absent	Absent
Cytochrome aa_3	Absent	Absent	Absent
NADH: cytochrome *c* reductase	Low[c]	Absent	Low[c]
Cytochrome *b*	Low[c]	Absent	Low[c]
Oligomycin sensitivity of ATPase	No	No	No (or yes)
Mitochondrial protein synthesis	Yes	No	Yes
Petite phenocopy by CAP[d]	Yes	No	Not done

RS = respiration sufficient
[a]No mtDNA
[b]By prolonged growth in presence of ethidium bromide
[c]<0.1 of wild-type levels
[d]Formed reversibly by prolonged growth in presence of chloramphenieol.

capable of forming formylmethionyl puromycin; of incorporating leucine into protein in the presence of CHX by a reaction sensitive at least in part to CAP; and of losing some of their components on prolonged exposure to this inhibitor (fig. 3). We have therefore begun a search for the polypeptide products of protein synthesis in mutant mitochondria in order to compare them with their wild-type counterparts. Our hope is to be able to distinguish between three alternatives: (1) That the mutation has resulted in the loss or alteration of a polypeptide already identified previously as a mitochondrial translational product (tables 6 and 8); the pleiotropism would then be referrable, presumably, to a regulatory, or integrative function of this component. (2) That the mutation is in a gene affecting a membrane polypeptide not hitherto identified as a component of the three enzyme complexes known to harbor mitochondrial translational products. (3) That the entity affected by the mutation is a component of the protein synthesizing *machinery* (e.g., a mitochondrial rRNA

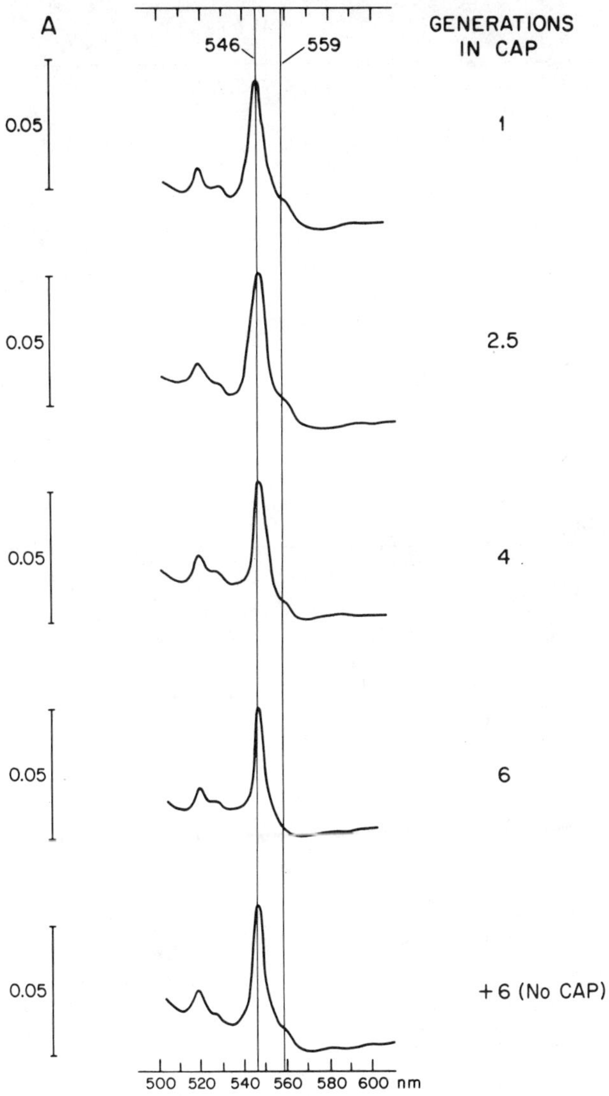

Fig. 3. Elimination of cytochromes *b* and *c* during growth in the presence of chloramphenicol. Cells of strain 73/1 were grown in YPGal medium containing 4 mg chloramphenicol/ml. Cells were chilled, harvested, and washed twice with cold distilled water before suspending in a concentration of 300 mg/ml in 30% glycerol, 50 *mM* K_2HPO_4 pH 7.4 and a pinch of dithionite added. Samples were frozen in liquid nitrogen prior to recording spectra. The spectrum at the bottom was recorded after incubation of the CAP-treated cells for six generations in fresh YPGal medium without inhibitor. The vertical lines at 546 and 559 nm correspond to the absorbance maxima of reduced cyt *c* and *b*, respectively.

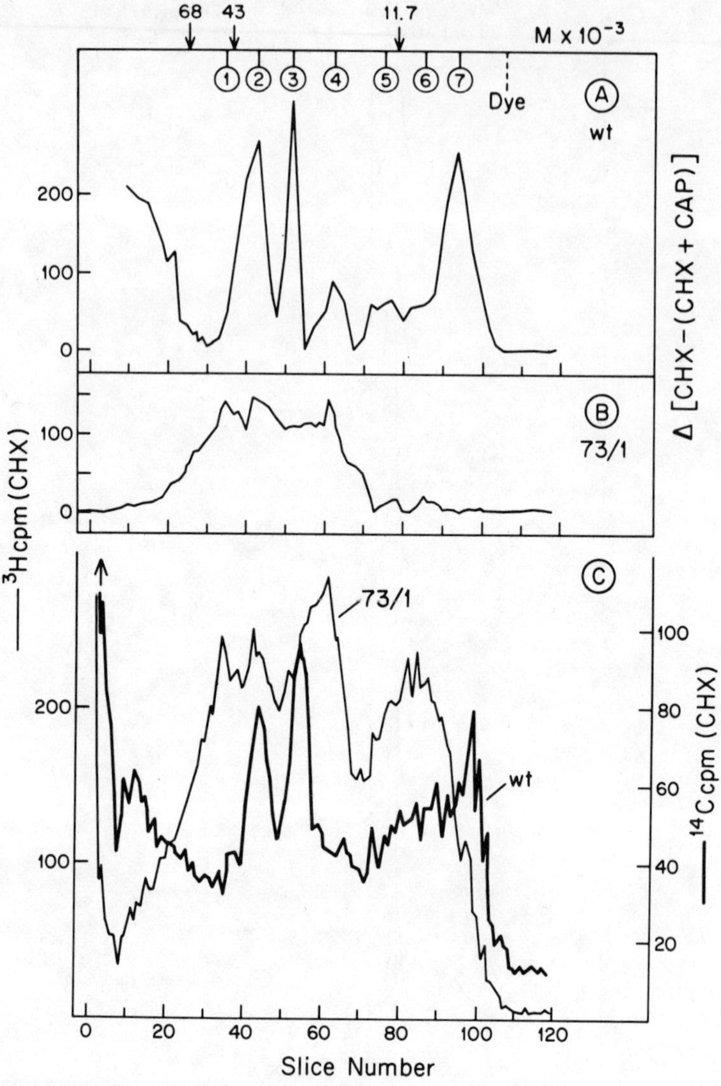

Fig. 4. Altered products of mitochondrial protein synthesis in strain 73/1. Incorporation of labeled leucine in the presence of CHX by whole cells; mitochondria were isolated, proteins solubilized with deoxycholate, dissociated and analyzed by electrophoresis in 10% acrylamide gels in the presence of sodium dodecyl sulfate; data are cpm found in 1 mm slices. *A*, CAP-sensitive incorporation by wild type (CHX alone minus CHX+CAP); *B*, Same as *A* but using strain 73/1; *C*, double-labeling experiment: wild type (wt) labeled with (^{14}C)-leucine and 73/1 with (^{3}H)-leucine, both in the presence of CHX, mixed and analyzed as described.

as in *poky Neurospora* [Kuriyama and Luck, 1974]); which, in turn, leads to changes in the quantitative aspects of translation or even the introduction of errors during this process.

As a preliminary step in this direction, Drs. Flury and Feldman have begun to examine—by means of double-labeling experiments—the total products of the CHX-resistant, CAP-sensitive incorporation on SDS-polyacrylamide gels. These experiments are hampered by the fact that both in the wild type, and especially in the mutant, not all the counts resistant to CHX are CAP-sensitive. Therefore, the patterns obtained are not as sharp as they might be (fig. 4). However, it is evident that the mutant does carry out mitochondrial protein synthesis and that its products appear quite different from those found in wild-type organelles. We are intrigued by three sets of observations. First, that the mutant pattern appears considerably more diffuse than the wild-type one; whether this effect is due to the "leakiness" already referred to or has a less trivial explanation (translational errors?) will have to be established. Proteolysis during isolation and analysis would appear to be ruled out since the pattern is retained when the two preparations, labeled with different isotopes, are mixed prior to isolation. Second, that, superimposed on this noise, there appears one major set of changes, a great enhancement of incorporation into a longer polypeptide(s) (No. 1) at the apparent expense of the smallest (No. 7). This behavior is reminiscent of the interconversions in size of a highly hydrophobic subunit (subunit 9, table 7) of the oligomycin-sensitive ATPase described by Tzagoloff and his collaborators (Tzagoloff and Akai, 1972; Sierra and Tzagoloff, 1973). Finally, the diazido derivative of EtdBr—after its photochemical conversion to the dinitrene—appears to be a highly selective affinity reagent for a single mitochondrial component. This component, which in the mutant has an apparent $M_r = 8{,}000$, is virtually absent in the mutant, and in its stead we find a large, labeled polypeptide in the $M_r = 40{,}000$ region. The same reactions can also be obtained with a soluble oligomycin-sensitive ATPase, and the product of both resembles subunit 9 in its electrophoretic and solubility properties.

Interactions during Mutagenesis

Rationale.—The intercalating dye EtdBr is an exquisitely sensitive probe for a variety of mitochondrial functions. When added in micromolar concentrations to cells of *S. cerevisiae* either in growth medium or buffer, it is capable of converting such a population quantitatively into progenitors of ρ^- clones (Slonimski et al., 1968; Mahler et al., 1971; Mahler and Perlman, 1972; Mahler, 1973b). Such exposure—even of a transient nature—is now known to result in the deletion of variable but large segments of their mtDNA including its totality (see above, the DNA of ρ^0 mutants). Exposure of cells to the agent under mutagenic conditions is known to (1) affect both the structure (Gitler et al., 1969; Azzi and Santato, 1971) and organization (Kellerman et al., 1969; Mahler et al., 1971) of the mitochondrial inner membrane; (2) inhibit the synthesis of mtRNA (South and Mahler, 1968; Fukuhara and Kujawa, 1970; Richardson, 1973); and (3) inhibit the synthesis of cytochromes aa_3, b, and c_1 in that order (Kellerman et al., 1969; Mahler et al., 1971; Mahler and Perlman, 1972). On the level of mtDNA itself such exposure is known to (4) lead to intercalation of the dye between adjacent base pairs, resulting in conformational distortion that produces unwinding of covalently circular, supercoiled molecules (Waring, 1965, 1970; Bauer and Vinograd, 1971; Paoletti and LePecq, 1971; Denhard and Kato, 1973; Pigram et al., 1973); (5) inhibit the replication of the molecule (Goldring et al., 1970; Perlman and Mahler, 1971); (6) transient exposure of mtDNA in yeast cells results in a double-strand scission somewhere near the center of the molecule producing fragments that after isolation exhibit a molecular weight about 12.5×10^6, and which (7) upon exposure of the cells to an energy source are further degraded to ever smaller pieces until they become undetectable as discrete entities (Goldring et al., 1970; Perlman and Mahler, 1971; Mahler, 1973b). Some phases of this sequence must be reversible, or at least repairable, for it is known that a number of agents or treatments, either before, during, or subsequent to, the exposure to EtdBr can overcome its mutagenic action or rather its expression (Mahler, 1973a). Additional details will be found in Dr. Perlman's presentation (Perlman, this volume).

We decided to probe this remarkable specificity by means of radioactive EtdBr and followed its uptake by, and distribution within, cells under mutagenic conditions. We found that (1) cells or spheroplasts were able to take up EtdBr against a concentration gradient; (2) $\geq 90\%$ of the total intracellular EtdBr was found in the mitochondria (these two observations, suggesting a much higher effective concentration of the molecule inside the mitochondria, already go far in accounting for its apparent specificity); (3) upon separation of mtDNA from nDNA by various techniques, EtdBr remained associated exclusively with the former; and (4) of the intramitochondrial EtdBr about 10% appeared to be bound to mtDNA in a novel, particularly strong, and probably covalent fashion. The evidence for this last point (Mahler and Bastos, 1974) is that (1) the [^{3}H]-EtdBr associated with mtDNA either on sucrose (fig. 5A) or CsCl gradients was not removed either by digestion with NaOH or precipitation with trichloracetic acid; (2) various treatments designed for—and shown in *in vitro* experiments to be capable of—the removal of intercalated EtdBr from DNA (whether nuclear *or* mitochondrial) such as solvent extraction, exposure to certain resins, chromatography on polylysine-Kieselguhr columns, all failed to do so on the product obtained from exposed cells, but did so readily when the labeled EtdBr was added to a protoplast lysate (fig. 5B); (3) the product was not formed by cells either of a DNA^{0} petite or of strains known to be resistant to the action of EtdBr (i.e., petite negative yeasts); (4) after complete enzymatic digestion the EtdBr-containing mtDNA was degraded to a product that, upon analysis by thin-layer chromatography or high-voltage electrophoresis, differed from the four standard bases and, based on its absorbance spectrum, contained EtdBr in a modified form.

Reconstruction of in vivo sequence with isolated mitochondria—We next asked (Bastos and Mahler, 1974a,b) whether isolated purified mitochondria were also capable of performing this reaction and found (fig. 5C) that they were indeed able to do so. These data show that unmodified mtDNA co-sediments with the DNA of bacteriophage T7 ($s = 32.0$ S; $M_r = 25 \times 10^6$ [Thomas and McHattie, 1967]) and that upon modification it also becomes degraded to molecules with a distribution about one-half that size. We also compared both the kinetics and the stoichiometry of the reaction which may be formulated as

$$\text{mtDNA} + m\text{EtdBr} \rightarrow \text{mtDNA}' \cdot \text{EtdBr}_m \qquad (1)$$

and found them to be virtually identical whether we used cells or mitochondria isolated from them. Incidentally, in the former case, except for an initial lag of 10 min, the kinetic course of mutagenesis also coincides with that of reaction (1). The value of m, which may be expressed as the number of nucleotides for each molecule of EtdBr in the modification product, equals 110 ± 10 for several different haploid wild-type strains, both *in vivo* and *in vitro*. The value of m for isogenic diploids is the same, but since these strains contain twice as much mtDNA the absolute *amount* per cell in diploids is twice that in haploids. Reaction (1) does not require any external

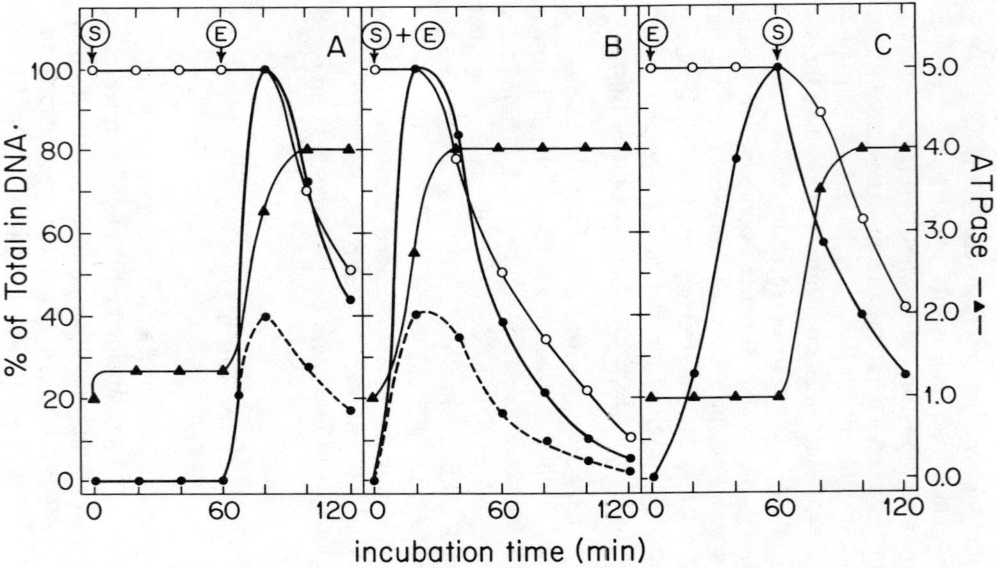

Fig. 5. Kinetics of mitochondrial reactions elicited by EtdBr. The designs for *A*, *B*, and *C* were similar except for the time of addition of the components succinate (S) and EtdBr (E), which were as indicated. The reactions involving modification and labeling of [^{14}C]-prelabeled DNA by [^{3}H]-EtdBr (5 μCi in 2 ml) and the measurement of the induction of ATPase are described in Mahler and Bastos (1974) and Bastos and Mahler (1974b); the EtdBr was provided by the labeled samples. [^{14}C]-DNA is represented by the open circles; [^{3}H]-EtdBr in DNA by the solid circles; and ATPase by the solid triangles. The values corresponding to the base line (or 100%) were 2.8×10^3 dpm for [^{14}C] in *A*, *B*, and *C*, and 7.1×10^3 for [^{3}H] in *C*, for the 2 ml of incubation mixture employed; ATPase activity was 4 μmoles P_i liberated $\times$ 10 $min^{-1} \times mg^{-1}$ protein. The dashed line in *A* and *B* sets the maximal level of EtdBr incorporated into DNA as determined in *C* as 100%, while the solid line connecting the solid circles uses the maximal level obtained in the respective experiment as 100%.

energy source, nor is it dependent on the intramitochondrial ATP present; it is unaffected by the specific inhibitors of the terminal ATP synthetase (or ATPase) oligomycin and Dio 9, but is inhibited completely by EDTA, and in part by lipophilic uncouplers of oxidative phosphorylation such as dinitrophenol or CCCP.

Although the modification product is stable to prolonged incubation ($\geq$2 h) of the mitochondria in the medium in which it was formed, the addition of an energy source (either ATP or succinate) results in its rapid degradation (fig. 6). Coincident with this degradation there is an *activation* of an endogenous ATPase, also shown in the figure. Therefore, isolated mitochondria are also able to catalyze the following reactions:

$$[\mathrm{EtdBr}_m \cdot \mathrm{DNA}'] + (\mathrm{ATP}) \xrightarrow{\text{nuclease(s)}} \text{fragments} + \ldots \qquad (2)$$

$$\mathrm{mtDNA} + m\mathrm{EtdBr} + (\mathrm{ATP}) \rightarrow \text{fragments} + \ldots \qquad (3) = (1) + (2)$$

$$n\mathrm{ATP} \xrightarrow{[\mathrm{EtdBr}_m \cdot \mathrm{DNA}']} n\mathrm{ADP} + n\mathrm{P_i} \qquad (4)$$

Results with a number of (1) respiratory inhibitors and uncouplers, and (2) mutants of defined phenotype, affecting either enzymes of the cellular system for damage repair and recombination or the mitochondrial ATPase, have convinced us that there exists a close and perhaps obligatory link between the two most important sets of mitochondrial functions: the ones concerned with energy coupling and those concerned with the genetic functions of mtDNA. Some of the results on which we base this inference are summarized in tables 12 and 13. These tables also include results with different modulators (Mahler, 1973a), especially euflavine (see also Perlman, this volume), and with petite negative strains. It is clear also (cf. table 13) that we now have available mutants that are blocked specifically in any one of the three component reactions (1), (2), and (4), and by their use can expect to gain some insight into the nature of the events and of the proteins responsible.

Attempts at further reconstitution.—Although the experiments just described suggest that isolated mitochondria are capable of performing the initial reactions responsible for mutagenesis *in vivo* and can be used to analyze their sequence, kinetics, and stoichiometry, an under-

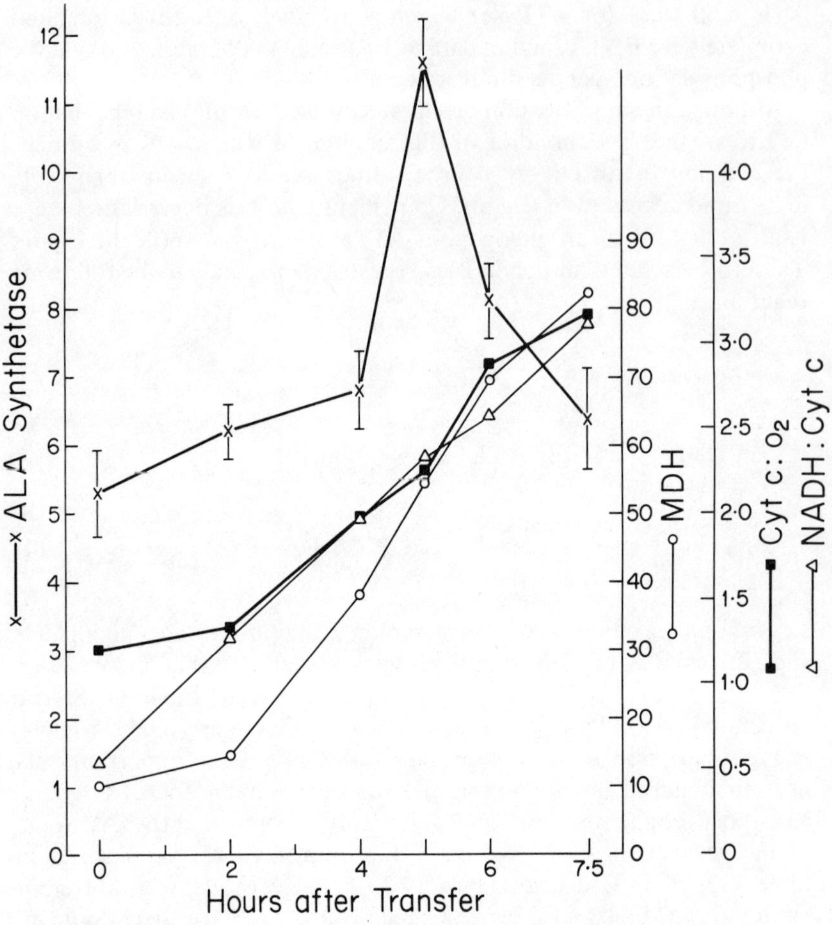

Fig. 6. Derepression of various mitochondrial activities. Cells of strain A364A were grown on a repressing medium (1% yeast extract, 2% peptone, 2% glucose) and harvested in exponential phase ($A_{600} = 0.4$), washed, concentrated 4 × and resuspended in a derepression medium consisting of YM1 with 0.25% glucose plus 3% ethanol. Samples were removed at the indicated times, mitochondria isolated (Mahler and Dawidowicz, 1973), and enzymatic activities determined as described by Perlman and Mahler (1974) except for ALA Synthetase, which was measured in triplicate, essentially by the method of Marver et al. (1966); means and standard deviations are indicated.

TABLE 12

EFFECTS OF VARIOUS INHIBITORS ON MITOCHONDRIAL REACTIONS

Agent or Treatment		Effects			
		On Reaction 1[a]	On Reaction 2		On Reaction 4
Class	Representative		With succinate	With ATP	With succinate
	EDTA	Complete inhibition	[b]	[b]	[b]
Inhibitors of ATP/ADP transport	Atractyloside	None	None	Inhibits	Inhibits
Inhibitors of ATPase	Oligomycin	None	Inhibits	Inhibits	Inhibits
	Dio 9	None	Inhibits	Inhibits	Inhibits
Uncouplers	Arsenate	n.d.	n.d.	Inhibits	[c]
	Dinitrophenol	Inhibits	Inhibits	Inhibits	n.d.
	CCCP, FCCCP, 1199	Inhibits	Inhibits	Inhibits	[c]
	Colicin K	Inhibits	Inhibits	Inhibits	[c]
Respiratory Inhibitors	Antimycin A	None	Inhibits	None	Inhibits
	Malonate	None	Inhibits	None	Inhibits
Modulators of Mutagenesis	Euflavine	None	n.d.	Complete inhibition	Complete inhibition
	Galactose/Glucose	Inhibits	None	n.d.	None
	Caffeine	None	None	n.d.	None
	Cycloheximide	None	None	n.d.	None

n.d. = not done

[a]See text for reactions.

[b]Since reaction 1 is required for reactions 2 and 4, its absence necessarily precludes any effects thereon; whenever we have checked whether reaction 4 shows any correlation with inhibitors of reaction 1, we have found it.

[c]Activation independent of, and competitive with, presence of covalent modification product.

TABLE 13

EFFECTS OF VARIOUS MUTATIONS ON MITOCHONDRIAL REACTIONS

Strain	Class	Phenotype	Effects (Relative to Wild-Type)		
			On Reaction 1[a]	On Reaction 2[a]	On Reaction 4[a]
O^R–4	*oli*-r	Oligomycin resistant ATPase	Normal	Normal in rate but oligomycin resistant	Normal in extent but oligomycin resistant
73/1	*res*$^-$	Respiration-deficient, but not ρ^-; oligomycin resistant ATPase	Lowered in extent but not in rate	Normal in rate but oligomycin resistant	Reduced in level and oligomycin resistant
uvs ρ5	*uvs* ρ	ρ^--mutation-prone for *uv*, resistant to EtdBr	Normal	Absent	Absent
uvs ρ72	*uvs* ρ	ρ^--mutation-prone for *uv* and EtdBr	Increased in extent	Accelerated	Increased in extent
2C4	*rec 4*	Recombination deficient, EtdBr resistant	Decreased in extent but not in rate	Normal	Normal
rad 6	*rad*	X-ray, uv-sensitive, EtdBr resistant	Very low	Very low	Very low
Hansenula wingei	petite$^-$	No stable mutants by EtdBr	Absent	Absent	Absent
Torulopsis utilis	petite$^-$	No stable mutants by EtdBr	Absent	Absent	Absent
Kluyveromyces lactis	petite$^-$	No stable mutants by EtdBr	Absent	Absent	Absent

[a]See text for reactions.

standing of the molecular events responsible can come only from investigations on better-defined systems. As a first step to accomplish this aim, we wanted to see whether we could obtain the degradation of the modification product containing [^{3}H]-EtdBr in a soluble system. As shown in table 14, this has been accomplished with a combination of the soluble oligomycin-sensitive ATPase purified by the method of Tzagoloff and Meagher (1971), and a crude mitochondrial extract eluted from a DNA-cellulose column, supplemented with ATP. The response to omission of these components and to the specific inhibitors of the ATPase (Dio 9 and oligomycin) as well as its resistance to the uncouplers (CCCP and Colicin K) are those expected for a functional rather than adventitious association between them. We are therefore now in a position to inquire into the nature and the properties of the mitochondrial protein(s) contained in the DNA-binding fraction and the response when the various components are isolated from mutant strains.

Interactions during Derepression

Rationale.—Facultatively anaerobic, petite positive yeasts such as *S. cerevisiae* are remarkably versatile organisms, capable of growth on a variety of carbon sources and using either glycolysis or respiration as their primary mode of carbon dissimilation and energy generation.

TABLE 14

CHARACTERISTICS OF RECONSTRUCTED SYSTEM FOR DEGRATION OF MTDNA · [^{3}H] ETDBR

System	Fraction of Counts Released (%)	Inhibition (%)
Complete[a]	70	
Omit		
ATPase	2	
DNA-binding protein	2	
All protein	2	
ATP	7	
Add		
CCCP (10^{-5} M)	52	16
Colicin K (10^{-6} M)	61	13
Oligomycin (20 μg/ml)	17	76
Dio 9 (20 μg/ml)	12	83

[a] 1 μCi of modification product, 0.5 mg of oligomycin-sensitive ATPase, 3.7 mg of DNA-binding protein in 2.0 ml of 0.1 *M* sorbitol, 100 *mM* KH_2PO_4, 100 *mM* NH_4Cl and 10 *mM* TES; incubated for 60 min at 30° C and TCA-precipitable counts determined.

Under appropriate conditions the transitions between these different modes can be shown to be both reversible and not to require duplication or even proliferation on the cellular level (for reviews and references see Linnane et al., 1972; Mahler, 1973a; Perlman and Mahler, 1974). These observations suggest that the yeast cell must contain whole batteries of regulatory devices capable of adjusting the intracellular milieu so as to utilize the changing extracellular one to its maximal advantage. To a first approximation, the systems for the catalysis of glycolysis and its accessories such as the glyoxylate cycle are localized outside—while those concerned with respiration are sequestered inside—the mitochondria. This consideration suggests that any such regulation must not only be exercised on the cell's mitochondria and all its other compartments, but also must make certain that the sequence and extent of any controlling features be imposed on these constituents in synchrony. It is perhaps in considerations such as these that one may search for the teleological significance of the retention of some few selected polypeptides encoded in mtDNA, to be expressed within the organelle when full mitochondrial expression becomes essential, while all its other constituents are produced elsewhere and only their integration and assembly into a fully functional organelle is performed locally. These transitions should therefore provide an excellent means for the investigation of possible integrative and regulatory interactions in mitochondrial biogenesis.

Relation to the cell cycle.—One of the most fundamental questions that may be posed is whether, in fact, as earlier experiments had already suggested, mitochondrial development could be dissociated completely from other cellular events. For instance, we (Grimes et al., 1974) knew already that mitochondrial mass and number per cell and, hence, mitochondrial duplication were completely independent of the cell cycle. What about mitochondrial *differentiation* to produce a fully functional respiratory organelle? To this end we have devised a simple system for the study of release from catabolite repression (derepression) in the absence of any significant increases in cell number or mass (Mahler, Bastos, et al., 1974; Mahler, Feldman, et al., 1974). It consists of repressed cells in exponential growth on 5% glucose, harvested and re-suspended after fourfold concentration in a simple derepression medium containing 0.25% glucose and 3% ethanol. Such cells (table 15, first column) show significant increases in all the parameters previously employed by us (Perlman and Mahler, 1974) to define derepression under more physiological conditions. However,

TABLE 15

DEREPRESSION OF ENZYMATIC ACTIVITIES IN CELL CYCLE MUTANTS

ACTIVITY	INCREASE (Strain, temp.)			
	w.t. 30°	w.t. 36°	*cdc 28*, 38°	*ts 135*, 36°
GDH	6.9	2.7	3.6	3.8
MDH	3.3	1.9	4.2	2.8
succ:c	10.0	2.0	14.0	4.4
NADH:c	5.0	4.1	7.5	6.6
$c:O_2$	2.2	1.9	1.3	1.5
$NADH:O_2$	5.0	7.5	5.2	5.7
Cell dry weight	1.5	2.0	1.10	1.15
Protein	1.0	1.3	1.27	1.20

All values represent $(\text{activity})_t/(\text{activity})_0$ where t = 7.5 h after transfer of mid-exponential (A_{600} = 0.4) cells from 5% glucose to 0.25% glucose-3% ethanol at A_{600} = 2.0. In general, increase first becomes detectable at t = 1 h, and is complete by t = 6h. *cdc 28* is a temperature-sensitive "start," *ts 135* a DNA initiation mutant (Hartwell et al., 1974); *ts 314*, a DNA synthesis mutant of somewhat different phenotype, gives a similar pattern.

except for an increase of cellular proteins and dry weight of <30% during the first 90 min, this form of derepression occurs in the complete absence of any augmentation of cell number, size, mass, or other evidence of proliferation; the number of cells with buds remains constant at <0.5%. The strain employed (A364A) was chosen on purpose since it is the wild type for a large number of temperature-sensitive mutants originally selected and characterized by Hartwell and his colleagues (Hartwell, 1967; Hartwell and McLaughlin, 1968; Hartwell et al., 1970) and used by us (Mahler and Dawidowicz, 1973; Feldman and Mahler, 1974) in some of our earlier and ongoing investigations. Of particular relevance is the availability of such mutants (*cdc* mutants) blocked, at the non-permissive temperature, in a number of functions that regulate the cell division cycle (Hartwell et al., 1974). Some of the *cdc* mutants are incapable of entering S phase by virtue of an inability to initiate DNA synthesis (Hartwell, 1973), while others are believed to even lack the ability to enter G_1 and hence are blocked at the very start of the cycle. Clearly such mutants should be of the greatest utility in determining whether derepression is dependent on the duplication of nDNA or of *any* other cellular function linked to the cell cycle. The data in the table show that the answer is clearly in the negative. Indications to this effect had already been obtained previously by the use of hydroxyurea

(Slater, 1973) as an inhibitor of DNA synthesis (Mahler, Bastos, et al., 1974). These results appear to rule out any model which postulates that derepression is in *any* way dependent on the replication of any part of the nuclear genome, for instance, to render it competent for transcription by unmasking from, or forming sequences not subject to, a block by a resident repressor (Barath and Küntzel, 1972). Since the content of mtDNA of such cells undergoes no substantial change during derepression, such a model would appear to be ruled out for mtDNA as well. Nevertheless, the presence of multiple copies of mtDNA per cell, and the relative inaccuracy of the methods employed so far can not exclude the possibility that derepression requires the prior duplication of a small segment of the molecule (Perlman and Mahler, 1971); perhaps specifying a positive control element. However, previous studies had already eliminated the possibility that a mitochondrial transcriptional or translational product was required for the derepression at least of those polypeptides translated on cytoribosomes (Perlman and Mahler, 1974).

Regulation of ALA synthetase levels.—Delta-aminolevulinate (ALA) synthetase is an interesting mitochondrial protein. As the enzyme catalyzing the first committed step in heme biosynthesis one would expect it to be subject to, and participate in, a variety of regulatory interactions, and this expectation appears to be borne out in a number of instances. In mammals activity of the enzyme controls the level of a number of cytochromes; it exhibits a very short half-life and its synthesis is subject to repression by heme and induction by a variety of agents (Meyer and Schmid, 1973; Tschudy and Bonkowsky, 1972). In yeast at least three sets of interactions have been reported: (1) several mutants appear to be pleiotropically deficient both in heme and lipid synthesis (Bard et al., 1974; Gollub et al., 1974); (2) although present in ρ^- mutants (which produce cytochrome *c*) the simultaneous presence of a recessive nuclear mutation—which by itself reduces but does not abolish heme synthesis in wild type—leads to a complete abolition of all heme and cytochrome synthesis (Sander, Mied, et al., 1973); (3) there is an indication of anomalous pattern of biosynthesis of the enzyme during derepression (Jayaraman et al., 1971). We have therefore begun an investigation of the biosynthesis of the enzyme during derepression under the conditions outlined above and its response to CHX and CAP. Our results so far are shown in figures 7 and 8.

The cyclical pattern of biogenesis exhibited by the enzyme is quite

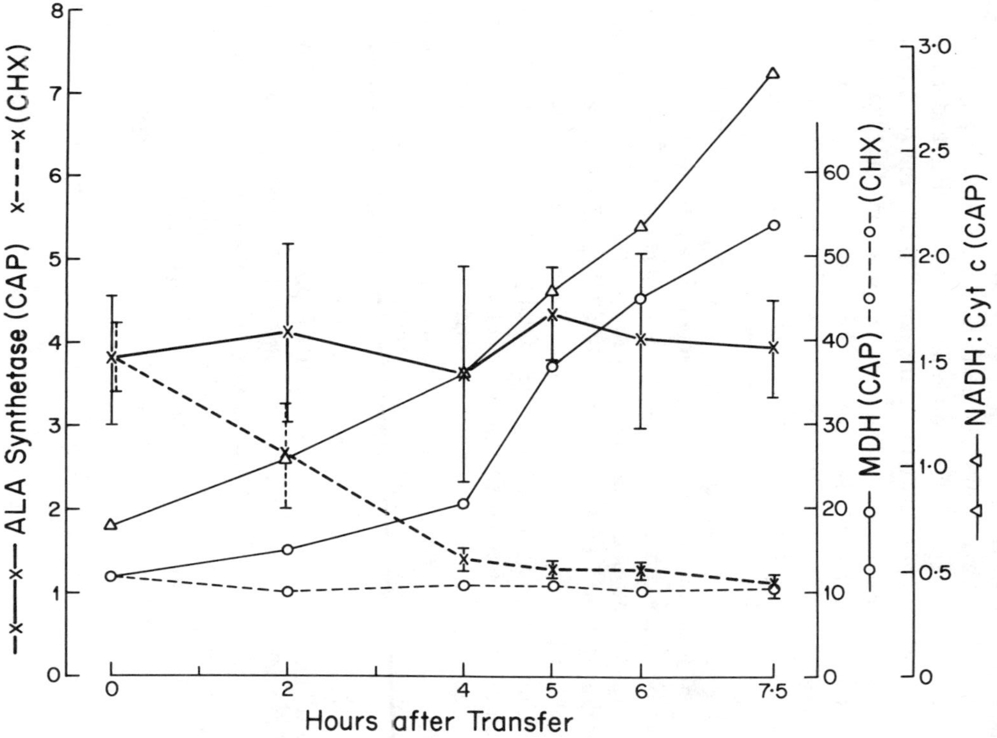

Fig. 7. Derepression of mitochondrial activities in the presence of inhibitors. Experimental procedure was similar to that described in the legend to fig. 6 except for the addition of CAP (4 mg/ml) or CHX (10 μg/ml) where indicated.

unusual and is not shared by any of the derepressible enzymes—whether located inside or outside the mitochondria—previously studied in our or other laboratories. The mitochondrial L-malate dehydrogenase (MDH), a matrix enzyme, and NADH:cytochome *c* reductase (NADH:*c*), a characteristic activity of the inner membrane, are shown as examples. The ALA synthetase pattern is, however, highly reproducible in both its qualitative and quantitative (timing and extent) aspects. The effects of the two inhibitors added at the time of initiation of derepression are demonstrated in figure 8. Addition of CAP prevents derepression while that of CHX leads to the destruction or inactivation of preexisting enzyme with an apparent half-life of 3 h. Furthermore (not shown), when CAP is present throughout the prior growth under repressing conditions but removed at the time of transfer, there is *no* effect on the base-line level, and the cyclic pattern is conserved but with a delay of one hour. When CAP is added at the time of maximal derepression (t = 5 h) there is no significant alteration in the kinetics of the decay of the enzyme; it drops to the base-line level within ≤2.5 h. However, addition of CHX produces an enhancement in both the rate and extent of this decay. Finally, when either of the inhibitors is added during the initial 2.5 h after transfer and then removed, there is no derepression of ALA synthetase, at least not during the subsequent 5 h, while the other enzymes *are* synthesized. Our interpretations may be summarized as follows: the enzyme exists in two forms, a base-line and a derepressible variety. The latter is synthesized only during the initial (fermentative) phase of derepression, prior to the period when respiration by newly elaborated mitochondria comes to be of importance. Its synthesis probably continues for a period longer than 5 h into derepression but becomes masked at that point by a rapid and specific process of inactivation or degradation. Both forms also appear to be subject to rapid degradation (turnover), but at different rates, when their synthesis in the cytoplasm is interrupted by CHX. The derepressible—but not the base-line enzyme—requires for its activity or integration also the presence of an entity synthesized by, or dependent on a product of, the mitochondrial translational system.

In some of these respects, at least, this behavior of ALA synthetase is strongly reminiscent of that of another enzyme also presumed to fulfill a key function during derepression, namely, the mitochondrial DNA-dependent RNA polymerase (South and Mahler, 1968). The uniquely cyclical nature in the levels of these two enzymes may go far in providing a regulatory device for the onset and particularly the *shut-down* of derepression: what will need to be done next is obviously to inquire into the possible mechanisms for turning the synthesis of the derepressible enzymes on and off.

SUMMARY

In summary then, we have attempted to show that the autonomous genetic system resident in the mitochondria of fungi and its constituents and products are tightly integrated and coordinated with the rest of the cell in expression and function. Its purpose is to provide, on demand, a limited number of polypeptides to a limited number of enzyme complexes of the inner membrane, polypeptides which are required for the proper elaboration and integration of a fully competent respiratory organelle. Such respiratory competence is not only dependent on products of mitochondrial genes but in turn influences the competence of these genes. Multiple reciprocal loops—at both the stoichiometric and regulatory plane—can be discerned at all levels of integration, starting at the individual protein, and continuing through the enzyme complex, the inner membrane, the organelle, and finally the whole cell. Mitochondrial biogenesis is thus a process that must be approached both conceptually and experimentally as a problem of cellular and not of sub-cellular dimensions.

ACKNOWLEDGEMENTS

H.R. Mahler is the recipient of a Research Career Award, KO6 GM 05060 from the Institute of General Medical Sciences. The experimental work described in this report was supported by Research Grant GM 12228 from the National Institute of General Medical Sciences, National Institutes of Health, U.S. Public Health Service. This is publication number 2630 from the Chemical Laboratories, Indiana University.

We wish to thank Prof. P. S. Perlman and Dr. F. Feldman for many fruitful discussions; Profs. Leland Hartwell, P. S. Perlman, P. P. Slonimski, and E. Moustacchi for providing us with the yeast strains used in this investigation; and Ms. K. Assimos for devoted and highly competent assistance in some of the derepression experiments.

LITERATURE CITED

Azzi, A., and M. Santato. 1971. Interaction of ethidium with the mitochondrial membrane: Cooperative binding and energy-linked changes. Biochem. Biophys. Res. Commun. 44:211-17.

Barath, Z., and H. Küntzel. 1972. Cooperation of mitochondrial and nuclear genes specifying the mitochondrial genetic apparatus in *Neurospora crassa*. Proc. Nat. Acad. Sci. USA. 69:1371-74.

Bard, M., R. A. Woods, and J. M. Haslam. 1974. Porphyrin mutants of *Saccharomyces cerevisiae:* Correlated lesions in sterol and fatty acid biosynthesis. Biochem. Biophys. Res. Commun. 56:324-30.

Barnett, W. E., and D. H. Brown. 1967. Mitochondrial transfer ribonucleic acids. Proc. Nat. Acad. Sci. USA. 57:452-58.

Bastos, R. N., and H. R. Mahler. 1974*a*. A synthesis of labeled ethidium bromide. Arch. Biochem. Biophys. 160:643-46.

Bastos, R. N., and H. R. Mahler. 1974*b*. Molecular mechanisms of mitochondrial genetic activity: Effects of ethidium bromide on the DNA and energetics of isolated mitochondria. J. Biol. Chem. 249:6617-27.

Bauer, W., and J. Vinograd. 1971. The use of intercalative dyes in the study of closed circular DNA. Prog. Molec. Subcell. Biol. 2:181-215.

Bernardi, G., F. Carnevali, A. Nicolaieff, G. Piperno, and G. Tecce. 1968. Separation and characterization of a satellite DNA from a yeast cytoplasmic "Petite" mutant. J. Mol. Biol. 37:493-505.

Boardman, N. K., A. W. Linnane, and S. M. Smillie (eds.). 1971. Autonomy and biogenesis of mitochondria and chloroplasts. American-Elsevier, New York.

Bolotin, M., D. Coen, J. Deutsch, B. Dujon, P. Netter, E. Petrochilo, and P. P. Slonimski. 1971. La recombinaison des mitochondries chez *Saccharomyces cerevisiae*. Bull. Inst. Pasteur 69:215-39.

Borst, P., and L. A. Grivell. 1971. Mitochondrial ribosomes. FEBS Letters 13:73-88.

Borst, P. 1972. Mitochondrial nucleic acids. Ann. Rev. Biochem. 41:333-76.

Carnevali, F., C. Falcone, L. Frontali, L. Leoni, G. Macino, and C. Palleschi. 1973. Informational content of mitochondrial DNA from a "low density" petite mutant of yeast. Biochem. Biophys. Res. Commun. 51:651-58.

Casey, J., M. Cohen, M. Rabinowitz, H. Fukuhara, and G. S. Getz. 1972. Hybridization of mitochondrial transfer RNA's with mitochondrial and nuclear DNA of grande (wild type) yeast. J. Mol. Biol. 63:431-40.

Clark-Walker, G. D., and A. W. Linnane. 1966. *In vivo* differentiation of yeast cytoplasmic and mitochondrial protein synthesis with antibiotics. Biochem. Biophys. Res. Commun. 25:8-13.

Coen, D., J. Deutsch, P. Netter, E. Petrochilo, and P. P. Slonimski. 1970. Mitochondrial genetics. I. Methodology and phenomenology. Symp. Soc. Exp. Biol. 24:449-96.

Cohen, M., J. Casey, M. Rabinowitz, and G. S. Getz. 1972. Hybridization of mitochondrial transfer RNA and mitochondrial DNA in petite mutants of yeast. J. Mol. Biol. 63:441-51.

Cooper, C. S., and C. J. Avers. 1974. Evidence of involvement of mitochondrial polysomes and messenger RNA in synthesis of organelle proteins, pp. 289-303.

In A. M. Kroon and C. Saccone (eds.), The biogenesis of mitochondria. Academic Press, New York.

Dawidowicz, K. 1972. Studies on the mitochondrial protein synthesis system in *Saccharomyces cerevisiae.* Ph.D. thesis, Indiana University.

Dawidowicz, K., and H. R. Mahler. 1973. Synthesis of mitochondrial proteins, pp. 503-22. *In* F. T. Kenny, B. A. Hamkalo, G. Favelukes, and J. T. August (eds.), Gene expression and its regulation. Plenum Press, New York.

Denhardt, D. T., and A. C. Kato. 1973. Comparison of the effect of ultraviolet radiation and ethidium bromide intercalation on the conformation of superhelical ϕX174 replicative form DNA. J. Mol. Biol. 77:479–94.

Deutsch, J., P. Dujon, P. Netter, E. Petrochilo, P. P. Slonimski, and M. Bolotin-Fukuhara. 1974. Mitochondrial genetics. VI. The *petite* mutation in *Saccharomyces cerevisiae:* Interrelations between the loss of the ρ^+ factor and the loss of the drug resistance mitochondrial genetic markers. Genetics 76:195–219.

Ebner, E., L. Mennucci, and G. Schatz. 1973. Mitochondrial assembly in respiration-deficient mutants of *Saccharomyces cerevisiae.* I. Effect of nuclear mutations on mitochondrial protein synthesis. J. Biol. Chem. 248:5360–68.

Ebner, E., T. L. Mason, and G. Schatz. 1973. Mitochondrial assembly in respiration-deficient mutants of *Saccharomyces cerevisiae.* II. Effect of nuclear and extrachromosomal mutations on the formation of cytochrome *c* oxidase. J. Biol. Chem. 248:5369–78.

Eccleshall, T. R., and R. S. Criddle. 1974. The DNA-dependent RNA polymerases from yeast mitochondria, pp. 31-46. *In* A. M. Kroon and C. Saccone (eds.), The biogenesis of mitochondria. Academic Press, New York.

Ephrussi, B., H. Hottinguer, and J. Tavlitzki. 1949. Action de l'Acriflavine sur les levures. II. Etude genetique du mutant "petite colonies". Ann. Inst. Pasteur 76:419–50.

Ephrussi, B. 1950. The interplay of heredity and environment in the synthesis of respiratory enzymes in yeast. Harvey Lectures 46:45–67.

Epler, J. L. 1969. The mitochondrial and cytoplasmic transfer ribonucleic acids of *Neurospora crassa.* Biochemistry 8:2285–90.

Epler, J. L., L. R. Shugart, and W. E. Barnett. 1970. N-formylmethionyl transfer ribonucleic acid in mitochondria from *Neurospora.* Biochemistry 9:3575–79.

Faye, G., H. Fukuhara, C. Grandchamp, J. Lazowska, F. Michel, J. Casey, G. S. Getz, J. Locker, M. Rabinowitz, M. Bolotin-Fukuhara, D. Coen, J. Deutsch, B. Dujon, P. Netter, and P. P. Slonimski. 1973. Mitochondrial nucleic acids in the petite colonie mutants: Deletions and repetitions of genes. Biochimie 55:779–92.

Faye, G., C. Kujawa, and H. Fukuhara. 1974. Physical and genetic organization of *petite* and *grande* yeast mitochondrial DNA. IV. *In vivo* transcription products of mitochondrial DNA and localization of 23S ribosomal RNA in *petite* mutants of *Saccharomyces cerevisiae.* J. Mol. Biol. 88:185–203.

Feldman, F., and H. R. Mahler. 1974. Mitochondrial biogenesis: Retention of terminal formyl methionine in membrane proteins and regulation of their synthesis. J. Biol. Chem. 249:3702-9.

Fukuhara, H. 1968. Informational role of yeast mitochondrial DNA studied by hybridization with different classes of RNA, pp. 303-25. *In* E. C. Slater, J. M. Tager,

S. Papa, E. Quagliariello (eds.), Biochemical aspects of the biogenesis of mitochondria. Adriatica Editrice, Bari.

Fukuhara, H., and C. Kujawa. 1970. Selective inhibition of the *in vivo* transcription of mitochondrial DNA by ethidium bromide and by acriflavin. Biochem. Biophys. Res. Commun. 41:1002–8.

Fukuhara, H., G. Faye, F. Michel, J. Lazowska, J. Deutsch, M. Bolotin-Fukuhara, and P. P. Slonimski. 1974. Physical and genetic organization of petite and grande yeast mitochondrial DNA. I. Studies by RNA-DNA hybridization. Mol. Gen. Genet. 130:215–38.

Getz, G. S. 1972. Organelle biogenesis, pp. 386–438. *In* C. F. Fox and A. Keith (eds.), Membrane molecular biology. Sinauer Assoc. Inc., Stamford, Conn.

Gitler, C., B. Rubalcava, and A. Caswell. 1969. Fluorescence changes of ethidium bromide on binding to erythrocyte and mitochondrial membranes. Biochim. Biophys. Acta 193:479–81.

Goffeau, A., Landry, Y., F. Foury, M. Briquet, and A.-M. Colson. 1973. Oligomycin resistance of mitochondrial adenosine triphosphatase in a pleiotropic chromosomal mutant of a "petite-negative" yeast, *Schizosaccharomyces pombe*. J. Biol. Chem. 248:7097–105.

Goldring, E. S., L. I. Grossman, D. Krupnick, D. R. Cryer, and J. Marmur. 1970. The petite mutation in yeast: Loss of mitochondrial deoxyribonucleic acid during induction of petites with ethidium bromide. J. Mol. Biol. 52:323–35.

Goldring, E. S., L. I. Grossman, and J. Marmur. 1971. Petite mutation in yeast. II. Isolation of mutants containing mitochondrial deoxyribonucleic acid of reduced size. J. Bact. 107:377–81.

Gollub, E. G., P. Trocha, P. K. Liu, and D. B. Sprinson. 1974. Yeast mutants requiring ergosterol as only lipid supplement. Biochem. Biophys. Res. Commun. 56:471–77.

Gordon, P., and M. Rabinowitz. 1973. Evidence for deletion and changed sequence in the mitochondrial deoxyribonucleic acid of a spontaneously generated petite mutant of *Saccharomyces cerevisiae*. Biochemistry 12:116–23.

Gordon, P., J. Casey, and M. Rabinowitz. 1974. Characterization of mitochondrial deoxyribonucleic acid from a series of petite yeast strains by deoxyribonucleic acid-deoxyribonucleic acid hybridization. Biochemistry 13:1067–75.

Grandi, M., A. Helms, and H. Küntzel. 1971. Fusidic acid resistance of mitochondrial G Factor. Biochem. Biophys. Res. Commun. 44:864–71.

Grimes, G. W., H. R. Mahler, and P. S. Perlman. 1974. Regulation of mitochondria and their genomes. I. Effects of nuclear gene dosage. J. Cell Biol. 61:565–74.

Grivell, L. A., P. Netter, P. Borst, and P. P. Slonimski. 1973. Mitochondrial antibiotic resistance in yeast: Ribosomal mutants resistant to chloramphenicol, erythromycin, and spiramycin. Biochim. Biophys. Acta 312:358–67.

Halbreich, A., and M. Rabinowitz. 1971. Isolation of *Saccharomyces cerevisiae* mitochondrial formyltetrahydrofolic acid: Methionyl-tRNA transformylase and the hybridization of mitochondrial fMet-tRNA with mitochondrial DNA. Proc. Nat. Acad. Sci. USA. 68:294–98.

Hartwell, L. H. 1967. Macromolecule synthesis in temperature-sensitive mutants of yeast. J. Bact. 93:1662–70.

Hartwell, L. H., and C. S. McLaughlin. 1968. Mutants of yeast with temperature-sensitive isoleucyl-tRNA synthetases. Proc. Nat. Acad. Sci. USA. 59:422–28.

Hartwell, L. H. 1970. Periodic density fluctuation during the yeast cell cycle and the selection of synchronous cultures. J. Bact. 104:1280–85.

Hartwell, L. H., H. T. Hutchison, T. M. Holland, and C. S. McLaughlin. 1970. The effect of cycloheximide upon polyribosome stability in two yeast mutants defective respectively in the initiation. Mol. Gen. Genet. 106:347–61.

Hartwell, L. H. 1973. Three additional genes required for deoxyribonucleic acid synthesis in *Saccharomyces cerevisiae*. J. Bact. 115:966–74.

Hartwell, L. H., J. Culotti, J. R. Pringle, and B. J. Reid. 1974. Genetic control of the cell division cycle in yeast. Science 183:46–51.

Haselkorn, R., and L. B. Rothman-Denes. 1973. Protein synthesis. Ann. Rev. Biochem. 42:397–438.

Hollenberg, C. P., P. Borst, and E. F. J. van Bruggen. 1972. Mitochondrial DNA from cytoplasmic petite mutants of yeast. Biochim. Biophys. Acta 277:35–43.

Hollenberg, C. P., P. Borst, R. A. Flavell, C. F. van Kreijl, E. F. J. van Bruggen, and A. C. Arnberg. 1972. The unusual properties of mtDNA from a "low-density" petite mutant of yeast. Biochim. Biophys. Acta 277:44–58.

Huang, M., D. R. Biggs, G. D. Clark-Walker, and A. W. Linnane. 1966. Chloramphenicol inhibition of the formation of particulate mitochondrial enzymes of *Saccharomyces cerevisiae*. Biochim. Biophys. Acta 114:434–36.

Hutchison, H. T., L. H. Hartwell, and C. S. McLaughlin. 1969. Temperature-sensitive yeast mutant defective in ribonucleic acid production. J. Bact. 99:807–14.

Jayaraman, J., G. Padmanaban, K. Malathi, and P. S. Sarma. 1971. Haem synthesis during mitochondriogenesis in yeast. Biochem. J. 121:531–35.

Kellerman, G. H., D. R. Biggs, and A. W. Linnane. 1969. Biogenesis of mitochondria. XI. A comparison of the effects of growth-limiting oxygen tension, intercalating agents, and antibiotics on the obligate aerobe *Candida parapsilosis*. J. Cell Biol. 42:378–84.

Kroon, A. M., and A. J. Arendzen. 1973. The inhibition of mitochondrial biogenesis by antibiotics, pp. 135–55. *In* G. S. Van den Bergh, P. Borst, and E. C. Slater (eds.), Mitochondria biomembranes. North Holland, Amsterdam.

Kroon, A. M., E. Agsteribbe, and H. de Vries. 1972. Protein synthesis in mitochondria and chloroplasts, pp. 539–82. *In* L. Bosch (ed.), The mechanism of protein synthesis and its regulation. North Holland, Amsterdam.

Kroon, A. M., A. J. Arendzen, and H. de Vries. 1974. On the sensitivity of mammalian mitochondrial protein synthesis to inhibition by the macrolide antibiotics, pp. 395–402. *In* A. M. Kroon and C. Saccone (eds.), The biogenesis of mitochondria. Academic Press, New York.

Kroon, A. M., and C. Saccone (eds.). 1974. The biogenesis of mitochondria. Academic Press, New York.

Küntzel, H., and H. Noll. 1967. Mitochondrial and cytoplasmic polysomes from *Neurospora crassa*. Nature 215:1340–45.

Küntzel, H., and K. P. Schäfer. 1971. Mitochondrial RNA polymerase from *Neurospora crassa*. Nature New Biol. 231:265–69.

Küntzel, H., Z. Barath, I. Ali, J. Kind, and H.-H. Althaus. 1973a. Virus-like particles in an extranuclear mutant of *Neurospora crassa.* Proc. Nat. Acad. Sci. USA. 70:1574–78.

Küntzel, H., Z. Barath, I. Ali, J. Kind, H.-H. Althaus, and H. C. Blossey. 1973b. Expression of the mito-genome in wild type and in an extranuclear mutant of *Neurospora crassa*, pp. 195–210. *In* E. K. F. Bautz, P. Karlson, and H. Kersten (eds.), Regulation of transcription and translation of eukaryotes. Springer-Verlag, New York.

Kuriyama, Y., and D. J. L. Luck. 1974. Methylation and processing of mitochondrial ribosomal RNAs in *Poky* and wild-type *Neurospora crassa.* J. Mol. Biol. 83:253–66.

Kuriyama, Y., and D. J. L. Luck. 1974. Synthesis and processing of mitochondrial ribosomal RNA in wild type and poky strain of *Neurospora crassa*, pp. 117–33. *In* A. M. Kroon and C. Saccone (eds.), The biogenesis of mitochondria. Academic Press, New York.

Lamb, A. J., and W. Rojanapo. 1973. Preferential transcription of dG and dC rich mitochondrial DNA in cytoplasmic petite mutants of *S. cerevisiae.* Biochem. Biophys. Res. Commun. 55:765–72.

Lazowska, J., F. Michel, G. Faye, H. Fukuhara, and P. P. Slonimski. 1974. Physical and genetic organization of *petite* and *grande* yeast mitochondrial DNA. II. DNA-DNA hybridization studies and buoyant density determinations. J. Mol. Biol. 85:393–410.

Lederman, M., and G. Attardi. 1973. Expression of the mitochondrial genome in HeLa cells. XVI. Electrophoretic properties of the products of *in vivo* and *in vitro* mitochondrial protein synthesis. J. Mol. Biol. 78:275–83.

Linnane, A. W., G. W. Saunders, E. B. Gingold, and H. B. Lukins. 1968. The biogenesis of mitochondria. V. Cytoplasmic inheritance of erythromycin resistance in *Saccharomyces cerevisiae.* Proc. Nat. Acad. Sci. USA. 59:903–10.

Linnane, A. W., and J. M. Haslam. 1970. The biogenesis of yeast mitochondria. Current Topics in Cellular Regulation 2:101–72.

Linnane, A. W., J. M. Haslam, H. B. Lukins, and P. Nagley. 1972. The biogenesis of mitochondria in microorganisms. Ann. Rev. Microbiol. 26:163–98.

Locker, J., M. Rabinowitz, and G. S. Getz. 1974. Tandem inverted repeats in mitochondrial DNA of petite mutants of *Saccharomyces cerevisiae.* Proc. Nat. Acad. Sci. USA. 71:1366–70.

Luck, D. J. L., and E. Reich. 1964. DNA in mitochondria of *Neurospora crassa.* Proc. Nat. Acad. Sci. USA. 52:931–38.

Mackler, B., H. C. Douglas, S. Will, D. C. Hawthorne, and H. R. Mahler. 1965. Biochemical correlates of respiratory deficiencies. IV. Composition and properties of respiratory particles from mutant yeast. Biochemistry 4:2016–20.

Mahler, H. R., and P. S. Perlman. 1971. Mitochondriogenesis analyzed by blocks on mitochondrial translation and transcription. Biochemistry 10:2979–90.

Mahler, H. R., B. D. Mehrotra, and P. S. Perlman. 1971. Formation of yeast mitochondria. V. Ethidium bromide as a probe for the functions of mitochondrial RNA. Prog. Molec. Subcell. Biol. 2:274–96.

Mahler, H. R., and P. S. Perlman. 1972. Effects of mutagenic treatment by ethidium bromide on cellular and mitochondrial phenotype. Arch. Biochem. Biophys. 148:115–29.

Mahler, H. R., D. Dawidowicz, and F. Feldman. 1972. Formate as a specific label for mitochondrial translational products. J. Biol. Chem. 247:7439-42.

Mahler, H. R. 1973a. Biogenetic autonomy in mitochondria. CRC Crit. Rev. Biochem. 1:381-460.

Mahler, H. R. 1973b. Structural requirements for mitochondrial mutagenesis. J. Supramol. Struct. 1:449-60.

Mahler, H. R. 1973c. Genetic autonomy of mitochondrial DNA, pp. 181-208. *In* B. Hamkalo and J. Papaconstantinou (eds.), Molecular cytogenetics. Plenum Press, New York.

Mahler, H. R., and K. Dawidowicz. 1973. Autonomy of mitochondria of *Saccharomyces cerevisiae* in their production of messenger RNA. Proc. Natl. Acad. Sci. USA. 70:111-14.

Mahler, H. R., and R. N. Bastos. 1974. A novel reaction of mitochondrial DNA with ethidium bromide. FEBS Letters 39:27-34.

Mahler, H. R., R. N. Bastos, F. Feldman, U. Flury, C. C. Lin, P. S. Perlman, and S. H. Phan. 1974. Biogenetic autonomy of mitochondria and its limits. *In* A. A. Tzagoloff (ed.), Membrane biogenesis: Mitochondria, chloroplasts, and bacteria. Plenum Press, New York (in press).

Mahler, H. R., F. Feldman, S. H. Phan, P. Hamill, and K. Dawidowicz. 1974. Initiation, identification and integration of mitochondrial proteins, pp. 423-41. *In* A. M. Kroon and C. Saccone (eds.), The biogenesis of mitochondria. Academic Press, New York.

Mahler, H. R. and R. A. Raff. 1974. The evolutionary origin of the mitochondrion: A non-symbiotic model. Int. Rev. Cytol., in press.

Mehrotra, B. D., and H. R. Mahler. 1968. Characterization of some unusual DNAs from the mitochondria from certain "petite" strains of *Saccharomyces cerevisiae.* Arch. Biochem. Biophys. 128:685-703.

Marver, H. S., D. P. Tschundy, M. G. Perlroth, and A. Collins. 1966. δ-Aminolevulinic acid synthetase. I. Studies in liner homogenates. J. Biol. Chem. 241:2803-9.

Meyer, U. A., and R. Schmid. 1973. Hereditary hepatic porphyrias. Fed. Proc. 32:1649-55.

Michaelis, G., D. Douglass, M. J. Tsai, and R. S. Criddle. 1971. Mitochondrial DNA and suppressivenes of petite mutants in *Saccharomyces cerevisiae.* Biochem. Genetics 5:487-95.

Michel, E., J. Lazowska, G. Faye, H. Fukuhara, and P. P. Slonimski. 1974. Physical and genetic organization of *petite* and *grande* yeast mitochondrial DNA. III. High resolution melting and reassociation studies. J. Mol. Biol. 85:411-31.

Miller, P. L. (ed.). 1970. Control of organelle development. Academic Press, New York.

Mounolou, J. P., H. Jakob, and P. P. Slonimski. 1966. Mitochondrial DNA from yeast "petite" mutants: Specific changes in buoyant density corresponding to different cytoplasmic mutations. Biochem. Biophys. Res. Commun. 24:218-24.

Moustacchi, E. 1972. Determination of the degree of suppressivity of *Saccharomyces cerevisiae* strain RDIA. Biochim. Biophys. Acta 277:59-60.

Mitchell, H. K., and M. B. Mitchell. 1952. A case of "maternal" inheritance in *Neurospora crassa.* Proc. Nat. Acad. Sci. USA. 38:442-49.

Nagley, P., and A. W. Linnane. 1970. Mitochondrial DNA deficient petite mutants of yeast. Biochem. Biophys. Res. Commun. 39:989–96.

Nagley, P., and A. W. Linnane. 1972. Biogenesis of mitochondria. XXI. Studies on the nature of the mitochondrial genome in yeast: The degenerative effects of ethidium bromide on mitochondrial genetic information in a respiratory competent strain. J. Mol. Biol. 66:181–93.

Nikolaev, N., D. Schlessinger, and P. K. Wellauer. 1974. 30 S preribosomal RNA of *Escherichia coli* and products of cleavage by ribonuclease III: Length and molecular weight. J. Mol. Biol. 86:741–47.

Paoletti, J., and J.-B. Le Pecq. 1971. Resonance energy transfer between ethidium bromide molecules bound to nucleic acids. J. Mol. Biol. 59:43–62.

Perlman, P. S., and H. R. Mahler. 1970. Formation of yeast mitochondria. III. Biochemical properties of mitochondria isolated from a cytoplasmic petite mutant. J. Bioenergetics 1:113–38.

Perlman, P. S., and H. R. Mahler. 1971. Molecular consequences of ethidium bromide mutagenesis. Nature New Biol. 231:12–16.

Perlman, P. S., and H. R. Mahler. 1974. Derepression of mitochondria and their enzymes in yeast: Regulatory aspects. Arch. Biochem. Biophys. 162:248–71.

Pigram, W., W. Fuller, and M. E. Davies. 1973. Unwinding the DNA helix by intercalation. J. Mol. Biol. 80:361–65.

Poyton, R. O., and G. Schatz. 1975. Cytochrome *c* oxidase from bakers' yeast. III. Physical characterization of isolated subunits and chemical evidence for two different classes of polypeptides. J. Biol. Chem. 250:752–61.

Preer, J. R., Jr. 1971. Extrachromosomal inheritance: Hereditary symbionts, mitochondria, chloroplasts. Ann. Rev. Gen. 5:361–406.

Rabinowitz, M., J. Casey, P. Gordon, J. Locker, H.-J. Hsu, and G. S. Getz. 1974. Characterization of yeast grande and petite mitochondrial DNA by hybridization and physical techniques, pp. 89–105. *In* A. M. Kroon and C. Saccone (eds.), The biogenesis of mitochondria. Academic Press, New York.

Raff, R. A. and H. R. Mahler. 1972. The non-symbiotic origin of mitochondria. Science 177:575–82.

Raff, R. A., and H. R. Mahler. 1974. Symbiosis. Symp. Soc. Exp. Biol. 29 (in press).

Reijnders, L., and P. Borst. 1972. The number of 4 S RNA genes on yeast mitochondrial DNA. Biochem. Biophys. Res. Comm. 47:126–33.

Reijnders, L., C. M. Kleisen, L. A. Grivell, and P. Borst. 1972. Hybridization studies with yeast mitochondrial RNAs. Biochim. Biophys. Acta 272:396–407.

Richardson, J. P. 1973. Mechanism of ethidium bromide inhibition of RNA polymerase. J. Mol. Biol. 78:703–14.

Rifkin, M. R., D. D. Wood, and D. J. L. Luck. 1967. Ribosomal RNA and ribosomes from mitochondria of *Neurospora crassa*. Proc. Nat. Acad. Sci. USA. 58:1025–32.

Ross, E., E. Ebner, R. O. Poyton, T. L. Mason, B. Ono, and G. Schatz. 1974. The biosynthesis of mitochondrial cytochromes, pp. 477–89. *In* A. M. Kroon and C. Saccone (eds.), The biogenesis of mitochondria. Academic Press, New York.

Sager, R. 1972. Cytoplasmic genes and organelles. Academic Press, New York.

Sanders, J. P. M., R. A. Flavell, P. Borst, and J. N. M. Mol. 1973. Nature of the base sequence conserved in the mitochondrial DNA of a low-density petite. Biochim. Biophys. Acta 312:441-57.

Sanders, H. K., P. A. Mied, M. Briquet, J. Hernandez-Rodriguez, R. F. Gottal, and J. R. Mattoon. 1973. Regulation of mitochondrial biogenesis: Yeast mutants deficient in synthesis of δ-aminolevulinic acid. J. Mol. Biol. 80:17-39.

Schäfer, K. P., and H. Küntzel. 1972. Mitochondrial genes in *Neurospora:* A single cistron for ribosomal RNA. Biochem. Biophys. Res. Commun. 46:1312-19.

Schatz, G. 1970. Biogenesis of mitochondria, pp. 251-314. *In* E. Racker (ed.), Membranes of mitochondria and chloroplasts. Van Nostrand-Reinhold, New York.

Schatz, G., and T. Mason. 1974. The biosynthesis of mitochondrial proteins. Ann. Rev. Biochem. 43:51-87.

Scragg, A. H. 1974. A mitochondrial DNA-directed RNA polymerase from yeast mitochondria, pp. 47-57. *In* A. M. Kroon and C. Saccone (eds.), The biogenesis of mitochondria. Academic Press, New York.

Sebald, W., W. Machleidt, and J. Otto. 1973. Products of mitochondrial protein synthesis in *Neurospora crassa.* Eur. J. Biochem. 38:311-24.

Shakespeare, P. G., and H. R. Mahler. 1971. Purification and some properties of cytochrome *c* oxidase from the yeast *Saccharomyces cervisiae.* J. Biol. Chem. 246:7649-55.

Shannon, C., R. Enns, L. Whellis, K. Burchiel, and R. S. Criddle. 1973. Alterations in mitochondrial adenosine triphosphatase activity resulting from mutation of mitochondrial deoxyribonucleic acid. J. Biol. Chem. 248:3004-11.

Sherman, F., and P. P. Slonimski. 1964. Respiration-deficient mutants of yeast. II. Biochemistry. Biochim. Biophys. Acta 90:1-15.

Sierra, M. F., and A. Tzagoloff. 1973. Assembly of the mitochondrial membrane system: Purification of a mitochondrial product of the ATPase. Proc. Nat. Acad. Sci. USA. 70:3155-59.

Slater, E. C. 1973. The mechanism of action of the respiratory inhibitor, antimycin. Biochim. Biophys. Acta 301:129-54.

Slonimski, P. P., G. Perrodin, and J. H. Croft. 1968. Ethidium bromide induced mutation of yeast mitochondria: Complete transformation of cells into respiratory deficient nonchromosomal "petites". Biochem. Biophys. Res. Commun. 30:232-39.

Somlo, M., P. R. Avner, J. Cosson, B. Dujon, and M. Krupa. 1974. Oligomycin sensitivity of ATPase studied as a function of mitochondrial biogenesis, using mitochondrially determined oligomycin-resistant mutants of *Saccharomyces cerevisiae.* Eur. J. Biochem. 42:439-45.

South, D. J., and H. R. Mahler. 1966. RNA synthesis in yeast mitochondria: A derepressible activity. Nature 218:1226-32.

Tewari, K. K., J. Jayaraman, and H. R. Mahler. 1965. Separation and characterization of mitochondrial DNA from yeast. Biochem. Biophys. Res. Commun. 21:141-48.

Thomas, C. A., Jr., and L. A. MacHattie. 1967. The anatomy of viral DNA molecules. Ann. Rev. Biochem. 36:485-518.

Thomas, D. Y., and D. Wilkie. 1968a. Recombination of mitochondrial drug-resistance factors in *Saccharomyces cerevisiae.* Biochem. Biophys. Res. Commun. 30:368-72.

Thomas, D. Y., and D. Wilkie. 1968b. Inhibition of mitochondrial synthesis in yeast by erythromycin: Cytoplasmic and nuclear factors controlling resistance. Genet. Res. Camb. 11:33–41.

Tschudy, D. P., and H. L. Bonkowsky. 1972. Experimental porphyria. Fed. Proc. 31:147–59.

Tzagoloff, A., and P. Meagher. 1971. Assembly of the mitochondrial membrane system. V. Properties of a dispersed preparation of the rutamycin-sensitive adenosine triphosphatase of yeast mitochondria. J. Biol. Chem. 246:7328–36.

Tzagoloff, A., and A. Akai. 1972. Assembly of the mitochondrial membrane system. VIII. Properties of the products of mitochondrial protein synthesis in yeast. J. Biol. Chem. 274:6517–32.

Tzagoloff, A., M. S. Rubin, and M. F. Sierra. 1973. Biosynthesis of mitochondrial enzymes. Biochim. Biophys. Acta 301:71–104.

Tzagoloff, A., A. Akai, and M. S. Rubin. 1974. Mitochondrial products of yeast ATPase and cytochrome oxidase, pp. 405–21. *In* A. M. Kroon and C. Saccone (eds.), The biogenesis of mitochondria. Academic Press, New York.

Tzagoloff, A. (ed.) 1974. Membrane biogenesis: Mitochondria, chloroplasts, and bacteria. Plenum Press, New York.

Waring, M. J. 1965. Complex formation between ethidium bromide and nucleic acids. J. Mol. Biol. 13:269–82.

Waring, M. 1970. Variation of the supercoils in closed circular DNA by binding of antibiotics and drugs: Evidence for molecular models involving intercalation. J. Mol. Biol. 54:247–79.

Weiss, H., and B. Ziganke. 1974a. Cytochrome *b* in *Neurospora crassa*. Eur. J. Biochem. 41:63–71.

Weiss, H., and B. Ziganke. 1974b. Biogenesis of cytochrome *b* in *Neurospora crassa*, pp. 491–500. *In* A. M. Kroon and C. Saccone (eds.), The biogenesis of mitochondria. Academic Press, New York.

Werner, S. 1974. Isolation and characterization of a mitochondrially synthesized precursor protein of cytochrome oxidase. Eur. J. Biochem. 43:39–48.

Wilkie, D., G. W. Saunders, and A. W. Linnane. 1967. Inhibition of repiratory enzyme synthesis in yeast by chloramphenicol: Relationship between chloramphenicol tolerance and resistance to other antibacterial antibiotics. Genet. Res. Camb. 10:199–203.

Wintersberger, E. 1973. Transcription in mitochondria, pp. 179–93. *In* E. K. F. Bautz, P. Karlson, and H. Kersten (eds.), Regulation of transcription and translation in eukaryotes. Springer-Verlag, New York.

DAVID E. GRIFFITHS

Utilization of Mutations in the Analysis of Yeast Mitochondrial Oxidative Phosphorylation

3

The mitochondrial oxidative phosphorylation system is an integrated multienzyme system found on the inner membrane of the mitochondrion. In its simplest configuration it consists of a redox chain of electron and proton carriers concerned with the transport of reducing equivalents from citric acid cycle intermediates for the reduction of molecular oxygen. Associated with this system is the ATP-synthetase complex (oligomycin-sensitive ATPase) that catalyzes an ATPase reaction and a P_i-ATP exchange reaction, which are representative reactions of the terminal steps of oxidative phosphorylation. However, the mechanism of oxidative phosphorylation and the mode of interaction of the electron transport system with the ATP-synthetase complex is still a matter of controversy embodied in the chemical, conformational, and chemiosmotic hypotheses (Greville, 1969).

The known components of the electron transport chain and the ATP-synthetase complex, which have been the subject of intense investigations in the laboratories of Green, Racker, Tzagoloff, and Hatefi, have been elucidated by isolation and characterization of the individual components and correlation of their physical properties with their function in the integrated membrane-bound complex.

The electron transport chain can be isolated as four functional complexes: Complex I. NADH-ubiquinone reductase (Hatefi et al., 1962a); Complex II. succinate-ubiquinone reductase (Ziegler and Doeg, 1959); Complex III. ubiquinone-cytochrome *c* reductase (Hatefi et

Department of Molecular Sciences, University of Warwick, Coventry CV4 7AL, England.

al., 1962b); and Complex IV. cytochrome *c* oxidase (Griffiths and Wharton, 1961). The biogenesis of some of these complexes has been discussed in a recent review by Schatz and Mason (1974) and by Mahler (this volume). The available evidence indicates that three subunits of the cytochrome oxidase complex are synthesized on mitoribosomes and thus are assumed to be coded for by mitochondrial DNA. In addition, a component of cytochrome *b* has been shown to be synthesized in mitochondria. Cytochrome c_1 apoprotein is synthesized on cytoribosomes; however, the formation of holocytochrome c_1 requires mitochondrial protein synthesis, which indicates the necessity for combination of the apoprotein with at least one mitochondrially synthesized polypeptide before a functional cytochrome c_1 can be formed. Thus, these conclusions, which result from biogenesis experiments in the laboratories of Schatz, Tzagoloff, and Sebald and Weiss, indicate that at least one subunit associated with complex III and complex IV is synthesized on cytoribosomes.

The correlation of these results with the properties of the classical mitochondrial mutation, the cytoplasmic "petite" mutation of *Saccharomyces cerevisiae,* which results in respiration-deficient mitochondria lacking cytochromes aa_3, *b,* and c_1 and an energy transfer system, provides supporting evidence that mitoribosome-synthesized components of complex III and complex IV are products of the mitochondrial genome. However, it should be pointed out that no mitochondrial gene or genetic locus has yet been identified with any mitoribosome-synthesized component of these two complexes and that there is no evidence available at present to indicate the biogenetic origin of other components of complex III such as the non-heme iron protein and the antimycin binding site known to be present in this complex. Similarly there is no evidence to indicate the presence of mitochondrially synthesized (mitochondrially coded) components in complex I or complex II. However, the experimental evidence favoring a cytoribosome site of synthesis is based on studies of mitochondria of yeast cytoplasmic petite mutants and mitochondria of cells grown in the presence of inhibitors of mitochondrial protein synthesis. The retention of succinate dehydrogenase activity and NADH dehydrogenase activity in such mitochondria is certainly evidence in favor of the cytoribosomal synthesis (nuclear coding) of *these* enzymes, but caution must be exercised with respect to the biogenetic origin of other subunits of complexes I and II. A systematic investigation of all the components of these two complexes, particularly the non-heme iron components, and their presence or absence in cytoplasmic petite mutants lacking mtDNA would be of particular interest in this context.

Cytoplasmic petite mutants and inhibitors of protein synthesis have been extensively used by Linnane and coworkers to study mitochondrial permease or translocase systems. Perkins et al. (1972) found that most mitochondrial translocase systems were present in petite mitochondria and in mitochondria from yeast grown in the presence of chloramphenicol, indicating that the components of these systems are cytoribosome synthesized (nuclear coded). The ADP translocase system, in contrast, was modified in both experimental systems, and it was concluded that the ADP translocase contained a subunit coded for by mtDNA. However, recent studies by Kolarov and Klingenberg (1974) are not in agreement with this conclusion.

The ATP-synthetase complex (oligomycin-sensitive ATPase complex) has been intensively investigated by Tzagoloff et al. (1973). The complex has been shown to consist of ten subunits, four of which are tightly associated with the mitochondrial inner membrane and are synthesized on mitoribosomes (Tzagoloff and Meagher, 1971). Synthesis of these four subunits has been associated with the development of oligomycin sensitivity, and, as will be discussed later, these four mitoribosome-synthesized subunits contain the sensitivity sites for oligomycin, venturicidin, dicyclohexylcarbodiimide (DCCD), and triethyltin salts (TET). The remaining six subunits of the ATP-synthetase complex are synthesized on cytoribosomes (nuclear coded); five of these are associated with the F_1-ATPase, and the other is the oligomycin sensitivity-conferring protein, OSCP. (This name is a misnomer, for there is no evidence that OSCP interacts with oligomycin.)

The whole area of mitochondrial biogenesis has been extensively reviewed recently (Ashwell and Work, 1970; Linnane and Haslam, 1970; Linnane et al., 1972); and a critical evaluation that points out some of the limitations of the experimental techniques used has been recently published (Schatz and Mason, 1974). The following major features emerge:

1. Mitochondrial protein complexes involved in energy conservation contain subunits synthesized on cytoribosomes (nuclear coded) and subunits synthesized on mitoribosomes (mtDNA coded).
2. Three subunits of complex IV (cytochrome oxidase) and two subunits of complex III (ubiquinone-cytochrome *c* reductase) are mitoribosome synthesized (mtDNA coded).

3. Four subunits of the ATP-synthetase complex are mitoribosome synthesized (mtDNA coded).
4. There is tentative evidence that the ADP translocase complex contains a mitoribosome-synthesized component.
5. Nuclear and mitochondrial mutations result in modification of components of the oxidative phosphorylation system.
6. If, as is generally assumed, mtDNA codes for mitochondrial proteins, it should be possible to isolate mitochondrial mutants in which the primary structure of a mitochondrial protein is altered.
7. Study of mitochondrial mutants may lead to the isolation and characterization of previously unknown components of the energy conservation system.
8. Study of nuclear and mitochondrial mutants that lead to a modification of the mitochondrial energy conservation systems may give information as to the mechanism of oxidative phosphorylation and/or the structural organization of the component complexes.

Biochemical genetic studies of yeast mitochondria relating to features 3, 4, 5, 6, 7, and 8 above that are under way in our laboratory are the subject of this paper.

UTILIZATION OF MITOCHONDRIAL DRUG-RESISTANT MUTANTS FOR ANALYSIS OF OXIDATIVE PHOSPHORYLATION

A systematic investigation of the biochemical genetics of the mitochondrial oxidative phosphorylation complex in *Saccharomyces cerevisiae* has been under way in our laboratory for the past five years utilizing mitochondrial drug-resistant mutants. The rationale of these investigations has been discussed previously (Griffiths et al., 1972) and is based on the following premises:

1. Drugs (inhibitors, uncouplers, ionophores) that affect mitochondrial energy conservation reactions have specific inhibitor sites associated with specific protein subunits of the oxidative phosphorylation complex.
2. Ideally the drug should be metabolically inert, and its mode

of action should be predominantly, if not exclusively, on the mitochondrial inner membrane.

3. The drug should have little or no effect when glucose is the carbon source during the fermentative phase of growth but should specifically inhibit the secondary aerobic growth phase during which mitochondrial oxidation of ethanol occurs. Similarly the drug should specifically inhibit growth when an oxidizable substrate such as glycerol or ethanol is utilized as the sole carbon source.

4. Drug-resistant mutants should exhibit modified sensitivity to the drug at the mitochondrial and sub-mitochondrial level and, ideally, in a purified membrane-free enzyme preparation. Demonstration that the mutation is cytoplasmically inherited and located on mtDNA is good *a priori* evidence that a mitochondrial gene product that is a component of the oxidative phosphorylation complex has been modified. The demonstration of mitochondrial inheritance of the resistance mutation is of key importance in assessing whether we are dealing with modification of a drug-receptor site on the mitochondrial inner membrane as opposed to a permeability or detoxification phenomenon.

The points of interaction of inhibitors, uncouplers, and ionophores on energy conservation are illustrated in figure 1. The various inhibitors have different sites of action that can be differentiated in biochemical studies. A large number of drugs that affect mitochondrial energy conservation systems have been tested for possible use in the isolation of drug-resistant mutants (Griffiths et al., 1972; Griffiths, 1972; Lancashire, 1974). The drugs that show promise as specific inhibitors of mitochondrial energy conservation processes are listed in table 1; uncoupling agents, ATPase inhibitors, adenine nucleotide translocase inhibitors, and ionophores satisfy the general criteria outlined above. Many of these compounds have been examined in detail, and specific drug-resistant mutants whose properties are described below have been isolated. Many of these mutants have been shown to be mitochondrial mutants and provide useful new markers for mapping of the mitochondrial genome.

OLIGOMYCIN-RESISTANT MUTANTS

Mutants resistant to high levels of oligomycin and rutamycin were isolated from drug plates containing oligomycin at 2.5, 5.0, and 10.0

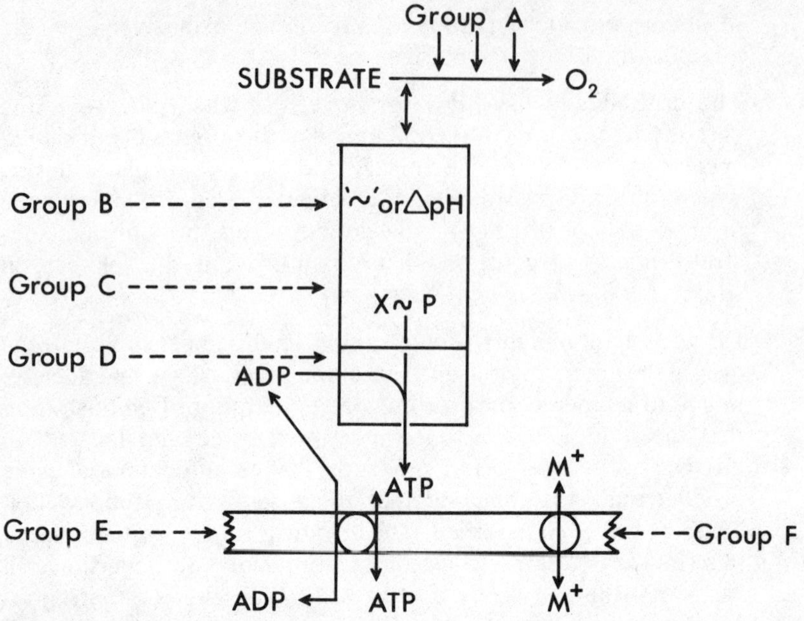

Fig. 1. Points of interaction of various inhibitors with the oxidative phosphorylation complex. Group *A:* inhibitors of electron transport; group *B:* uncoupling agents; group *C:* inhibitors of oxidative phosphorylation; group *D:* inhibitors of F_1-ATPase; group *E:* inhibitors of adenine nucleotide translocase; group *F:* ionophores.

μg/ml after the plates had been spread with approximately 5×10^6 cells previously ultraviolet-irradiated to reduce cell survival to 10–20%. The mutants isolated were subdivided into two classes on the basis of their cross-resistance to a number of inhibitors and uncouplers of mitochondrial energy conservation reactions (Avner and Griffiths, 1970; Avner and Griffiths 1973a; Griffiths et al., 1972). Class I mutants show cross-resistance to aurovertin, Dio-9, venturicidin, triethyltin, DCCD, bis-hexafluoroacetonyl acetone (1799), uncoupling agents, antimycin A, and protein synthesis inhibitors such as chloramphenicol, mikamycin, erythromycin, and cycloheximide. In contrast, Class II mutants are *specifically* resistant to oligomycin and structurally related antibiotics (rutamycin, ossamycin, peliomycin) and show no cross-resistance to venturicidin, triethyltin salts, DCCD, aurovertin, or uncoupling agents (table 2).

Genetic analysis of Class I mutants revealed a complex involvement of both nuclear gene and a cytoplasmic element (Avner & Griffiths,

TABLE I

INHIBITORS OF OXIDATIVE PHOSPHORYLATION AND THEIR
INHIBITORY EFFECTS ON GROWTH OF SACCHAROMYCES CEREVISIAE

Group	Inhibitor	Growth Inhibition in S. cerevisiae			Mutants Isolated
		Aerobic	Glycolytic	Differential	
Group A (Electron transport)	Antimycin A	Yes	No	Good	Yes
Group B (Uncouplers)	Dinitrophenol	Yes	Yes	Poor	Yes
	CCCP	Yes	Yes	Fair	Yes
	TTFB	Yes	Yes	Fair	Yes
	"1779"	Yes	No	Good	Yes
Group C (ATPase inhibitors)	Oligomycin	Yes	No	Good	Yes
	Rutamycin	Yes	No	Good	Yes
	Ossamycin	Yes	No	Good	Yes
	Peliomycin	Yes	No	Good	Yes
	Venturicidin	Yes	No	Good	Yes
	Trialkyltins	Yes	Yes	Good	Yes
	DCCD	Yes	Yes	Fair	Yes
	Botrycidin	Yes	No	Good	Yes
	Tetradifon	No	No	...	No
Group D (F_1 ATPase)	Aurovertin	Yes	No	Good	Yes
	Dio 9	Yes	No	Good	Yes?
	2-phenyl-isatogen	Yes	No	Good	No
Group E (ADP/ATP translocase)	Atractyloside	No	No	...	No
	Bongkrekic acid	Yes	No	Good	Yes
	Rhodamine 6G	Yes	No	Good	Yes
Group F (Ionophores)	Valinomycin	Yes	No	Good	Yes
	Gramicidin	No	No	...	No
	Nigericin	Yes	No	Good	Yes?
	Crown Polyethers	No	No	...	No
	Alamethicin	No	No	...	No

The results shown represent a screening of many laboratory stock yeast strains for their sensitivity toward growth inhibition by various inhibitors (Lancashire, 1974) when growing glycolytically (glucose carbon source) and aerobically (glycerol or ethanol carbon source).

Resistance was usually assayed by the "Drop-out" procedure described by Wilkie and Lee, (1968), using a range of plates containing various concentrations of inhibitor. In cases where both glycolytic and aerobic growth are inhibited the degree of specificity toward aerobic growth inhibition compared with glycolytic (called the glycolytic-aerobic differential) is given.

TABLE 2

CLASSIFICATION OF OLIGOMYCIN-, TRIETHYLTIN SULPHATE-, AND VENTURICIDIN-RESISTANT MUTANTS OF SACCHAROMYCES CEREVISIAE

Class	Levels of Resistance × Parental Strain						
	Oligomycin	Rutamycin	Venturicidin	Triethyltin	DCCD	"1799"	CAP, Antimycin A, CCCP, TTFB, Erythromycin, Cycloheximide
A. Oligomycin-resistant mutants							
1	>50	100	>50	10–20	>10	>20	2–5
2	>50	100	1	1	1	1	1
B. Triethyltin-resistant mutants							
1	>20		50	20		8–12	3–5
2A	1		10	20		1	1
2B	1		10	20		8	1
C. Venturicidin-resistant mutants							
1	>20		>50	20			3–5
2 (V,O)	>20		>100	1			1
2 (V,T)	1		10	>20		>20	1

1973b). Similar mutants have been isolated in studies of triethyltin-resistant, venturicidin-resistant, uncoupler-resistant, and ionophore-resistant strains (Griffiths, 1972), and a similar class of mutants have been described by Rank and Bech-Hansen (1973). No evidence has been obtained for a modification of a component in the ATP-synthetase complex in these mutants. The available evidence indicates that modification of mitochondrial membrane permeability to drugs may be involved, but no clear indication of specific biochemical changes in cellular membranes in these mutants is available at present.

All of the Class II mutants examined exhibited typical cytoplasmic inheritance, and the resistance determinants are located on mtDNA as judged by the criteria listed in table 3 (Avner and Griffiths, 1973b). Further genetic analysis of Class II oligomycin-resistant mutants (OLY^R) by recombination analysis and mapping studies have established two specific loci, OLI and OLII, on the mitochondrial genome for oligomycin resistants (Avner et al., 1973). The OLI and OLII loci are unlinked or very weakly linked, and the evidence indicates that the mutations are located on independent cistrons on mtDNA. Biochemical studies (Griffiths et al., 1972; Griffiths, 1972; Griffiths

TABLE 3

CRITERIA FOR MITOCHONDRIAL INHERITANCE AND GENETIC CHARACTERISTICS OF OLIGOMYCIN-RESISTANT MUTANTS OF SACCHAROMYCES CEREVISIAE

Criteria	Class I OL^R	Class II OL^R
1. Crosses of $OL^R \rho^+ \times OL^S \rho^+$ should show mitotic recombination; i.e., diploid zygotes give rise to resistant, sensitive, and mixed colonies.	Anomalous behavior; all colonies mixed. Resistance allele transmitted at very low rates.	Yes
2. Crosses of $OL^R \rho^+ \times OL^S \rho^-$ give only resistant zygotic progeny.	Colonies mixed; low transmission rates of resistance unaltered.	Yes
3. ρ^- strains derived from OL^R haploid may no longer carry the resistance allele.	No loss of resistance seen in any of the strains tested.	Yes (in some strains)
4. Meiotic products of an $OL^R \rho^+ \times OL^S \rho^+$ cross shown 4:0 segregation.	No; 2:2 segregation observed.	Yes
5. Linkage to other mitochondrial markers.	No	Yes

and Houghton, 1974) show that the ATPase and P_i-ATP exchange reactions of Class II mutants are markedly more resistant than these enzyme activities in the parental strains. These differences are demonstrable at the mitochondrial, sub-mitochondrial, and solubilized ATPase complex level and are consistant with the modification of one or more of the mitochondrially synthesized components of the oligomycin-sensitive ATPase complex. Current studies in this laboratory are concerned with the isolation of mitochondrially synthesized subunits of the ATPase complex and the comparison of mutant and parental strain subunits by peptide mapping and amino acid sequencing in order to establish a correlation between a mitochondrial mutation and a mitochondrial gene product.

TRIETHYLTIN-RESISTANT MUTANTS

Studies with triethyltin salts, specific inhibitors of mitochondrial oxidative phosphorylation, have led to the isolation of a series of triethyltin sulphate resistant mutants that fall into two general classes as found in the case of oligomycin-resistant mutants (table 2). A series of Class I resistant mutants similar to those isolated in studies of oligomycin mutants have been isolated. (Lancashire and Griffiths, 1971; Lancashire and Griffiths, 1975a). Class II mutants are specifically resistant to triethyltin and venturicidin and in some cases are also cross-resistant to the uncoupling agent 1799. Class II mutants are cytoplasmic, and the resistance determinant is deleted by the action of ethidium bromide during petite induction. Recombination studies indicate that the triethyltin mutations (TET^R) at locus TI are not allelic with the other mitochondrial mutations at loci OLI and OLII. This indicates that the binding or inhibitory sites of oligomycin and triethyltin are not identical and that the triethyltin binding site is located on a mitochondrial gene product different from those involved in oligomycin binding.

Interaction and cooperative effects between different binding sites on the mitochondrial inner membrane have been demonstrated in studies of the effect of the insertion of the TET^R phenotype into mitochondrial OLY^R mutants and provide an experimental basis for complementation studies at the ATP-synthetase level (Lancashire and Griffiths, 1975a).

VENTURICIDIN-RESISTANT MUTANTS

Venturicidin is another potent inhibitor of oxidative phosphorylation (Walter et al., 1967), and a series of venturicidin-resistant mutants

(VEN^R) similar to oligomycin-resistant and triethyltin-resistant mutants have been isolated (Griffiths et al., 1975). Class II mutants of *in vivo* phenotype $VEN^R OLY^R$ and $VEN^R TET^R$ are mitochondrial mutants. $VEN^R OLY^R$ mutants show a high degree of resistance to venturicidin and oligomycin at the whole cell and mitochondrial ATPase levels but, in contrast, little or no resistance at the mitochondrial level is observed with $VEN^R TET^R$ mutants. Mapping studies (Lancashire and Griffiths, 1975b) show that $VEN^R OLY^R$ mutants map at a new mitochondrial locus that is closely linked to the OLI locus. This locus has been termed OLIII. $VEN^R TET^R$ mutants map at the same locus as TET^R mutants and are probably identical to TET^R mutants of the same phenotype ($TET^R VEN^R$) isolated previously (Lancashire and Griffiths, 1975a).

Venturicidin resistance or sensitivity can thus be correlated with two binding sites on mitochondrial ATPase, one of which is common to the oligomycin binding site and the other to the triethyltin binding site. It has been postulated from biochemical genetic studies on OLY^R, TET^R, and VEN^R mutants that drug resistance involves modification of drug receptor sites on the mitochondrial ATPase (ATP synthetase) complex. These sites are composed of mitochondrially coded and mitochondrially synthesized components of the ATPase complex, and, to date, four mitochondrial loci have been established by genetic studies, OLI, OLII, OLIII, and TI (VI) (Lancashire and Griffiths, 1974b; Avner and Griffiths, 1973; Avner et al., 1973; Lancashire and Griffiths, 1975a; Lancashire et al., 1974). There is sufficient genetic evidence to indicate that loci OLI and OLIII are closely linked and that loci OLI, OLII, and TI (VI) are located on independent cistrons on mtDNA.

Binding of oligomycin to three attachment points on the ATPase complex, represented by loci OLI, OLII, and OLIII, and binding of venturicidin to two attachment points, represented by loci OLIII and TI (VI) can explain the biochemical genetic data. In the case of the TI (VI) locus the mutation probably does not modify a macrolide ring attachment point, but a modification involving the D-rhamnose residue of venturicidin is a strong possibility. This hypothesis can be tested directly in the case of venturicidin by analysis of the *in vivo* and *in vitro* sensitivity to the aglycone of venturicidin, which should lose the capability of binding to the TI locus due to loss of the rhamnose residue. As expected, venturicidin aglycone behaves as an analogue of oligomycin and not as venturicidin in its interaction with OLY^R, $VEN^R OLY^R$, and $VEN^R TET^R$ mutants. These findings provide strong evidence for separate interaction sites for oligomycin, triethyltin, and venturicidin and that we are dealing with modification of drug receptor sites on the ATPase complex.

The relationship of these mutations to the mitochondrially synthesized components of the ATP synthetase complex described by Tzagoloff and Meagher (1973) is under investigation, and should lead to the correlation of a mitochondrial mutation with modification of a mitochondrial gene product.

GENETIC ANALYSIS OF OLIGOMYCIN-, TRIETHYLTIN-, AND VENTURICIDIN-RESISTANT MUTANTS

The isolation of specific mitochondrial drug-resistant mutants that map at different sites on mtDNA provides additional loci for detailed

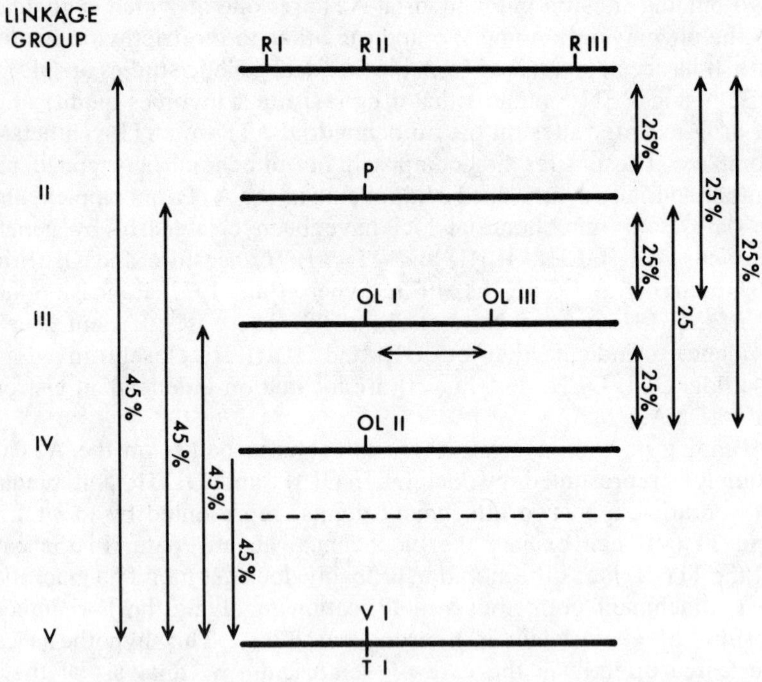

Fig. 2. Mitochondrial genetic map showing location of oligomycin-, triethyltin-, and venturicidin-resistant mutations. The values shown are total recombination frequencies observed in homosexual ($\omega^+ \times \omega^+$ and $\omega^- \times \omega^-$) genetic crosses (Lancashire and Griffiths, 1974). This summary also includes recombination data of other workers involving loci RI, RII, and RIII (Bolotin et al., 1971) and PI (Wolf et al., 1973).

mapping of the mitochondrial genome. Current information in this area is summarized in figure 2 and is a representation of mitochondrial loci in terms of *recombinational linkage groups* (Lancashire and Griffiths, 1975b). Recombinants between any of the ribosomal loci RI, RII, and RIII and any of the other loci, OLI, OLII, and PI, occur at a frequency of about 20–25%. This also applies to recombination between pairs of loci OLI, PI, and OLII, and would suggest that in this system 25% recombination is the maximum value obtainable between any pair of markers showing lack of close linkage. However, it is known from covariance data that linkage groups, 1, 2, 3, and 4 are not unlinked in the classical meaning of the word, i.e., on separate chromosomes. How, therefore, do we explain the 25% maximum recombination frequencies observed?

Models analogous to those applied in phage genetics (Visconti and Delbruck, 1953) are now being proposed and tested (see Birky, this volume) and are the source of controversy between several laboratories. The finding of free recombination by the ribosomal loci and the loci OLI, OLII, and PI has hindered identification of their relative positions on the mtDNA, but sufficient evidence is available to indicate that these loci are located at different sites.

Similar recombination studies involving TET^R mutants and VEN^R mutants have produced a completely different behavior pattern (Lancashire and Griffiths, 1975a,b), and the following results have been obtained:

1. The VEN^R and TET^R genes reassort randomly with the other known mitochondrial loci, RI, RII, RIII, OLI, OLII, and PI.

2. Recombination values of about 45% were observed between VEN^R and TET^R mutations and other mitochondrial loci.

3. The VEN^R and TET^R mutation is lost by treatment with ethidium bromide as observed for the mutations at loci RI, RII, RIII, OLI, OLII, OLIII, and PI but at a slower rate than that characteristic of the other markers.

4. Loss of the VEN^R and TET^R mutations following ethidium bromide treatment is *always* accompanied by loss of the ρ^+ state, i.e., the cell types ρ^+T^0 or ρ^+V^0 do not exist.

5. The TET^R mutation has several properties usually associated with episomes, such as effects on transmission and recombination of host genes (W. E. Lancashire, D. E. Griffiths, and P. P.

Slonimski, unpublished data) and determination of drug resistance.

These observations indicate that the TET^R and VEN^R mutations are located on a *separate* DNA molecule from that which is associated with the other mitochondrial loci. The fact that loss of the VEN^R and TET^R mutations is always accompanied by loss of the ρ^+ state indicates that the DNA molecule containing the former two sites is either directly responsible for maintenance of the ρ^+ state or indirectly responsible via its necessity for the stability of the mitochondrial DNA. This separate DNA molecule could even be a second mitochondrial DNA molecule that forms part of the mitochondrial genome. Provision would have to be made in such a model for the many genetic interactions that occur between the two sets of markers. A model that is under investigation proposes that VEN^R and TET^R are located on a small circular DNA that is a type of episome capable of being incorporated into the mitochondrial genome.

A possible candidate for the location of the TET and VEN markers is the small 2-micron circular DNA that has been identified in yeast by several workers (Guerineau et al., 1971; Clark-Walker, 1972; Clark-Walker, 1973; Zeman and Lusena, 1974). This DNA, known as omicron DNA, has a buoyant density similar to that of nuclear DNA, but is not a gene amplification product and cannot be related to "informosomal" DNA since highly purified nuclei do not contain the small circles. Preliminary investigations into the possibility that these small circles are responsible for the TET and VEN resistance determinants indicate no positive correlation between loss or retention of these markers and loss or retention of small circles (Zanders and Griffiths, unpublished data).

MUTANTS RELEVANT TO THE ADENINE NUCLEOTIDE TRANSLOCASE SYSTEM

The adenine nucleotide translocase is specifically inhibited by atractyloside (ATR) and bongkrekic acid (BA), and recent studies by Gear (1974) suggest that the lipophilic dye Rhodamine 6-G is also an inhibitor of the translocase system. Several classes of BA-resistant mutants have been isolated in this laboratory, but genetic analysis is not yet complete:

1. Class I mutants show a wide range of cross-resistance to other mitochondrial drugs.

2. Class II mutants are specifically resistant to BA. These mutants are characterized by a slow rate of growth and are petite negative. Preliminary genetic analysis indicates that these are nuclear mutants.

In addition, a third class of BA-resistant mutant that is cross-resistant to triethyltin, venturicidin, and "1799" but not cross-resistant to oligomycin has been isolated. The relationship of these to cytoplasmic (mitochondrial) triethyltin-resistant mutants described previously is under investigation. Further studies of mitochondrial triethyltin-resistant mutants have shown that they are cross-resistant to BA. Modified response to BA is also shown in *in vitro* translocase assays, whereas no modification of response to atractyloside was observed (Cain et al., 1974). These results indicate that the interaction sites of BA and ATR are not identical and that the ATP synthetase and translocase share a common mitochondrially synthesized subunit or that the ATP synthetase contains a mitochondrially synthesized subunit that is a regulatory subunit of ADP translocase. Some interesting results are also emerging with Rhodamine-6G-resistant mutants, and cytoplasmic mutants have been isolated that are cross-resistant to triethyltin and venturicidin. Biochemical studies of Rhodamine-6G mutants are in progress, and demonstration of modified response to Rhodamine-6G in the translocase assay will provide further evidence in support of the contention that the T and V loci represent modification of a mitochondrial gene product concerned in the full expression of the adenine nucleotide translocase system (Cain et al., 1974).

UNCOUPLER-RESISTANT MUTANTS

Uncoupler-resistant mutants have been isolated in this laboratory by selection against TTFB (tetrachloro trifluoromethyl benzimidazole). A class of mutants was selected that showed no cross-resistance against other mitochondrial drugs and inhibitors but showed cross-resistance against other uncoupling agents, even though of a different molecular structure (Griffiths, 1972). These mutants were shown to be cytoplasmic mutants, but as the resistance allele was not lost on petite induction with ethidium bromide, there is no evidence at present that this mutation involves modification of mtDNA. It was inferred from these studies that uncoupler resistance involved modification of a cytoplasmically determined binding site that was able to bind uncoupler molecules of different chemical structure. Supporting evidence has now been obtained by Hanstein and Hatefi (1974), who have described a mitochondrial uncoupler binding site that is able to discriminate between uncoupler ligands and ligands that have a similar chemical structure but are not uncouplers. In addition, binding studies with radioactive uncouplers have shown that the mitochondria from uncoupler-resistant strains have a modified binding site, as evidenced by a decrease in the affinity constant for the uncoupler (Hanstein, Skipton, and Griffiths, unpublished data). It also has been demonstrated recently that the uncoupler binding site is present in the ATP synthetase complex (Y. Hatefi, personal communication). The relationship of the component(s) involved in the uncoupler binding site to the mitochondrially determined binding sites for oligomycin, venturicidin, and triethyltin is the subject of active investigation.

IONOPHORE-RESISTANT MUTANTS

Studies with ionophores (Wright, Lancashire, and Griffiths, unpublished data) have led to the demonstration that valinomycin and nigericin are two potent inhibitors that appear to have a specific mode of action on mitochondrial metabolism. A specific valinomycin-resistant mutant has been isolated (Griffiths, 1972) and shown to be a mitochondrial mutant. The implications of this finding are that mitochondrial membranes contain a specific receptor site for ionophores such as valinomycin or that a co-transport system necessary for expression of valinomycin activity is modified in the mitochondrial valinomycin-resistant mutant. These possibilities are under active investigation in this laboratory, and the recent finding that valinomycin is a potent inhibitor of protein synthesis (Herzberg et al., 1974) is also being considered as an indication of the target site for valinomycin.

SUMMARY AND FUTURE TRENDS

This paper has described the use of drug-resistant mutants to analyze components of oxidative phosphorylation. Different mitochondrial gene products have been shown to be involved in the binding sites for oligomycin, venturicidin, and triethyltin; this being evidence not previously available from biochemical studies. These findings correlate well with known biochemical evidence and with studies of mitochondrial biogenesis carried out by Tzagoloff and co-workers. The correlation of a mitochondrial mutation with modification of a mitochondrial gene product should be achieved in the near future in studies of components of the ATP synthetase complex. In addition, the demonstration of new mitochondrial loci is of great value in mitochondrial genetics in "mapping" of the mitochondrial genome, and has opened up a wider range of possibilities in mitochondrial genetic studies. The possible involvement of "small circles," omicron-DNA, in mitochondrial gene expression is an intriguing possibility that has emerged as a result of studies on triethyltin- and venturicidin-resistant mutants. Of equal significance, in terms of mitochondrial membrane function, are the studies on uncoupler- and ionophore-resistant mutants that indicate clearly that mitochondrial membranes contain *protein* components that are involved in the mode of action of uncouplers and ionophores.

It is apparent that biochemical genetic studies of mitochondrial oxidative phosphorylation have elicited new parameters of this complex membrane-bound system and are providing the framework for further investigation of the genetics of energy conservation processes.

ACKNOWLEDGEMENTS

The work described in this paper has been carried out over the past five years by my research colleagues and graduate students including P. R. Avner, J. M. Broughall, K. Cain, E. J. Griffiths, W. E. Lancashire, P. Meadows, J. Mottley, and E. Zanders. The work was supported by grants from the Medical Research Council, the Science Research Council, the Tin Research Institute, and N.A.T.O. Grant 572.

REFERENCES

Ashwell, M., and T. S. Work. 1970. The biogenesis of mitochondria. Ann. Rev. Biochem. 39:251–90.

Avner, P. R., and D. E. Griffiths. 1970. Oligomycin resistant mutants in yeast. FEBS Letters 10:202–7.

Avner, P. R., and D. E. Griffiths. 1973a. Isolation and characterization of oligomycin-resistant mutants of *Saccharomyces cerevisiae*. Eur. J. Biochem. 32:301–11.

Avner, P. R., and D. E. Griffiths. 1973b. Genetic analysis of oligomycin resistant mutants of *Saccharomyces cerevisiae*. Eur. J. Biochem. 32:312–21.

Avner, P. R., D. Coen, B. Dujon, and P. P. Slonimski. 1973. Allelism and mapping studies of oligomycin resistant mutants in *S. cerevisiae*. Molec. Gen. Genet. 125:9–52.

Bolotin, M., D. Coen, J. Deutsch, B. Dujon, P. Netter, E. Petrochilo, and P. P. Slonimski. 1971. La recombinaison des mitochondres chez *S. cerevisiae*. Bull. Inst. Pasteur 69:215–39.

Cain, K., W. E. Lancashire, and D. E. Griffiths. 1974. Is the ADP-ATP translocase system influenced by mitochondrial genes? Biochem. Soc. Trans. 2:215–18.

Clark-Walker, G. D. 1972. Isolation of a circular DNA from a mitochondrial fraction from yeast. Proc. Nat. Acad. Sci. USA. 69:388-92.

Clark-Walker, G. D. 1973. Size distribution of circular DNA from petite-mutant yeast lacking ρ DNA. Eur. J. Biochem. 32:263-67.

Gear, A. R. L. 1974. Rhodamine 6G: A potent inhibitor of mitochondrial oxidative phosphorylation. J. Biol. Chem. 249:3628-37.

Greville, G. D. 1969. A scrutiny of Mitchell's chemiosmotic hypothesis of respiratory chain and plotosynthetic phosphorylation, pp. 1-72. *In* D. R. Sanadi (ed.), Current topics in bioenergetics. Vol. 3. Academic Press, New York.

Griffiths, D. E., and D. C. Wharton. 1961. Purification and properties of cytochrome oxidase. J. Biol. Chem. 236:1850-56.

Griffiths, D. E., P. R. Avner, W. E. Lancashire, and J. K. Turner. 1972. Studies of energy linked reactions: Isolation and properties of mitochondrial oligomycin resistant, trialkyltin resistant, and uncoupler resistant mutants of yeast, pp. 505-21. *In* G. F. Azzone, E. Carafoli, A. L. Lehninger, E. Quagliariello, and N. Siliprandi (eds.), The biochemistry and biophysics of mitochondrial membranes. Academic Press, New York.

Griffiths, D. E. 1972. The use of mutants in bioenergetics, pp. 95-104. *In* S. G. Van den Bergh, P. Borst, and E. C. Slater (eds.), Mitochondria biogenesis and bioenergetics. North Holland, Amsterdam.

Griffiths, D. E., and R. L. Houghton. 1974. Modified mitochondrial ATPase of oligomycin-resistant mutants of *Sacchromyces cerevisiae.* Eur. J. Biochem. 46:157-67.

Griffiths, D. E., R. L. Houghton, W. E. Lancashire, and P. M. Meadows. 1975. Isolation and properties of mitochondrial venturicidin-resistant mutants of *Saccharomyces cerevisiae.* Eur. J. Biochem. 51:393-402.

Guerineau, M., C. Grandchamp, C. Paoletti, and P. P. Slonimski. 1971. Characterization of a new class of circular DNA molecules in yeast. Biochem. Biophys. Res. Commun. 42:550-57.

Hanstein, W. G., and Y. Hatefi. 1974. Characterisation and localisation of mitochondrial uncoupler binding sites with an uncoupler capable of photoaffinity labelling. J. Biol. Chem. 249:1356-1362.

Hatefi, Y., A. G. Haavik, and D. E. Griffiths. 1962a. Preparation and properties of mitochondrial DPNH-coenzyme Q reductase. J. Biol. Chem. 237:1676-80.

Hatefi, Y., A. G. Haavik and D. E. Griffiths. 1962b. Reduced coenzyme Q (QH_2)-cytochrome *c* reductase. J. Biol. Chem. 237:1681-85.

Herzeberg, M., H. Breitbat, and H. Atlan. 1974. Interactions between membrane functions and protein synthesis in reticulocytes. Effects of valinomycin and dicyclohexyl-18-crown-6. Eur. J. Biochem.45:161-70.

Kolarov, J., and M. Klingenberg. 1974. The adenine nucleotide translocator in genetically and physiologically modified yeast mitochondria. FEBS Letters 45:320-24.

Lancashire, W. E. 1974. A genetic approach to oxidative phosphorylation. Ph.D. thesis, University of Warwick.

Lancashire, W. E., and D. E. Griffiths, 1971. Biocide resistance in yeast: Isolation and general properties of trialkytin resistant mutants. FEBS Letters 17:209-14.

Lancashire, W. E., and D. E. Griffiths. 1975a. Isolation, characterisation, and genetic analysis of trialkyltin resistant mutants of *Saccharomyces cerevisiae*. Eur. J. Biochem. 51:377–92.

Lancashire, W. E., and D. E. Griffiths. 1975b. Genetic analysis of venturicidin resistant mutants. Eur. J. Biochem. 51:403–13.

Lancashire, W. E., R. L. Houghton, and D. E. Griffiths. 1974. Two mitochondrial genes specifying venturicidin resistance in yeast. Biochem. Soc. Trans. 2:213–15.

Linnane, A. W., and J. M. Haslam. 1970. The biogenesis of yeast mitochondria. pp. 101–72. *In* B. L. Horecker, and E. R. Stadtman (eds.), Current topics in cellular regulation. Vol. 2. Academic Press. New York and London.

Perkins, M., J. M. Haslam, and A. W. Linnane. 1972. The effects of physiological and genetic manipulation on the anion transport systems of yeast mitochondria. FEBS Letters 25:271–74.

Rank, G. H., and N. T. Bech-Hansen. 1973. Single nuclear gene inherited cross-resistance and collateral sensitivity to 17 inhibitors of mitochondrial function in *S. cerevisiae*. Mol. Gen. Genet. 126:93–102.

Schatz, G., and T. Mason. 1974. Biosynthesis of mitochondrial proteins. Ann. Rev. Biochem. 43:51–87.

Tzagoloff, A., M. S. Rubin, and M. F. Sierra. 1973. Biosynthesis of mitochondrial enzymes. Biochim. Biophys. Acta. 301:71–104.

Tzagoloff, A., and P. Meagher. 1971. Properties of a dispersed preparation of the rutamycin-sensitive ATPase of yeast mitochondria. J. Biol. Chem. 246:7328–36.

Visconti, N., and M. Delbruck. 1953. The mechanism of recombination in phage. Genetics 38:5–33.

Walter, P., H. A. Lardy, and D. Johnson. 1967. Inhibition of phosphoryl transfer reactions in mitochondria by peliomycin, ossamycin, and venturicidin. J. Biol. Chem. 242:5014–18.

Wolf, K., B. Dujon, and P. P. Slonimski. 1973. Multifactorial mitochondrial crosses involving a mutation conferring paromomycin-resistance in *S. cerevisiae*. Mol. Gen. Genet. 125:53-90.

Zeman, L., and C. V. Lusena. 1974. Closed circular DNA associated with yeast mitochondria. FEBS Letters. 38:171–74.

Ziegler, D. M., and K. A. Doeg. 1959. The isolation and properties of a soluble succinic coenzyme Q reductase from beef heart mitochondria. Biochem. Biophys. Res. Commun. 1:344–49.

PHILIP S. PERLMAN

Cytoplasmic Petite Mutants in Yeast: A Model for the Study of Reiterated Genetic Sequences

4

INTRODUCTION

In the last decade great advances have been made in the analysis of the role of cytoplasmic genes in the biogenesis of mitochondria (Borst, 1972; Sager, 1972; Mahler, 1973a). As was made clear in the preceding papers, most workers in that area use cultured mammalian cells or fungi as the experimental system. Since results with those two groups of organisms are in good general agreement, it is possible to take advantage of organism-specific phenomena as an aid to the analysis. Since baker's yeast, *Saccharomyces cerevisiae,* is a facultative anaerobe, unlike many other yeasts, it has been possible to obtain respiratory-deficient mutants, ones that would be lethal in most eucaryotes. Such mutants have been very useful in delineating the role of cytoplasmic and nuclear genes in the biosynthesis of the cellular respiratory apparatus. However, the intensive analysis of the cytoplasmic respiratory-deficient mutants (known as petites or ρ^-) in recent years has revealed some unexpected information, information that demonstrates the usefulness of these mutants in the study of other matters of general interest to biologists and in particular geneticists.

In the space available I would like to develop the idea that cytoplasmic respiratory-deficient mutants may be useful for studying the mechanism of gene duplication and the genetic behavior of repeated deoxyribonucleic acid sequences (cf. Faye et al., 1973). To accomplish this, first I will review the earlier literature on these mutants; then I will describe

Department of Genetics, Ohio State University, Columbus, Ohio 43210 the evidence on the structure of mutant mitochondrial DNA (mtDNA) that shows that it contains internal repeats as well as other structural anomalies; then I will present some of our studies of the genetic behavior of such altered genomes; and finally, I will summarize our current degree of understanding of the mutagenic events leading to the generation of such repeats.

THE PETITE MUTATION

The first systematic study of cytoplasmic petite mutants was undertaken by Ephrussi and his students and was first reported in a series of papers commencing in 1949 (Ephrussi, Hottinguer, and Chimines, 1949; Ephrussi, Hottinguer, and Tavlitzki, 1949; Tavlitzki, 1949; Slonimski, 1949; Slonimski and Ephrussi, 1949; Ephrussi, L'Heritier, and Hottinguer, 1949; Ephrussi and Hottinguer, 1951). In that work they showed that euflavine is capable of converting virtually all respiratory-sufficient (hereafter referred to as wild-type or ρ^+) cells in a growing yeast culture to respiratory deficiency. At the time it was quite a remarkable finding that a chemical agent could induce a specific mutant syndrome with such efficiency. It is probably true that this particular response is quite unique to yeast since many of the specific responses to mutagens (see below) do not occur in the short term in organisms where the respiratory-deficient phenotype would be lethal in the long term.

The mutation, called ρ^-, was shown to be quite stable in that it never reverts to wild-type. The results of crosses with wild-type cells revealed a non-Mendelian mode of inheritance; most of the diploid progeny were wild-type, and when they were induced to undergo meiosis, all tetrads were found to contain four wild-type and no petite spores. When two petites are mated, all of the progeny are petite; thus all cytoplasmic petite mutants fall into a single complementation group and are deficient in the same cytoplasmic genetic element, which was termed "ρ." Additionally they reported that the mutation arises spontaneously with a rate that varied from strain to strain.

Since Ephrussi's germinal papers in this area, other workers have confirmed and extended these findings and have shown that many other agents or physical treatments are capable of inducing the cytoplasmic petite mutation (reviewed by Nagai et al., 1961). Although many intercalating dyes, such as euflavine and ethidium bromide (EtdBr), are effective, others, such as proflavine, are not (Mahler, 1973b). Additionally some agents that bind to DNA but not by intercalation (such as berenil) are also effective (Mahler and Perlman, 1973). Irradiation with ultraviolet light, but not X-rays (Raut and Simpson, 1955), growth at elevated temperatures (Sherman, 1959), and treatment with various inorganic ions (Lindegren et al., 1958) are other means of generating this mutation in yeast. The only common feature of these treatments that seems reasonable in light of current knowledge is interference with replication of the cytoplasmic genetic element. One could envision that growth of yeast cells in the presence of (or following treatment with) an agent that selectively interferes with cytoplasmic but not nuclear gene duplication would result in progeny cells (buds) that failed to receive a cytoplasmic genome from the parent cell. Certain agents, such as euflavine, do inhibit mtDNA replication and convert buds but not (immediately) mother cells into petite mutants (Ephrussi and Hottinguer, 1951); other treatments, especially with ethidium bromide, convert even the mother cell to respiratory deficiency (Slonimski et al., 1968). Thus, it appears that there may be several mutagenic mechanisms that differ from one another in at least one step. More will be said later in this paper about the primary mutagenic events caused by EtdBr.

When physiological properties of petite mutants were studied, it was found that all mutants have the same alteration (table 1). That

TABLE 1

THE PETITE PHENOTYPE

	Absent in petites	*Present in petites*
Growth characteristics	No growth on non-fermentable carbon sources Petite diploids cannot undergo meiosis	Grows on glucose at a slightly reduced rate Yields small (petite) colonies on solid glucose media
Respiratory enzymes	Cytochromes c_1, b, and aa_3 Cyt c oxidase; NADH:cyt c reductase; Succ:cyt c reductase	Cytochrome c Succinate and NADH dehydrogenase; TCA cycle enzymes; F_1-ATPase (oligo-insensitive)
Structural aspects	No inner-membrane cristae	Recognizable organelles with inner and outer membranes.
Macromolecular metabolism	Mitochondrial ribosomes and mitochondrial protein synthesis	MtDNA (in some mutants) MtDNA and RNA polymerases Mitochondrial transcripts in strains where mtDNA is present

is, they are respiration-deficient and lack all respiratory pigments except for cytochrome *c* (which is present at a slightly elevated level) (Slonimski and Ephrussi, 1949). All lack activities associated with the missing cytochromes (b, c_1 and aa_3) but possess low levels of certain mitochondrial dehydrogenases (Mackler et al., 1964, Mahler et al., 1971). Little ultrastructural analysis has been undertaken, but generally it is agreed that petite mitochondria are abnormal; they are present and recognizable in petite cells, but they lack the infoldings of the mitochondrial inner membrane (cristae) that are typical of wild-type mitochondria (Yotsuyanagi, 1962; Smith et al., 1969; Perlman and Mahler, 1970b). Recent studies have revealed some internal structure in petite mitochondria, but it is unlike anything seen in the parent cells grown under the same conditions (Grimes et al., unpublished data). A recent study by Perlman and Mahler (1970b) demonstrated that Krebs cycle and other enzymes present in petite cells retain the same subcellular localization as is found in wild-type cells; in this sense, it is reasonable to conclude that mitochondria in petite cells are rather similar to those in wild-type cells except that they do not respire.

It has been suggested that petite cells are an ideal experimental system for studying functions of mitochondria that are independent of oxidation and phosphorylation (Perlman and Mahler, 1970b). The localization of specific enzymes and enzyme systems in the mutant mitochondria is consistent with the notion that they are still functional in some respects. The recent demonstration that petite cells will not grow when mitochondrial ATP translocation is inhibited further supports this idea (Subik et al., 1972).

GENETIC HETEROGENEITY OF PETITE MUTANTS: SUPPRESSIVENESS

The first evidence suggesting that petites are a heterogeneous class of mutants appeared in 1955 (Ephrussi et al., 1955). It was reported that certain mutants when crossed to a wild-type produced diploids in which the wild-type cytoplasmic genetic element (ρ^+) has been lost (fig. 1). Such mutants were said to have suppressed ρ^+ and were termed "suppressive petites" or petites that possess a "suppressive cytoplasmic genetic element." In work that followed it was shown that suppressiveness is a quantitative character in that different petites can suppress ρ^+ to different extents (Ephrussi and Grandchamp, 1965). No physiological differences between suppressive and neutral petites have been detected yet. As will be discussed below, certain petites

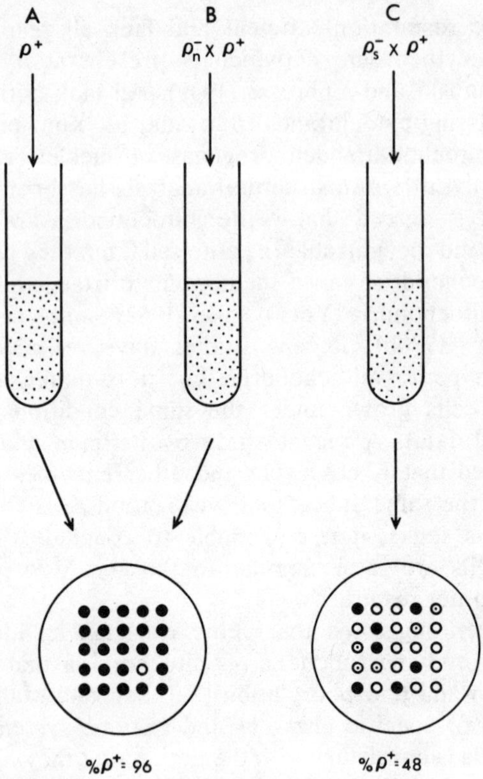

Fig. 1. Genetic analysis of petite mutants. *A*. Control: The percentage of spontaneous petites in the wild-type tester culture is measured by plating cells on solid glucose medium. The resulting clones are tested for respiratory ability by the tetrazolium overlay procedure of Ogur et al. (1957). Wild-type colonies are denoted by shaded circles and petite ones by unshaded circles. *B*. Neutral petite (ρ_N^-) × wild-type: Cultures of both haploid strains (with complementing nuclear auxotrophic requirements and of opposite cellular mating types) are mixed in liquid medium. Mating requires from 2–4 hours in most cases; after that time the mating mixture is washed and plated onto minimal medium to select for prototrophic diploids. The resulting clones (called zygotic clones) are then scored for the percentage ρ^+ as above. If the percentage of ρ^- zygotic clones equals the percentage of petites in the tester culture, then the petite is neutral; if it exceeds the percentage of haploid petites, then the mutant is suppressive. *C*. Suppressive petite (ρ_S^-) × wild-type: The cross is performed and zygotic clones analyzed as above. In such a cross three classes of clones are usually obtained, pure ρ^+, pure ρ^-, and mixed (partly shaded circles). The latter class is usually scored as ρ^+. In the figure, 12 of 25 clones are scored as ρ^+, and so the petite is 50% suppressive using the equation, suppressiveness = 100(X − Y/1 − Y) where X is the percentage of petite zygotic clones and Y is the percentage of petites in the wild-type tester (Sherman and Ephrussi, 1962).

retain some elements of mitochondrial macromolecular biosynthesis (table 1), but those differences do not result in an alteration of the respiration-deficient phenotype.

The mechanism of suppressiveness is still unknown, although it is likely that the mechanism is a genetic one rather than physiological. There appear to be interactions between mutant and wild-type cytoplasmic genetic elements which occur in the zygote that result in the inactivation of the wild-type element. The suppressiveness of a particular petite is reproducible when experimental conditions are held constant; however, when growth conditions are altered, the percentage of ρ^+ zygotic clones can be increased or decreased (Ephrussi et al., 1966). Presumably, the changed conditions influence the efficiency of the events that inactivate wild-type. The degree of suppressiveness of a petite also can be influenced by the genotype of the wild-type tester used in the cross (Bech-Hansen and Rank, 1973b; Michaelis et al., 1971; Perlman, unpublished data); however, no one has ever reported a tester strain that can increase or decrease the suppressiveness of a particular petite by a very large amount. It has been observed that the level of suppressiveness of a petite may change upon subcloning (Ephrussi and Grandchamp, 1965; Saunders et al., 1971; Nagley and Linnane, 1972; Bech-Hansen and Rank, 1973a; Perlman, unpublished data). Certain of those cases may be explained most simply by postulating that the petite cell has given rise to a bud that failed to receive the suppressive factor. Treatments that convert ρ^+ to ρ^- also are capable of converting a suppressive petite into a neutral one (Uchida, 1972; Kao, 1973; Perlman et al., unpublished data). Some cases in which an increase in suppressiveness was found are not understood so easily; this problem has not been studied extensively yet in the light of current knowledge and should be reexamined.

I will return to the nature of the postulated interactions that suppress ρ^+ in later sections.

MITOCHONDRIAL DNA IN PETITES: DELETIONS AND REITERATIONS

The discovery of mtDNA in the early 1960s (Nass and Nass, 1963a,b; Luck and Reich, 1964; Corneo et al., 1966; Tewari et al., 1966) was soon followed by the demonstration of mtDNA in petite mutants of yeast (Mounolou et al., 1966; Mehrotra and Mahler, 1968). A careful analysis of the properties of petite mtDNA can give some insight mutagenic mechanism; in addition, some novel properties of xperimental system are revealed.

ld-type mtDNA has a molecular weight of roughly 50×10^6 daltons. The genome is believed to be circular with a contour length of 25 μm (Hollenberg et al., 1969, 1970). Existing evidence suggests that all informational sequences in fungal mtDNA are present in one copy per 50×10^6 daltons (Schäfer and Küntzel, 1972). In a cell, roughly 15% of the total DNA is mitochondrial, and a haploid cell contains roughly 44 genomes of 50×10^6 daltons each (Fukuhara, 1969; Grimes et al., 1974). The molecule has an unusually high content of A+T (roughly 81%), and detailed thermal denaturation studies have shown that the distribution of bases along its length is non-random; that is, there are regions of very high A+T interspersed with regions of much lower A+T (Bernardi et al., 1970, 1972; Bernardi and Timasheff, 1970; Piperno et al., 1972). Recently, Prunell and Bernardi (1974) have suggested that the A+T-rich regions, constituting roughly one-half of the genome, serve as spacers that separate the information-containing higher G+C regions. They noted that such an arrangement of genes and spacers is probably a eucaryotic characteristic of mtDNA since it resembles the arrangement of certain genes in eucaryotic nuclei (Brown et al., 1972; Brown and Sugimoto, 1972; Clarkson et al., 1973a,b). The molecular weight estimates are based chiefly on electron microscopic analysis of molecules released onto grids from isolated organelles (Hollenberg et al., 1970); to date no one has been able to isolate 25 μm linear or circular molecules in reasonable yield although molecules of roughly 10–15 μm are obtained quite readily (Blamire et al., 1972; Mahler, unpublished data). Thus, it is clear that the mtDNA of yeast is larger than mtDNA from higher animals, which has a contour length of about 5 μm (see Borst, 1972). The use of renaturation kinetics to estimate the genetic complexity of the mitochondrial genome in yeast has yielded equivocal results, results that probably are due to the very unusual base composition and arrangement present in the molecule (Hollenberg et al., 1969; Faye et al., 1973; Christiansen et al., 1974; Michel et al., 1974).

Many petite mutants have been tested for the presence of mtDNA and the following pertinent observations have been made:

1. Petites sometimes lack mtDNA entirely, and all such petites are neutral (Nagley and Linnane, 1970). This finding is a rather recent one, and no systematic analysis of the relative efficiencies of various mutagens in inducing this class of petite has been undertaken under carefully controlled conditions.

2. When mtDNA is present in a petite, it is often physically different from wild-type mtDNA. Petites with increased, decreased, or unchanged buoyant density (average base composition) have been reported (Mounolou et al., 1966; Mehrotra and Mahler, 1968; Carnevali et al., 1969; Michaelis et al., 1971; Hollenberg et al., 1972; Bernardi et al., 1970; Lazowska et al., 1974); in addition, an individual wild-type strain can yield all of these types (Lazowska et al., 1974). The earliest reported cases were ones where the petite mtDNA was enriched in A+T relative to the parent strain; ones with unchanged base composition were also found, and only recently have rare ones with increased content of G+C been obtained. Although all suppressive petites possess mtDNA, no correlation between its base composition and suppressiveness has been noted even after a careful study (Michaelis et al., 1971).

3. Linnane and his colleagues have suggested that all mtDNA-containing petites (referred to later as "senseful petites") are suppressive; that is, all truly neutral petites are those lacking mtDNA (Nagley and Linnane, 1970, 1972; Nagley et al., 1973). Although this hypothesis is attractive for a variety of reasons, it is not entirely correct. Several laboratories have isolated and characterized senseful petites that are neutral, although such mutants may be rather rare (Hollenberg, Borst, et al., 1972; Moustacchi, 1972; Slonimski, personal communication; Perlman, unpublished data). In several such instances the neutral senseful petites have been shown to contain a normal (wild-type) amount of mtDNA per cell (Mahler and Perlman, unpublished data). It is thus reasonable to assign suppressiveness to some feature of petite mtDNA, but one that is not obligatory and that can be present to a greater or lesser extent.

4. In recent years mutations in wild-type mtDNA, other than ρ^-, have been characterized. These are believed to be base substitution mutations, and each was detected because it confers resistance to one of a number of antibiotics, such as chloramphenicol (C), erythromycin (E), spiramycin, oligomycin (O), paromomycin (P), and mikamycin, each of which interferes with mitochondrial biogenesis or function in cells growing on a non-fermentable carbon source, where mitochondrial function is required for prolonged growth (Thomas and Wilkie, 1968; Linnane et al., 1968; Coen et al., 1970; Wakabayashi and Gunge, 1970; Stuart, 1970; Howell et al., 1974). It has been shown that petites retaining any marker can be isolated (Gingold et al., 1969; Rank, 1970; Coen et al., 1970; Bolotin et al., 1971; Uchida and Suda, 1973; Suda and Uchida, 1974; Deutsch et al., 1974; Perlman et al.,

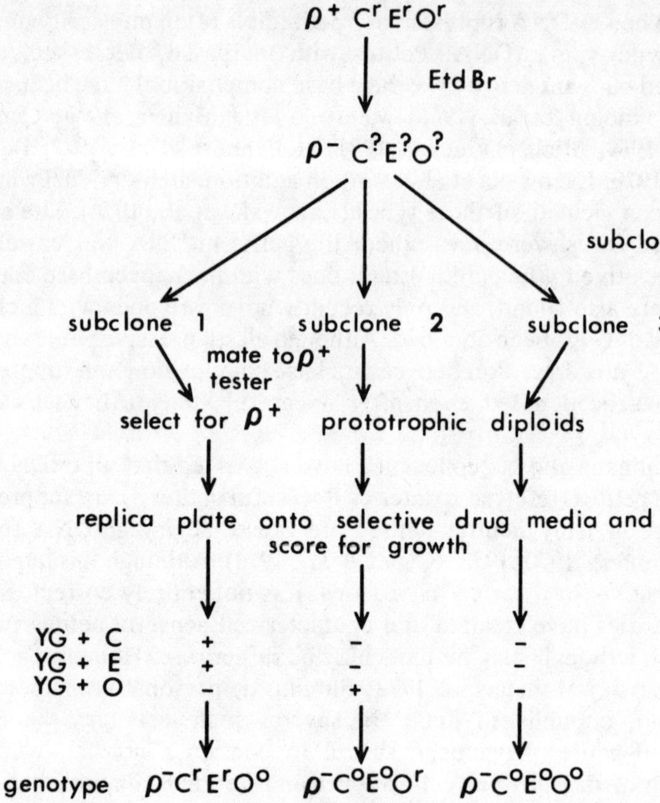

Fig. 2. Analysis of mitochondrial genotypes in petites. This procedure is essentially as described by Deutsch et al. (1974). Where two markers are transmitted by a petite (subclone 1), it is necessary to test for growth on selective medium containing both drugs before the genotype can be established with certainty. This test is suitable for screening both neutral and suppressive petites.

unpublished data; Slonimski, personal communication). Since petites cannot grow on media selective for the drug-resistance phenotype, it is not possible to score them directly for markers that might be present in their mtDNA. Genotypes of petites then are determined by measuring the ability of a petite to transmit a marker to the wild-type progeny of a cross with a wild-type tester. This procedure is illustrated in figure 2. It is, of course, possible that a marker might be present but in a form that interferes with its transmission to wild-type progeny. Studies of genetically defined petites show that any of the possible genotypes may be found in a suppressive or a neutral petite and in petites of different degrees of suppressiveness. Furthermore, petites that have lost all known markers may be suppressive. Thus, it appears that there is no single specific region of mtDNA that must be retained for suppressiveness to be expressed.

5. The origin of sequences contained in petite mtDNA has been studied in several laboratories. An early hypothesis was that petite mtDNA arises during an aberrant round of mtDNA replication in a wild-type in which "nonsense" sequences are generated with concomitant deletion of sequences necessary for the wild-type phenotype (Carnevali et al., 1969). The hypothesis of nonsense regions being present was rather attractive at first since many of the petites characterized first have mtDNA with a very unusual base composition, some being nearly pure A+T, which is, in an informational sense, nonsense. It is unlikely that ongoing mtDNA replication is required for petite induction since many of the most effective mutagens actually are potent inhibitors of just that process.

Recent studies using DNA-DNA hybridization procedures have searched for the presence of sequences in petite mtDNA that are not found in wild-type mtDNA, that is, for the hypothetical nonsense regions. If a petite genome possesses only sequences found in the wild-type mtDNA, then wild-type mtDNA should be capable of removing all of the petite mtDNA from solution; on the other hand, if the petite contains, say, 50% nonsense sequences, then only 50% of it would anneal to wild-type mtDNA. When experiments of this sort were carried out, it was found that some petites do have only wild-type sequences and so could be viewed as simple deletion mutants; with certain other petites the data showed that some new sequences were present (Gordon and Rabinowitz, 1973; Lazowska et al., 1974). Sanders et al. (1973) have argued that such results may be due to artifact (caused by the unusual nature of petite mtDNA), and, for the moment, theirs seems to be a persuasive argument.

6. It has been possible to estimate the amount of wild-type information retained in an individual petite using the techniques of DNA-DNA hybridization and reannealing kinetics. Although the amount retained varies among a set of mutants, it is clear that mutants rarely retain more than 60% and often retain less than 10% (Sanders et al., 1973; Lazowska et al., 1974; Michel et al., 1974). These data are quite strong for comparing different petites using a single method of analysis, but it is a bit disconcerting that the techniques of DNA-DNA hybridization and reannealing kinetics do not yield quantitatively equivalent

results for each petite tested. In any event, if a mutation were a simple deletion event, then the molecular weight of the mutant genome should equal the amount of retained information; that is, a petite with 1% retained information should have mtDNA molecules of 0.25 μm in length, and one with 10% should yield 2.5 μm molecules. On the contrary, petites have mtDNA of at least 10 μ in length (and probably larger) regardless of the amount of retained information (Faye et al., 1973; Locker et al., 1974; Mahler and Perlman, unpublished data). This finding requires that the molecules be intramolecularly reiterated. Some direct physical evidence supports the above conclusion. In several instances repeating units of a size anticipated by the known kinetic complexity of the mtDNA in question have been observed using the denaturation mapping procedure of Inman (1967)(Faye et al., 1973; Locker et al., 1974). Those studies revealed the presence of head-to-tail repeats as well as unusual inverted regions. Additional studies of that sort should provide an extensive catalog of the types of sequence alterations that may be found in petite mutants and that, in turn, will give valuable insight to the primary mutagenic process. I will return to this point later.

7. The nature of the retained sequences has been studied using DNA-RNA hybridization procedures; in those studies petite mtDNA was scored for the presence or absence of wild-type genes for stable RNA species (Faye et al., 1973), and a number of tRNA species (Cohen et al., 1972; Cohen and Rabinowitz, 1972) may be retained in individual petites. When an individual petite is tested for retention of a set of genes for stable RNAs, it is rare for all to be present. As was found with genetic markers, a wide variety of combinations of retained and lost sequences is possible. In several cases retention of specific RNA genes has been correlated with retention of specific antibiotic-resistance markers (Fukuhara et al., 1974). These results further demonstrate that at least some of the informational content of petite mtDNA is derived from wild-type sequences. It has also been shown that petites are active in mitochondrial RNA synthesis using the petite mtDNA as template and that the products bind as well to wild-type as to mutant mtDNA (Fukuhara et al., 1969; Fauman et al., 1973). In one case active leucyl tRNA has been identified in extracts of petite mitochondria (Faye et al., 1973); thus, the retained sequences are quite normal, at least over a span of 100 or more bases.

8. The DNA-RNA hybridization studies cited above have revealed additional information about the nature of petite mtDNA. It appears

that individual retained sequences may be differentially reiterated in petite mtDNA. To use a specific example, Faye et al. (1973) have reported that petite strain D21 has retained genes for both leucyl and isoleucyl tRNA. However, the mutant has more copies of leucyl tRNA than does wild-type but has roughly the same number of isoleucyl tRNA genes. This is best explained by proposing that reiteration of wild-type sequences occurred during the mutagenic events, but that not all retained sequences necessarily are reiterated to the same extent. A similar conclusion can be drawn from a consideration of the transmission genetics of markers contained in petite mtDNA (see below).

The above paragraphs have summarized the current status of the analysis of petite mtDNA. It is clear that essentially single-copy wild-type sequences become reiterated and rearranged during the mutagenic process. It is widely believed that gene duplication was an important factor in the evolution of eucaryotic genomes; additionally, reiterated sequences are a common feature of eucaryotic genomes. Yet the molecular processes that would produce such alterations (other than non-reciprocal recombination) and that control such events have not been characterized despite our detailed knowledge of the arrangement of some such repeated sequences. I suggest that some new insights to those important processes may be obtained from a detailed analysis of the molecular biology of petite induction. Before proceeding with a review of what is already known about petite induction, I would like to return to the question of suppressiveness.

A MOLECULAR MECHANISM FOR SUPPRESSIVENESS

A number of possible mechanisms for suppressiveness have been discussed in the literature and recently reviewed by Mahler (1973a). All of them focus on the nature of events occurring in the zygotes formed by the fusion of a suppressive petite cell with a wild-type one. Some workers have suggested that suppressive petite mitochondrial genomes, by right of their apparent simplicity or avidity for membrane replication sites, would have a replicative advantage over wild-type mtDNA so that they would be transmitted preferentially to the diploid progeny of the zygote (Slonimski et al., 1968; Carnevali et al., 1969; Rank and Person, 1969). A similar hypothesis suggests that petite mtDNA (or mitochondria) are transmitted preferentially to buds because of some intrinsic difference between the two types of mtDNA (mitochondria) in the zygote. Alternatively, the same result would be obtained if wild-type mtDNA were preferentially destroyed due to the presence of petite mitochondria; such destruction could be mediated only by petite mtDNA or a transcript of it since petites are not active in mitochondrial protein synthesis. And finally, it has been proposed that the spreading of hypothetical nonsense sequences in petite mtDNA to wild-type mtDNA by recombination might inactivate wild-type mtDNA (Coen et al., 1970).

My colleague Dr. Birky and I have recently proposed a molecular mechanism for suppressiveness that we feel is more compatible with the above-reviewed genetic and biochemical observations than is any of the previous models; in a sense, ours is a modification of the last two models (Perlman and Birky, 1974). We propose that the interaction occurring in zygotes that suppresses ρ^+ is recombination; we further propose that it is the nature of the repeating units in petite mtDNA and not hypothetical nonsense sequences that determines whether a particular petite is suppressive or neutral and that determines differences in level of suppressiveness among various petites.

Basically we suggest that petites whose repeating units contain internal deletions will result in the conversion of ρ^+ strands to ρ^- ones in a gene conversion-like process that requires the formation of a hybrid region (heteroduplex) containing one ρ^+ and one ρ^- strand in an otherwise canonical recombinational exchange. In figure 3 we present four possible arrangements of repeating units in mtDNA that happen to have retained the markers C^r and E^r. All four possess homology with wild-type (due to retained sequences), and so recombination with wild-type mtDNA should occur provided that the necessary organellar interactions can occur between physiologically different mitochondria. The petite mtDNA does recombine with wild-type (Coen et al., 1970; Bolotin et al., 1971; Saunders et al., 1971; Deutsch et al., 1974) or even with another DNA-containing petite (Michaelis et al., 1973), often with spectacular efficiency (see table 2).

Arrangement A, in which a region of wild-type information has been retained intact and simply repeated in a head-to-tail fashion (as is usually the case with reiterated sequences in eucaryotes), should behave as a simple deletion mutant. Reciprocal recombination could occur within a repeat; the act of recombination should be consummated with no adverse effect on wild-type mtDNA yielding one wild-type and one petite recombinant. Such a petite should be neutral since

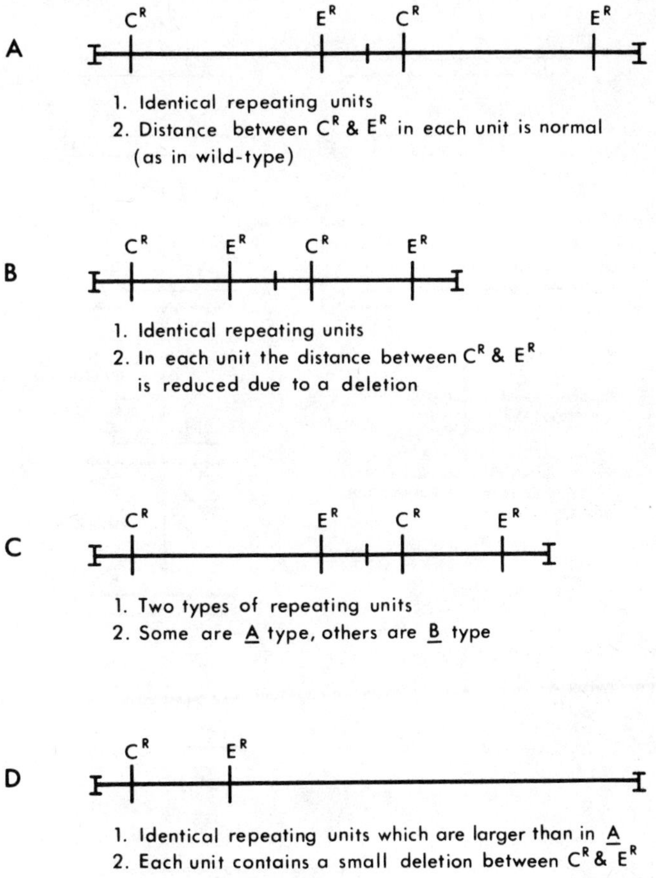
A
C^R E^R C^R E^R
1. Identical repeating units
2. Distance between C^R & E^R in each unit is normal (as in wild-type)
B
C^R E^R C^R E^R
1. Identical repeating units
2. In each unit the distance between C^R & E^R is reduced due to a deletion
C
C^R E^R C^R E^R
1. Two types of repeating units
2. Some are A type, others are B type
D
C^R E^R
1. Identical repeating units which are larger than in A
2. Each unit contains a small deletion between C^R & E^R

Fig. 3. Some possible arrangements of mitochondrial sequences in petite mtDNA. In the figure are depicted one or two repeating units in molecules that repeat the basic unit(s) as shown. Distinguishing features in each case are indicated in the figure and the predicted behavior in crosses with wild-type is discussed in the text. For clarity we have not shown the spiramycin locus between C^r and E^r; Deutsch et al. (1974) have shown that such petites usually retain that locus as well.

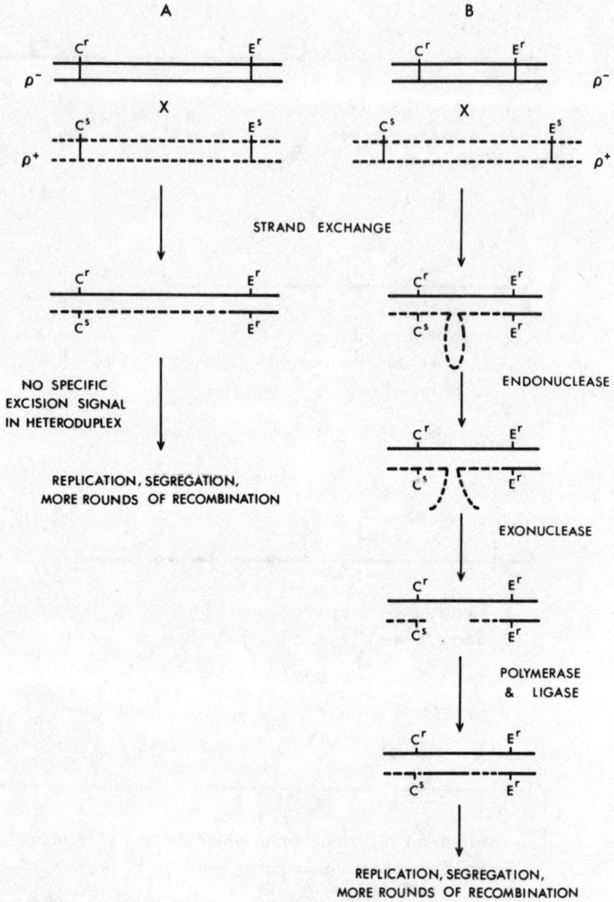

Fig. 4. The role of internal deletions in suppressiveness. The elements of this model are shown as a series of interactions of duplex DNA molecules, each line representing a single strand. Petite strands are solid lines, and wild-type ones are broken lines. Only a portion of each circular genome is shown; the distances between markers in petite A is as in wild-type, while in petite B it is reduced due to an internal deletion (see fig. 3). In crosses A and B only one of the two (possible) reciprocal recombinants is shown. Refer to the text for additional description of the events shown in the figure.

it is not capable of inactivating the wild-type mtDNA present in the zygote by recombination.

Petite B differs from A in that there is an internal small deletion within each repeating unit. Recombination will still be initiated within a repeating unit, but now a heteroduplex would be formed in which base-pairing would be disrupted at the position of the internal deletion (see fig. 4). Such a heteroduplex region would contain a region of single-stranded DNA in the wild-type strand; that single-stranded loop could serve as a substrate for a single-strand specific endonuclease. Once the strand were broken, a process not unlike pyrimidine dimer excision/repair could ensue, in which a region of wild-type DNA (the loop) is excised, the resulting gap is replaced by polymerase action (except for the loop whose bases are not present on the petite strand used as template) and the remaining nick is sealed by ligase. The result of this process (besides exchange of genetic information) is the net loss of bases from the wild-type mtDNA; if the lost bases are essential for respiratory sufficiency, then the act of recombination will have converted a wild-type mitochondrial genome into a petite one. According to this scheme, each round of recombination would reduce the number of wild-type genomes remaining in the cell, and a petite that causes these events to occur each round would be highly suppressive. For the moment, we propose that recombination occurs preferentially *within* each repeat and not at the juncture of two repeats; this is likely since all petites have junctures but not all are highly suppressive.

Along these lines, it is possible to envision alternate arrangements of repeating units that would be of intermediate suppressiveness; two of these are indicated in fig. 3C and D. In petite C the mtDNA contains some repeats of type A and others of type B so that only a fraction of the rounds of recombination would result in a loss of wild-type information. In D the amount of retained information is larger than in cases A, B, and C, and an internal deletion is found in only one place; recombinational exchanges occurring distal to the internal deletion would be as in case A while those near the internal deletion would be as in case B.

We find that this model is consistent with all of the known evidence concerning petite genetics and mtDNA. It is especially compatible with the knowledge that virtually any retained region may be found in neutral and suppressive petites. Thus, it is not necessary to postulate the existence of specific suppressive regions nor is it necessary to rely on postulated "nonsense" regions for suppressiveness. Currently we are testing some of the predictions that follow from this model. What is lacking at the moment is a systematic study of repeating units in petites compared to the genetic behavior of each such petite. Technically such an analysis is feasible since the physical analysis of petite mtDNA has become quite sophisticated in recent years. The main difficulty lies in establishing criteria needed for selecting sets of petites appropriate to this study from the large number of petites that are available or that can be generated with ease.

Studies of the transmission genetics of petites suggest a means to this end (Perlman et al., unpublished data). In one experiment we isolated from a wild-type strain a set of 21 petite mutants all of which had retained the mitochondrial markers C^r and E^s but had lost those for O^s and P^s; thus their genotype is $\rho^- C^r E^s O^o P^o$ (for completeness, they have also retained the marker ω^-; cf. Birky, this volume). All of the petites are capable of transmitting their markers (recombining) to wild-type progeny in a cross to a wild-type tester strain of the genotype $\rho^+ \omega^- C^s E^r O^s P^s$; however, it was found that the petites transmit their markers to various extents. Table 2 summarizes the results of crosses using one petite member of each transmissional class that has been observed to date. It is clear that the level of transmission of the two markers is not constant among these petites; in one class (two other petites like this one were isolated independently) the transmission of E^s exceeds that of C^r. Two of the petites shown transmit both markers equally but to a greater extent than does the parent (IL16); it is these two that are essentially neutral while the poorer-transmitting classes are both suppressive. Since all of the strains used are isonuclear and derive their mitochondrial sequences from the same parent, it seems quite reasonable to ascribe the transmissional levels to variations in gene arrangement or dosage of the mtDNA. It is tempting to speculate that petites such as E10-5/1-3 show differential reiteration of E^s relative to C^r. Additional details about these petites are contained in the legend to table 2.

Thus, using the transmission genetics of petites to assess genetic differences within a large set of petites, it is possible to reduce such a set to a smaller number of distinctly different mutants. Some care must be taken in such studies to avoid accumulation of segregants lacking mtDNA in the course of subculturing the mutants; we have had, however, excellent success in maintaining these petites for over two years with no changes in transmission level or class that could not be remedied by subcloning. We have since shown that transmis-

TABLE 2

PETITE TRANSMISSIONAL CLASSES

Strain $\omega^- C^r E^s$	RANDOM PROGENY (%)				N	%C^r	%E^s	% ρ^- Zygotic clones
	C^rE^s	C^sE^r	C^rE^r	C^sE^s				
ρ^+ IL16	46.6	44	6.1	3.3	5,831	52.7	49.9	4
ρ^- E10-5/1-3 . . .	19.7	63.7	2.3	14.4	4,831	22	34.1	60
ρ^- E10-10/10-7/3	17	71.2	4.4	7.3	563	21.4	24.3	39
ρ^- B30-41/1-3 . .	64.9	28.1	2.6	4.4	2,600	67.5	69.3	8
ρ^- B30-13/7 . . .	91.4	6.8	1	0.9	4,561	92.4	92.3	5

Each strain was mated to strain 55-R5-3C/221 ($\rho^+\omega^- C^s E^r$).

Strain IL16 and its derived petites were grown to exponential phase in glucose medium (Perlman and Mahler, 1970); the tester strain, 55R5-3C/221 ($\rho^+\omega^- C^s E^r$) was grown in glycerol medium. Cells were mated in suspension in glucose medium at roughly 10^7/ml for 2-3 hours (Sena et al., 1973), washed, and plated on minimal glucose medium to select for prototrophic diploids. The genotypes of random progeny of the cross were determined by the replica plating of progeny subclones onto selective media (Coen et al., 1970). Suppressiveness (percentage of ρ^- zygotic clones minus the number in the $\rho^+ \times \rho^+$ cross) was measured by staining clones containing wild-type cells with triphenyl tetrazolium (Ogur et al., 1957). Each petite mutant shown in the table is an independent isolate; each has been in use in our laboratory since 1972, and none has shown any change in transmission or suppressiveness that was not restored by subcloning. Petites having an "E" in the strain designation were obtained by EtdBr mutagenesis, and those with the letter "B" are berenil induced (Mahler and Perlman, 1973); we will show elsewhere that these two mutagens induce, in part, different transmissional classes of petites (Perlman, unpublished data).

sional classes are found with petites having other combinations of mitochondrial markers, and currently we are extending this work. Thus, a systematic study of the physical properties of mtDNA in such petites should constitute a test of our suppressiveness model; it is important that petites that are largely homologous (identical genotype) and differ chiefly in suppressiveness be used for this analysis. In addition, the results of such a study will yield valuable insights into the genetic behavior of repeated sequences resembling those present in all eucaryotic nuclear genomes.

INDUCTION OF PETITES BY ETHIDIUM BROMIDE

With these insights into the great variety of petite mtDNAs we should return to the question posed earlier, namely, the nature of the mutagenic events giving rise to such rearrangements. In this section I want to develop the notion that repair processes are found in yeast mitochondria and that they may be responsible for generating multiply repeated genomes such as are found in petites (Mahler and Perlman, 1972a; Mahler, 1973c).

Studies of the molecular biology of petite induction were made possible by the discovery by Slonimski et al. (1968) of the great efficacy of the trypanocidal agent ethidium bromide (EtdBr) in inducing the petite mutation in non-growing yeast cells. Treatment of a starved culture of yeast with a low concentration of EtdBr for a brief period of time converts the cells to petite virtually quantitatively with no decrease in cellular viability (see fig. 9). Other agents in use to this end such as euflavine or ultraviolet irradiation either require several generations of growth for complete conversion or else kill most of the cells in the population.

A brief exposure to EtdBr is sufficient for the mutagenic conversion even when the cells are subsequently grown in the absence of the drug. Professor Mahler and I showed that the transient treatment with EtdBr does not inactivate preexisting respiratory enzymes (Mahler et al., 1971; Mahler and Perlman, 1972b). It does, however, restrict the ability of such treated cells to make any new respiratory enzymes during subsequent growth. These results made it clear to us that EtdBr has an effect on mitochondria, the expression of which is delayed and does not require the continued presence of the drug.

EFFECTS OF ETDBR ON PARENTAL STRANDS OF MTDNA

In order to understand the nature of the inactivation of the mitochondrial genome under these conditions, we undertook a study of the fate of parental mtDNA in EtdBr-treated cells (Mahler et al., 1971; Perlman and Mahler, 1971a). In our study we first labeled the parental mtDNA with ^{3}H- or ^{14}C-adenine, starved the cells, incubated with 50 μM EtdBr for 2.5 hours, and then grew the cells in fresh medium that lacked EtdBr. We used CsCl gradients to analyze the level of retention of parental (labeled) mtDNA in treated and control cells at the end of the labeling period (t_o), after starvation, during EtdBr treatment, and after 1, 3, and 4 generations of growth (table 3). The results showed that the mtDNA peak became broadened after the treatment with the drug (no growth) and that roughly 30% of the parental label was lost (failed to band in CsCl). During growth further loss of counts in the peak was observed resulting in retention of less than 6% of the parental mtDNA after three generations, with most of the loss occurring during the first two generations; in controls parental mtDNA was retained quantitatively. Subsequent analysis of similarly treated samples on neutral sucrose gradients demonstrated a reduction in fragment size during the treatment followed by a further reduction in size during growth (fig. 5; Mahler and Perlman, 1972, and unpublished data). Thus exposure to a mutagenically effective dose of EtdBr produces some immediate physical damage in mtDNA under non-growing conditions; the bulk of the degradation occurs

TABLE 3

FATE OF PARENTAL MtDNA IN RESPONSE TO EtdBr TREATMENT

Treatment	COUNTS REMAINING (%)			
	Control	EtdBr	Control	EtdBr
End of labeling	100	100	100	100
1 hr starvation	99	106	107	117
1-1/4 hr EtdBr-treated	104	76	102	116
2-1/2 hr EtdBr-treated	107	70	103	120
Grown after EtdBr				
1 generation	101	28	121	141
3 generations	99	$<6^a$	191	204
4 generations	107	$<1^a$	204	219

Two aliquots of actively derepressing cells (strain Fleischmann) were labeled for two hours, one with ^{3}H-adenine and the other with ^{14}C-adenine. The cells were washed and suspended in 0.1 M phosphate buffer at pH 6.5. Both samples were starved for one hour, after which EtdBr (20 μg/ml) was added to the sample; the other was incubated further but without EtdBr. After 1-1/4 hours of EtdBr treatment the population was 98% petite, and after 2-1/2 hours it was 99.9%. After 2-1/2 hours in buffer (+/− EtdBr) the cells were washed and suspended in fresh glucose medium containing no drug. After 1, 3, and 4 generations of growth, aliquots of EtdBr-treated and control cells were harvested. The EtdBr-treated population remained >97% petite throughout the experiment. Corresponding cell samples were mixed (e.g., control, 1 generation sample [^{14}C] plus EtdBr-treated 1 generation sample [^{3}H] and converted to spheroplasts using the procedure of Leon and Mahler [1968]). The samples then were run on preparative CsCl density gradients that were fractionated and analyzed as described in Perlman and Mahler, 1971.

[a] No peak was present. Counts above the baseline in the mtDNA region were totaled.

after removal of exogenous EtdBr. In a separate experiment it was shown that the maximum damage to mtDNA in starved cells has been inflicted in two hours and that no further loss of label or reduction in size occurred when the cells were incubated for an additional 22 hours at 30° in the presence of EtdBr. Similar results were reported by Goldring et al. (1970), but in that case EtdBr was added to a growing culture and was retained during subsequent growth; we have confirmed their findings using growing cells and have shown that the continued presence of EtdBr has no effect on the rate or extent of degradation of parental mtDNA.

Recently, Mahler and Bastos have evaluated the nature of the delayed EtdBr effect and have obtained a rather unexpected result. They synthesized ^{3}H-labeled EtdBr (Bastos and Mahler, 1973) and showed that it binds covalently to mtDNA during the treatment in buffer (Mahler and Bastos, 1974a). The amount of bound EtdBr reaches a maximum at about the time when mutagenesis is extensive; the kinetics of binding and mutagenesis are quite similar. The binding reaction itself reduces the mtDNA to large fragments of roughly 12.5 $\times 10^6$ daltons that probably correspond to the fragments obtained by Mahler and Perlman (1972) in whole cells treated with EtdBr in buffer. Thus, covalently bound EtdBr is most likely the basis of the long-term effects of a transient treatment with EtdBr.

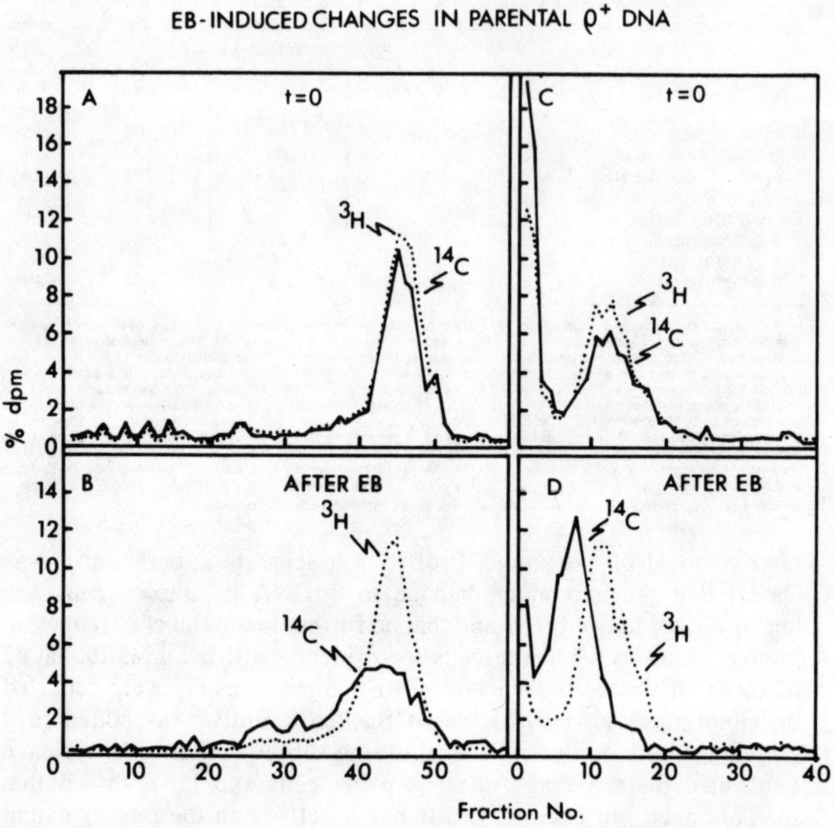

Fig. 5. Fate of parental mtDNA strands during EtdBr treatment. The experiment was as in the legend to table 3 except that a haploid strain (IL-8-8C) grown on lactate was used. Cells were labeled in the presence of cycloheximide to obtain preferential synthesis of mtDNA (Grossman et al., 1969). In the figure are shown mtDNA profiles of untreated cells (t = 0) and control (^{3}H) or EtdBr-treated (^{14}C) cells after one hour of starvation plus 2 1/2 hours of treatment with 25 μM EtdBr. DNA was analyzed in CsCl gradients (panels *A* and *B*) and neutral sucrose gradients (panels *C* and *D*). In the CsCl gradients density increases to the left; the nuclear peak, if labeled, would center at about fraction 20. In the sucrose gradients sedimentation was from the left; the nuclear peak would be well resolved from the mitochondrial one (between fractions 25 and 30). Consult figure 5 in Mahler and Perlman (1972a) for further details; this figure is from Mahler and Perlman (1972a), with permission.

The same workers subsequently showed that this reaction could be obtained *in vitro* using an isolated mitochondrial preparation. Their studies implicate the mitochondrial membrane in the mutagenic process and ATP in the delayed extensive fragmentation events. The binding reaction does not require ATP but is somewhat sensitive to the energetic status of the mitochondrial membrane; when ATP is added (or generated by the oxidation of succinate), the bound EtdBr is released as the parental mtDNA is degraded, presumably by an ATP requiring nuclease. They have studied this reaction by measuring an EtdBr-stimulated ATPase activity in mitochondria; all of their results to date indicate that the ATPase is F_1 (Mahler and Bastos, 1974b).

Figure 6 is a schematic summary of the studies of the EtdBr interaction with mtDNA and of the fate of parental strands during treatment and during the first few generations of growth after treatment (or upon addition of ATP to isolated mitochondria). Other properties of these fragments as well as their proposed role in the generation of petite mitochondrial genomes will be discussed in subsequent sections.

It is now well established that EtdBr can induce DNA-containing petites and that such petites contain gene rearrangements described in earlier sections. It would appear that high-molecular-weight petite genomes arise from the fragments at some time following treatment; in fact, it is likely that they arise from pieces smaller than the EtdBr-bound large fragments (that is, from the products of the EtdBr-stimulated ATPase / DNAse).

INSIGHTS TO EVENTS IN SINGLE ETDBR-TREATED CELLS

In all of our biochemical studies we have measured reactions occurring in a population of cells. In order to correlate biochemical events with genetic ones, it is necessary first to establish that the genetic events are not occurring in a restricted portion of the cell population; for, if they were rare, they could well be consequences of biochemical events that are quite distinct from the average measured in our studies. We set out to learn whether each EtdBr-treated cell is capable of yielding some senseful petite progeny and, in particular, more than one type of senseful petite (Givler and Perlman, unpublished data).

We generated two populations of EtdBr-treated cells (treated under growing conditions), one 17% ρ^- and the other 87% ρ^-. Roughly fifty clones, each derived from a single treated cell, were picked

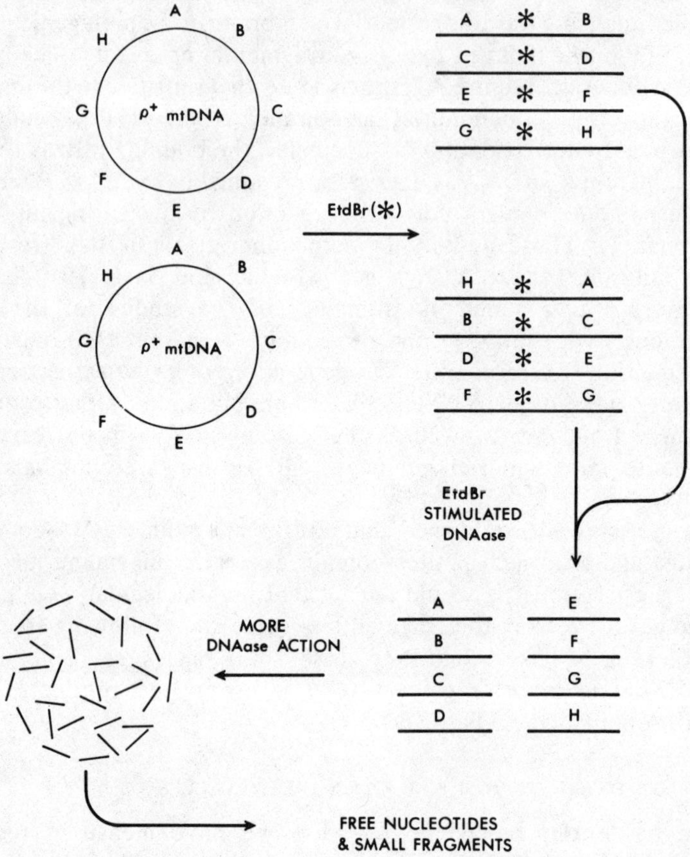

Fig. 6. EtdBr-induced fragmentation of ρ^+ mtDNA. In the figure are summarized several steps in the response of ρ^+ mtDNA to treatment with EtdBr. Regions of mtDNA are designated with the letters A through H. The circular parental genomes are broken randomly to form roughly one-quarter molecules during the EtdBr binding reaction. Nothing further happens until energy is applied to the system (by resumption of growth or by addition of ATP or succinate to isolated mitochondria); when energy is present, further degradation occurs, and bound EtdBr is released. It would appear that the initial fragmentation is by an endonuclease; it is not known how many nucleases are involved in subsequent events or whether there are exonucleases involved.

TABLE 4

RETENTION OF MITOCHONDRIAL MARKERS IN PETITES

	No. of Clones Tested	No. of Subclones Tested	% of Senseful Petites	PETITE GENOTYPE				
				$C^oE^oO^o$	$C^rE^rO^o$	$C^oE^oO^r$	$C^rE^rO^r$	$C^rE^oO^o$
Low dose of EtdBr[a]								
Pure ρ^-	14	703	6.7	656	25	19	3	0
Sectored	32	1,637	28.0	1,181	178	255	22	1
Total	46	2,340	21.4	1,837	203	274	25	1
High dose of EtdBr[b]								
Pure ρ^-	41	2,050	4.4	1,960	20	70	0	0
Sectored	17	851	10.2	764	45	39	2	1
Total	58	2,901	6.1	2,724	65	109	2	1

Wild-type strain (32)5-1/6 ($\rho^+\omega^+C^rE^rO^r$) was treated under growing conditions (repressed) with 25 μM EtdBr. After 10 and 30 minutes of treatment, cells were washed and plated on glucose solid medium. A number of primary clones from each dose was suspended and subcloned, and 50 petite subclones from each primary clones were tested for retention of each marker by the replica-cross procedure of Deutsch et al. (1974) using strain D6 ($\rho^+\omega^+C^sE^sO^s$) as tester. Petites that transmitted more than one marker by this test were further tested to be certain that they were not mixtures of two single marker-retaining mutants. In particular, the $C^rE^rO^r$ mutants have been shown to yield $\rho^+C^rE^rO^r$ diploid progeny in crosses; these rare petites are currently under further investigation.

[a] 17% pure ρ^-; 24% pure ρ^+; 59% sectored.

[b] 87% pure ρ^-; 0.7% pure ρ^+; 12.3% sectored.

and subcloned; from each clone tested, fifty petite subclones were scored for retention of the mitochondrial markers C^r, E^r, and O^r, and the results are summarized in table 4. In this analysis we included clones that were entirely petite as well as some that were a mixture of wild-type and petite progeny (called "sectored" or "petite abcédées" in the literature). The pure petite clones are derived from cells whose level of mutagenesis probably exceeded that of those that gave rise to sectored clones; the latter were not due to the plating of aggregates and contained from 0.4 to 60% wild-type cells. Since sectored colonies are quite rare when one plates untreated wild-type cells on similar media, it is clear that sectored clones are derived from EtdBr-affected cells and so should be included in this analysis.

For the low-EtdBr-dose sample (17% pure ρ^- clones) half of the pure petite clones yielded at least one marker-containing petite in the sample tested; all of the sectored clones contained marker-containing petites with a mean of 32%. Many individual treated cells yielded daughters with different combinations of retained markers; of the 42 clones that yielded senseful petites, 22 yielded two different genotypes and 16 yielded three. In a related experiment we found that a single EtdBr-treated cell is capable of yielding daughters that are members of different transmissional classes for the same retained markers (Perlman et al., unpublished data).

For the high-EtdBr-dose sample, where most tested clones were pure petite, 35 of 58 contained senseful petites while the recovery of senseful petites was higher for the sectored clones. Of the 41 pure petite clones whose progeny were tested, 21 yielded marker-containing petites. Again, most treated cells yielded two or three distinct petite genotypes.

These data show a correlation between the ability of treated cells to yield some wild-type progeny and the absolute yield of senseful petites. In most such cases 20–30% of the petite progeny of the treated cell retained some parental information. This argues indirectly that the petite genomes were generated soon after growth resumed. Had there been a delay of, say, three generations before stable petite genomes were present in the progeny of treated cells, then one would expect no more than 1/16 senseful petites in each clone (arising in the fourth generation). It should be noted also that these values are an underestimate of senseful petites in each clone since wc have not tested for the retention of all mitochondrial DNA sequences, but only for those few associated with the three markers tested; at least two other linkage groups, comprising as much as one-half of the map, were not included in this experiment.

These results rule out the possibility that rare cells in the population are responsible for yielding senseful petites in response to EtdBr treatment. The high yield of senseful petites per clone and the heterogeneity of genotypes obtained strengthens our argument that the EtdBr-induced fragments of the parental mtDNA are the immediate precursors of the stable, reiterated petite genomes.

Before proceeding to some of our most recent work, I shall summarize the points that have been made so far. It has been shown that EtdBr treatment of wild-type yeast cells immediately reduces the molecular weight of mtDNA and predisposes those fragments to further size reduction during the next few generations (in an energy-requiring reaction). Virtually every treated cell is capable of yielding some petites containing mtDNA. The mtDNA of the resulting petites is higher in molecular weight than are the fragments, and it contains gene arrangements that did not exist in the untreated parent cells. It appears that the fragmented parental mtDNA is the precursor of the petite mtDNA. So, the events that must follow include an increase in molecular weight (reiteration) and a restoration of properties that will allow self-replication.

A MODEL FOR THE GENERATION OF REITERATED PETITE GENOMES

Given our further experience with this paradigm, we proposed that petite genomes are constructed by recombination of the fragments (or copies of them) within a treated cell (Mahler and Perlman, 1972). Figure 7 illustrates two variations of how this might occur, beginning with a large EtdBr-induced fragment (as was indicated in figure 6 as a product of the EtdBr binding reaction). In principle, smaller fragments generated by the EtdBr-stimulated ATPase could enter this scheme as well but would be expected to result in different types of petites (less retained information and more highly reiterated). If the distal portions of the fragment possess some homology, then a single reciprocal recombination event could circularize the linear primary fragment; such a circularized molecule should then be stable to exonuclease attack and should be no more sensitive to endonucleases than would be the equivalent portion of an intact wild-type genome. Since mtDNA contains many regions of essentially pure A+T interspersed among the G+C-rich (gene) regions (Prunell and Bernardi, 1974), it is tempting to suggest that most fragments of mtDNA would contain such A+T-rich regions and that they might serve as the region at which the postulated recombination event would occur. Provided that the fragment contains a replicative origin, it could replicate in one of two ways. If replication proceeds as in L-cells (Robberson et al., 1972; Berk and Clayton, 1974), *Tetrahymena* (Upholt and Borst, 1974), and trypanosome (Brack et al., 1972) mitochondria (through a Cairns-type intermediate), then closed-circular copies of the unit fragment would be formed (Case I). These molecules, which are all identical, could then recombine to form larger circular multiples of the unit fragment. A series of such events would generate a highly reiterated, high-molecular-weight petite genome. On the other hand, if replication proceeded by a rolling circle process (for which no precedent exists in organelles), then a long, reiterated tail would be formed attached to the original circle (Case II). The tail would then fold back on itself, pair and recombine, producing a petite genome as in Case I. Since each cell seems capable of generating at least some senseful petite progeny and since extensive destruction of parental strands occurs in the first few generations after treatment, it would appear that this process must occur fairly soon after growth resumes. Case II is patterned after a current model of ribosomal DNA gene amplification that occurs in *Xenopus* oocytes (Hourcade et al., 1973).

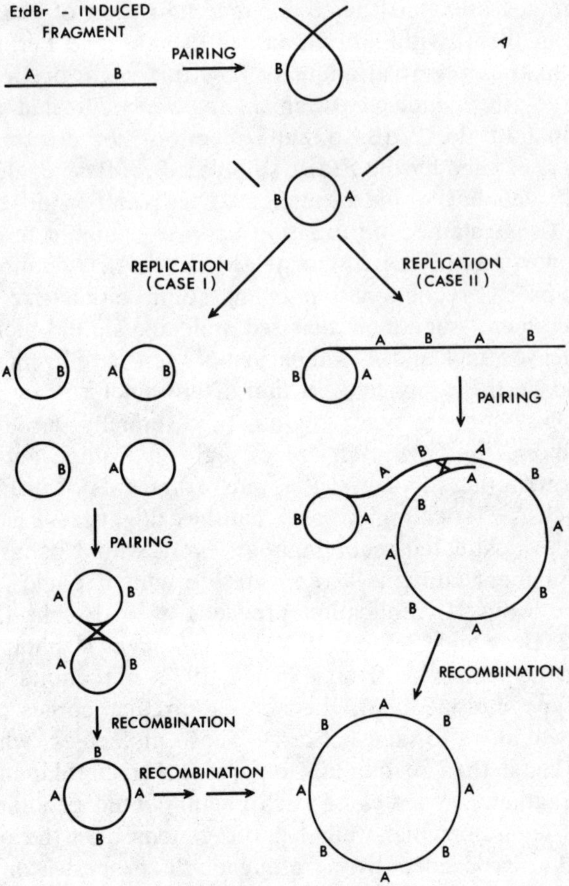

Fig. 7. A mechanism for reiteration. This figure (and the accompanying text) depicts one possible mechanism for generating reiterated petite genomes starting with an EtdBr-induced fragment of a ρ^+ genome. Several recombination events are required in both cases shown. Given the available evidence, it is likely that this process could be occurring many times in each treated cell (many fragments involved). Depending upon the number of initially circularized pieces in a cell, one could obtain non-identical repeating units (differential reiteration) only if replication is as in case I.

Although we have no direct physical evidence in support of the details of this model, we have made the following relevant observations:

1. EtdBr treatment itself inhibits mtDNA replication; mtDNA synthesis does resume within one generation but at an initially reduced rate (perhaps reflecting the reduced amount of parental template strands) (Perlman and Mahler, 1971a).
2. EtdBr-induced fragments can recombine with wild-type mtDNA (see below).
3. The ability of the fragments to recombine with wild-type is enhanced by treatments that stabilize the fragments (see below).
4. EtdBr-induced fragments are a necessary but not sufficient step for the loss of ρ^+. In other words, such cells can be caused to yield many wild-type progeny under appropriate conditions (reversal)(see below).

ETDBR-INDUCED FRAGMENTS CAN RECOMBINE

In my laboratory we have devised a procedure that measures the ability of mtDNA in EtdBr-treated cells to recombine with wild-type mtDNA in a cross (Kao, 1973; Perlman et al., unpublished data). In the broadest terms our experiments resemble marker rescue experiments as have been done in phage systems. Typically we treat cells in buffer or medium with EtdBr, remove exogenous EtdBr, and immediately mate the cells to a wild-type tester strain having a complementing nuclear and mitochondrial genotype. We then assess the level of transmission of markers in the treated parent to the diploid wild-type progeny of the cross and compare that to untreated control crosses. An experiment of this sort, using growing cells, is illustrated in figure 8. Here we found that the ability of transmitting each individual marker exceeds the percentage of wild-type cells (pure and sectored) in the treated population at all doses tested; marker transmission greatly exceeds the percentage of *pure* wild-type cells in the treated population. Thus, affected (inactivated and partly inactivated ρ^+) genomes still can recombine. When a similar experiment was done using starved cells, treated with EtdBr in buffer, the level of marker transmission was even higher; in table 5 we compare marker transmission levels in growing and starved cells comparably mutagenized by EtdBr at three dosages (20%, 10%, and 5% wild-type colonies). Cells that were unable to degrade the parental mtDNA extensively during the drug treatment (e.g., the starved cells) had a very high level of marker transmission compared with the growing cells. Similarly,

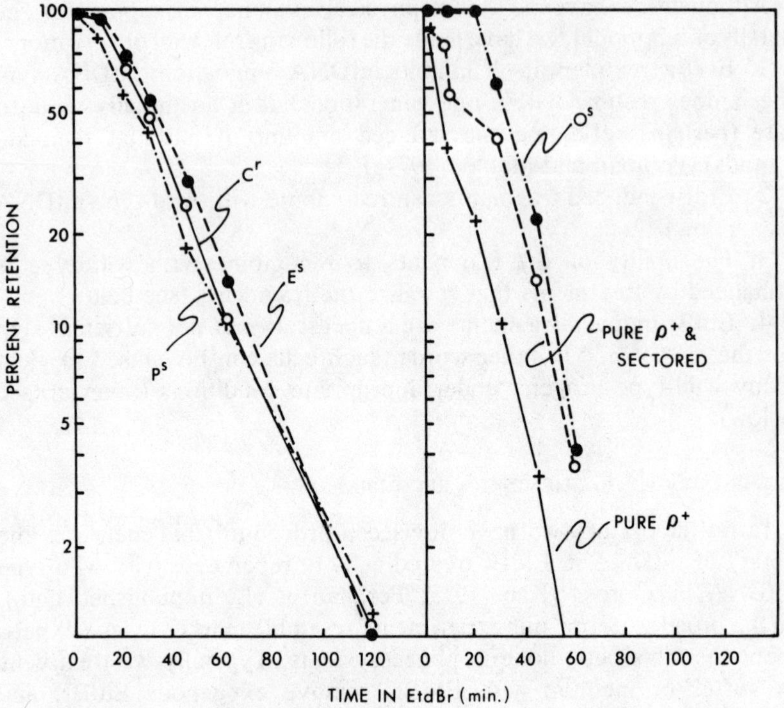

Fig. 8. Marker elimination by EtdBr treatment under growing conditions. Glucose-repressed cells of strain IL16-10B ($\rho^+ \omega^- C^r E^s O^s P^s$) were treated with 12.5 μM EtdBr in medium. After 0, 10, 20, 30, 45, 60, and 120 minutes of drug treatment, an aliquot was washed free of EtdBr and mated to strains appropriate for measuring transmission of all four markers of the IL16-10B mitochondrial genome. Treated cells were scored for percentage of ρ^+ after they had been washed; in the righthand panel are presented data where ρ^+ was scored as pure ρ^+ *and* sectored clones or as pure ρ^+ only. Marker retention is as defined in the legend to table 6.

when cells were treated with EtdBr in buffer containing an energy source (glucose), marker elimination was enhanced. We feel that these results are exactly as expected based on previously described biochemical studies and furthermore show that EtdBr-induced fragments can recombine.

Several other observations from experiments of this sort have been made that also extend the biochemical studies.

If treated cells are grown in medium (lacking exogenous EtdBr) for one or more generations prior to mating, the level of marker

TABLE 5

ETDBR TREATMENT ELIMINATES MARKERS MORE RAPIDLY IN GROWING CELLS THAN IN NON-GROWING CELLS

MUTAGENESIS			MARKER RETENTION			
% ρ^+ and sectored	% pure ρ^+ (non-growing)	% pure ρ^+ (growing)	Non-growing %E^s	Non-growing %O^s	Growing %E^s	Growing %O^s
20	6.7	5.2	56	52	32	28.0
10	0.9	2.5	35	30	18	15.0
5	0.7	0.6	50	40	10	8.5

Treated strain was IL16-10B ($\rho^+ \omega^- C^r E^s O^s$).

Tester strain was (32)5-1/6 ($\rho^+ \omega^+ C^r E^r O^r$).

These data are from the experiments shown in figs. 8 and 10. Refer to the legends of those figures for further details and to the text for a discussion of the data.

transmission falls well below the level shown by cells mated before growth had resumed (Kao, 1973). This decrease is probably a reflection of the extensive destruction of EtdBr-induced fragments that occurs when growth resumes and therefore shows that the fragments are necessary for the transmission of markers from the treated cell (tables 6 and 7).

In our experience, most senseful EtdBr-induced petites are somewhat suppressive. If the immediate products of EtdBr interaction with wild-type mtDNA were, in fact, stable petite genomes, then such treated cells should be suppressive. When mutagenized cells are crossed with a wild-type tester immediately after drug treatment, all diploid clones are wild-type; therefore, the immediate products of EtdBr treatment are not suppressive. When the cells have been grown for one to three generations (without exogenous EtdBr) and then mated, now a significant percentage of petite diploids (suppressed wild-types) is obtained (Kao, 1973). These preliminary results indicate, as did other observations, that the petite genomes arise within the first few generations after mutagenesis and are not the immediate products of the inactivation events.

Our marker elimination studies indicate that all known linkage groups are eliminated at about the same rate (fig. 8). This argues that there are no specific EtdBr-sensitive points on the mitochondrial genome and that the fragmentation process is random. If there are specific sensitive points, then it would be required that not all of them are broken in the initial fragmentation events; otherwise, multiple arrangements (transmissional classes) of the same retained region would not be found. The presence of multiple breakpoints is also supported,

TABLE 6

EFFECTS OF EUFLAVINE AND PRE-MATING GROWTH ON ETDBR-INDUCED MUTAGENESIS AND MARKER ELIMINATION

Drug Treatment	Growth	% ρ^+	RANDOM PROGENY (%)					RETENTION		
			C^rE^s	C^sE^r	C^rE^r	C^sE^s	N	%C^r	%E^s	%O^{s}*
A. None	None	96.6	64.9	29.2	2.3	3.6	470	100	100	100
(control)	1 Gen.	96.6	55.3	37.1	3.3	4.3	488	100	100	100
	3 Gen.	93.9	41.4	47.9	3.5	7.2	290	100	100	100
B. Euflavine	None	94.5	60.5	30.8	3.3	5.4	590	94.9	96	79
(no EtdBr)	1 Gen.	87.2	37.8	57.5	2.6	2.1	341	68.9	67	76
	3 Gen.	60.1	28.3	67	1.4	3.1	420	65	65	62
C. EtdBr	None	0.2	8	90.9	0.5	0.7	448	13	13	12.4
(no Eufl.)	1 Gen.	0.4	1.2	98.3	0.2	0.2	482	2.4	2.4	ND
	3 Gen.	0.3	0.5	99.5	0	0	427	1.1	1	0
D. EtdBr, then	None	22.5	37.6	53.9	3.1	5.4	386	60.6	63	65
Euflavine	1 Gen.	15.3	18	77	2.8	2.1	283	36	34	34
(Reversal)	3 Gen.	18.6	11.8	87.8	0.4	0.8	490	24.8	24	26

Treated strain was IL16-10B ($\rho^+\omega^-C^rE^sO^s$).

Tester strain was 55-R5-3C/221 ($\rho^+\omega^-C^sE^rO^s$).

*Tester strain was D243-4A-O^r ($\rho^+\omega^+C^sE^sO^r$).

Cells of IL16-10B growing exponentially in glucose medium were treated with 20 μM EtdBr for 30 min. They were washed and suspended in 0.1M phosphate buffer (pH 6.5) and incubated for 60 min at 30° C with (sample D) or without 10 μM euflavine (sample C). Control cells (no EtdBr treatment) also were incubated in buffer with (sample B) or without euflavine (sample A). All samples then were washed and placed in fresh glucose medium; an aliquot of each was mated immediately and after 1 and 3 generations of growth. Diploids were selected on minimal medium and the genotypes of random progeny scored by replica plating. The percentage of wild-type cells in each sample was measured by plating an aliquot of each sample before mating was begun. Marker retention (the last three columns on the right) is the level of marker transmission normalized to the appropriate control sample.

TABLE 7

EFFECTS OF ANTIMYCIN A AND PRE-MATING GROWTH ON EtdBr-INDUCED MUTAGENESIS AND MARKER ELIMINATION

Drug Treatment	Growth	%ρ^+	RANDOM PROGENY (%)					RETENTION	
			C^rE^s	C^sE^r	C^rE^r	C^sE^s	N	%C^r	%E^s
A. None	None	95	47.9	43.6	3.5	5.0	461	100	100
(control)	1 Gen.	93.1	45	47.6	1.1	6.3	271	100	100
	3 Gen.	93.7	41	48.2	3.6	7.2	307	100	100
B. Anti A	None	92	42.9	45	5.3	6.8	381	91	94
(no EtdBr)	1 Gen.	93.1	55	34.1	3.1	7.8	258	126	122
	3 Gen.	90.8	47.9	40.1	4.9	7.2	359	118	114
C. EtdBr	None	5.6	15	81.4	1.5	2.1	473	32	32
(no Anti A)	1 Gen.	2.4	11	87.1	0	1.9	365	23.9	25.2
	3 Gen.	2.8	5.4	93.6	0.2	0.9	466	12.6	13.1
D. EtdBr, then	None	23.7	29.2	64.4	2.1	4.4	128	60.9	63.5
Anti A	1 Gen.	17.2	27.3	66.1	2.6	4.1	389	64.4	61.2
(reversal)	3 Gen.	10.3	16.2	79.7	0.7	3.5	458	38	41

Treated strain was IL16-10B ($\rho^+\omega^-C^rE^s$).

Tester strain was 55R5-3C/221 ($\rho^+\omega^-C^sE^r$).

The experimental design was as in the legend to table 6 except that Anti A at 1 μg/ml was used in place of euflavine.

albeit weakly, by the recent studies of Locker et al. (1974) which showed that petites retaining the C^rE^r region of mtDNA may have different repeat intervals; a more thorough analysis of the number of different repeat intervals possible given a particular genotype may provide a map of the break points and that, in turn, may give insight to the nature of the EtdBr binding sites.

It has been reported by several groups and we have confirmed in table 4 that O^r petites are more readily obtained from a single batch of treated cells than are C^r ones (Uchida and Suda, 1973; Deutsch et al., 1974). But in the preceeding paragraph I have noted that these two markers are eliminated at the same rate. The explanation of this apparent discrepancy lies in a consideration of the design of the two types of experiments. In a marker elimination experiment (where markers are assayed for ability to recombine, at a time when stable petite genomes do not yet exist) one measures the effects of fragmentation on genetic transmission. In a petite screening experiment, one assays the relative success of stabilizing a particular fragment or the suitability of a fragment to serve as a substrate for the reiteration process. It should be concluded that the O^r region is reiterated more easily than is the C^r region. Extension of this analysis to other markers should provide insights to the base sequence requirements of the reiteration process.

REPAIR OF ETDBR-INDUCED DAMAGE

Returning to evidence in support of our repair hypothesis, Professor Mahler and I discovered in 1971 that the ability of EtdBr-mutagenized cells to yield wild-type progeny can be restored to a great extent by heating the cells briefly prior to the resumption of growth (Perlman and Mahler, 1971b). Since we had already shown that the mtDNA in such EtdBr cells had sustained physical damage (and now we know that it contained bound EtdBr) before we had heated the cells, we concluded that the generation of large fragments is not sufficient for the loss of ρ^+. We termed the process "reversal" and have since found four other procedures—namely, treatment before growth resumes with chloramphenicol, antimycin A (anti A), or euflavine, or else a brief period of growth in medium containing a non-fermentable carbon source (metabolic reversal)—which accomplish the same result (Mahler and Perlman, 1972; Mahler, 1973c). Several of these protocols are presented in figure 9; at the moment our procedures of choice are the use of anti A or euflavine. It should be noted that euflavine,

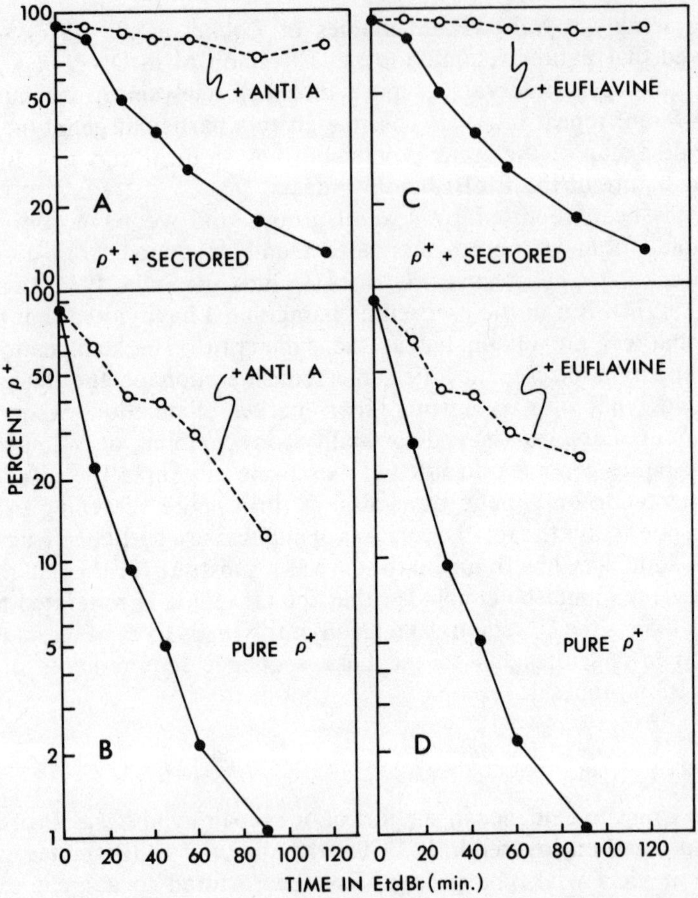

Fig. 9. Reversal of EtdBr mutagenesis. Repressed cells of strain IL16-10B were starved in 0.1 M phosphate buffer (pH 6.5) for one hour and then treated with 25 μM EtdBr. At the intervals noted, cells were plated on solid medium or placed in tubes containing euflavine at 10 μM or anti A at 1 μg/ml. The tubes were incubated for one hour at 30°, then diluted and plated. Colonies were then scored for the percentage of wild-type; panels A and C show data where ρ^+ was scored as pure ρ^+ *and* sectored clones, and panels B and D show data where only pure ρ^+ clones were scored as ρ^+. In experiments of this sort the EtdBr-treated cells are known to have sustained physical damage to mtDNA before treatment with the reversing agent; thus, it is clear that the treatment interferes with subsequent events and restores the ability of mutagenized cells to yield ρ^+ progeny. Similar data are obtained using the other reversal procedures cited in the text. In related experiments it has been shown that treatment of cells with EtdBr and anti A (or euflavine) results in protection of ρ^+ from EtdBr-induced loss; however, covalent binding of EtdBr and initial fragmentation still occur under such conditions.

a potent inducer of the petite mutation in growing cells, is not mutagenic to starved cells. If EtdBr-treated cells are grown in glucose medium for one generation prior to treatment with a reversing agent, the ability to be reversed by these agents is greatly reduced. This further points out the great importance of events occurring during the first few generations of growth following mutagenesis.

We have studied the effects of reversing treatments on the size and stability of mtDNA in EtdBr-treated cultures and find that the treatments primarily interfere with the ATP-dependent, extensive degradation of mtDNA fragments that occurs quite rapidly without reversing treatment (Mahler and Perlman, unpublished data). The treatments do not cause the release of bound EtdBr when added after EtdBr treatment, nor do they restore the large fragments to a higher-molecular-weight form. The reversing agents were then tested for their ability to interfere with the effects of EtdBr by adding both mutagen and reversing agent to wild-type cells simultaneously; when the percentage of ρ^+ cells was scored as a function of time in the drugs, it was found that the reversing agent had a protective effect. When those studies were extended biochemically, it was noted that protected cells still bind EtdBr to mtDNA and that the large fragments are still formed. This argues quite strongly that the protection/reversal is due to interference with events that would occur during the first few generations of growth. The reversing agents, like EtdBr, do not have to be present during growth, and they can be applied to non-growing cells transiently. We have interpreted the studies of reversal as implicating the mitochondrial membrane in this process of fragment degradation; since this argument is still a hypothesis, we refer the interested reader to the original papers for more details (Mahler and Perlman, 1972; Mahler, 1973c; Mahler and Bastos, 1974b; Bastos and Mahler, 1974).

We have recently evaluated the genetic correlates of these treatments as an extension of the experiments described above (tables 6 and 7). We found that brief exposure to euflavine or anti A of cells whose marker transmission has been reduced by EtdBr treatment increases the level of transmission markedly. Here thc level of marker transmission was increased to a greater extent than was the level of ρ^+; when cells were mated after reversal treatment *and* one or three generations of growth, we found that marker transmission and percentage ρ^+ decreased somewhat but the difference between the two parameters was retained. This change, which occurs during growth, is probably due to two factors, namely, residual inhibition of mtDNA

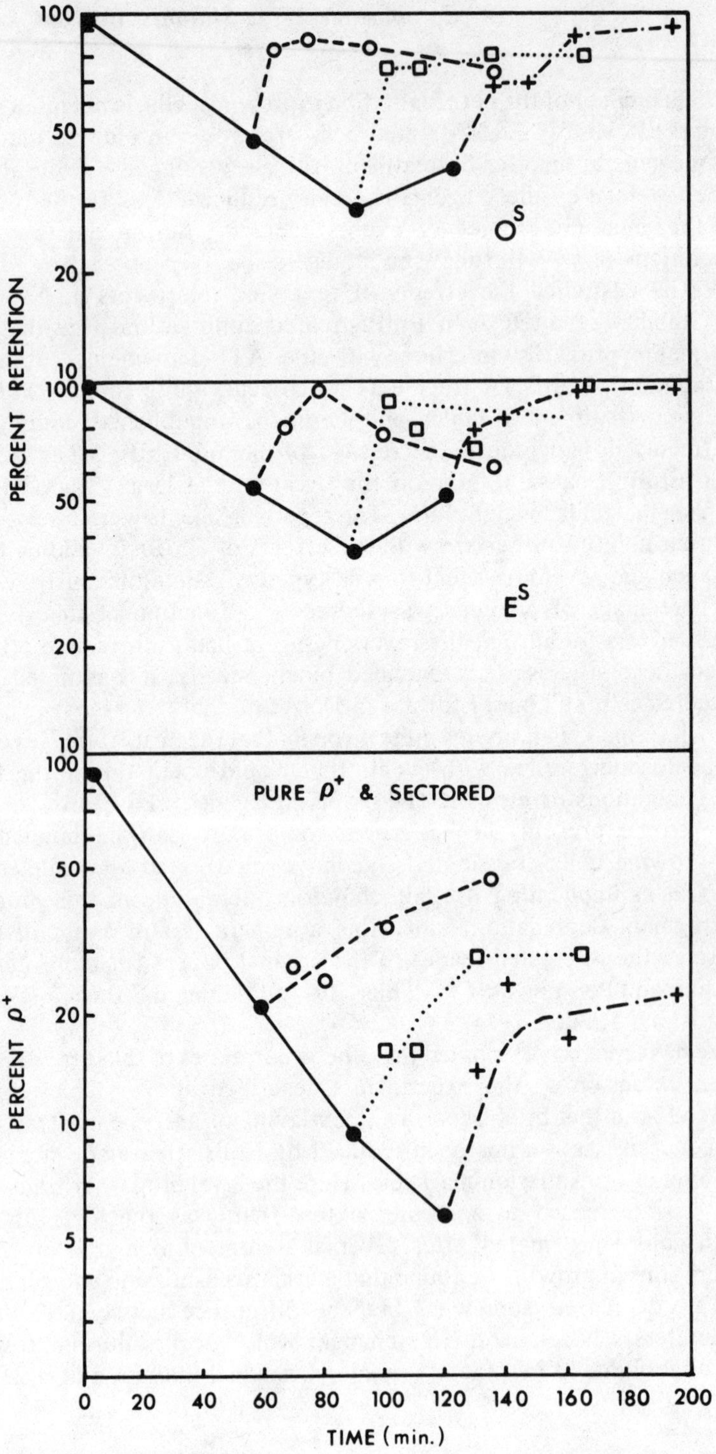

PERCENT RETENTION
PERCENT ρ+
O^S
E^S
PURE ρ+ & SECTORED
TIME (min.)
100
50
20
10
5
0
20
40
60
80
100
120
140
160
180
200

replication and partial further degradation of EtdBr-induced large fragments. In a separate experiment we studied the effects of anti A reversal of cells mutagenized and reversed under non-growing conditions (fig. 10). Here marker transmission was restored to almost the control level while ρ^+ was raised only partially. We feel that these studies provide strong support for the contention that EtdBr-induced large fragments can recombine and that stabilizing them enhances that ability as well as the ability of yielding wild-type progeny. The fact is that the treated cells still contain fragments and little if any intact mtDNA; thus the increased yield of wild-type cells would seem to be a consequence of the repair of the physical damage; similar events, but ones that were not fully successful, would give rise to altered gene arrangements in petites. Figure 11 indicates our current hypothesis for the mechanism of restoration of ρ^+ in treated cells. Clearly, the smaller the fragments, the less likely will be the restoration of ρ^+.

GENETIC CONTROL OF PETITE INDUCTION

There exist some mutants whose rate of EtdBr-induced petite formation is altered compared with their parent strain (table 8); some are EtdBr-resistant, and one is EtdBr-sensitive. They were originally isolated in a number of laboratories by selecting for increased rate of UV-induced petite mutation (Moustacchi, personal communication; Mahler and Perlman, 1972), increased UV-induced lethality (Cox and Parry, 1968; Moustacchi, 1971), nuclear recombination deficiency (Rodarte-Ramon and Mortimer, 1972), and mitochondrial respiratory deficiency (Flury et al., 1974), and were characterized for EtdBr sensitivity in our laboratory by A. Fraenkel and in Mahler's laboratory by P. Perlman and R. Bastos. In summary, EtdBr-resistant mutants

Fig. 10. Genetic studies of anti A reversal under nongrowing conditions. Repressed cells of strain IL16-10B ($\rho^+ \omega^- C^r E^s O^2$) were starved for one hour and then treated with 25 μM EtdBr. After 60 min of EtdBr treatment an aliquot of cells was washed, and part mated to cells of strain (32)5-1/6 ($\rho^+ \omega^+ C^r E^r O^r$); the rest of the washed cells were suspended in buffer containing anti A at 1 μg/ml. After 10, 20, 40, and 75 min of anti A exposure at 30° C, an aliquot of cells was washed and then mated. Each sample generated in this experiment was tested for the degree of mutagenesis by plating before mating. Samples of 90 and 120 min EtdBr-treated cells were reversed and mated as was the 60 min sample. Mutagenesis and marker retention results are presented in the figure. The effects of reversal treatment on marker transmission are immediate, while effects on percentage of ρ^+ are delayed slightly. Refer to figure 8, table 5, and the text for further discussion of this experiment.

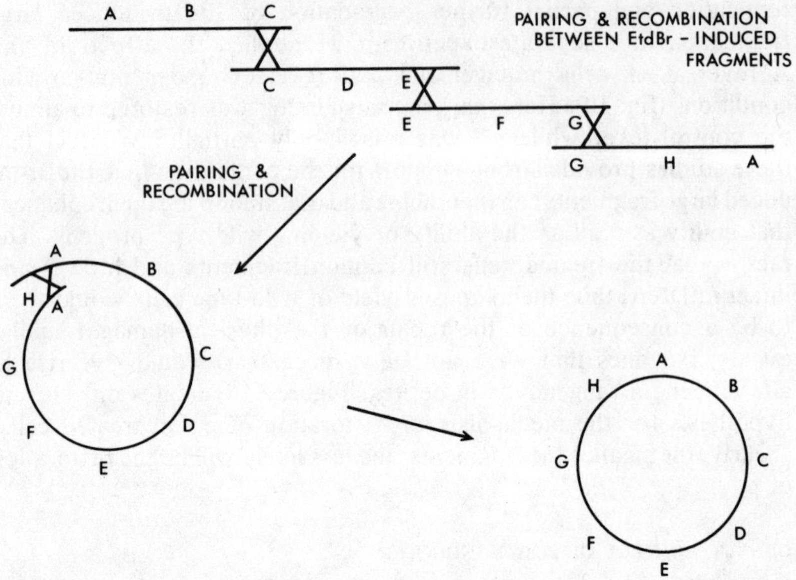

Fig. 11. The role of recombination in the recovery of ρ^+. In the figure is presented our current working hypothesis for the recovery of ρ^+ after EtdBr treatment. We propose that large overlapping fragments (derived from different mtDNA molecules in the treated cell) pair and recombine until a complete linear genome with redundant ends is obtained. That molecule could then recircularize to complete the reconstruction (repair) of a ρ^+ genome. Clearly, the larger the fragments, the fewer recombinational events are needed. Our observation that stabilization of the large fragments increases the efficiency of this process is compatible with this hypothesis. This model (and the one shown in figure 7) assumes that recombination of mtDNA molecules can occur in haploid cells; that is, it assumes that recombination does not require zygote formation or else that it is not restricted to zygotes. Some evidence suggesting that recombination occurs in unmated cells exists (Sena, 1971).

exist that show a variety of alterations in the reactions which we have implicated in the mutagenic process (table 8). All of the alterations are consistent with the measured EtdBr resistance; in one mutant or another the following altered responses were measured by Mahler and Bastos (1974b and unpublished data): reduced rate of EtdBr binding, reduced extent of binding, reduced rate or extent of EtdBr release (DNA fragmentation), and absence of EtdBr stimulation of ATPase. The mutant that is EtdBr-sensitive shows an increased rate and extent of EtdBr binding and an increased rate of DNA fragmentation correlated with an excessive extent of ATPase stimulation. In

TABLE 8

GENETIC CONTROL OF EtdBr-INDUCED REACTIONS WITH mtDNA

Strain Tested	Mutagenesis Rate	EtdBr Binding Reaction		Release of bound EtdBr	Stimulation of ATPase
		Rate	Extent		
Wild-type	Standard	Standard	Standard	Standard	3- to 4-fold increase
$\rho 5$	Slow	Standard	Standard	No release	No increase
$\rho 72$	Fast	Increased	Increased	Increased rate	5-fold increase
rec4$^-$	Slow	Standard	Reduced	Reduced rate and extent	3- to 4-fold increase
rad274	Slow	None	None		
rad284	Slow	Standard	Standard	Reduced rate	
73/1	Slow	Standard	Reduced	Standard	3- to 4-fold increase
Petite	Slow	Standard	Reduced		

The data summarized in this table were obtained by Bastos and Mahler (1974b and unpublished data). Mutants $\rho 5$ and $\rho 72$ were isolated by Moustacchi by scoring for increased rate of petite induction by UV irradiation. *Rec4*$^-$ (strain 2C4) was isolated by Rodarte-Ramon and Mortimer (1972) as a nuclear mitotic recombination-deficient mutant. *Rad274* and *rad284* were isolated by Cox and Parry (1968) as *uvs* mutants. 73/1 is a mitochondrial respiration-deficient mutant that is not a petite (Flury et al., 1974). All of the petites reported in table 2 show essentially the same response to EtdBr as shown in the table. Mutagenesis rates of strains designated *rec4*$^-$, *rad274*, and *rad284* were measured by A. Fraenkel; strains $\rho 5$ and $\rho 72$ were originally characterized by Mahler and Perlman (1972), and 73/1 and the petite strain by Perlman (unpublished data).

general, respiration-deficient mutants are resistant to EtdBr-induced marker elimination; although petites have not yet been studied in great detail, it appears that they show a reduced yield of bound EtdBr and are deficient in the fragmentation process. These results do not constitute a thorough study of the genetic control of these events, but they do argue quite strongly that the events are related intimately to the mutagenic process. Throughout our studies we were aware that many of the parameters measured, especially the biochemical ones, could be unrelated or side reactions; at this point, it is still conceivable that they are side reactions, but they are clearly related to the process under investigation.

SUMMARY

In this paper I have reviewed the literature on the petite mutation and discussed its use in studying the nature of mitochondrial genomes and mitochondrial genetics. In addition, I have discussed some more recent studies and summarized some of our recent work with the genetics of petites in order to develop the idea that these mutants may be of great use as a model system for studying the mechanism of generating reiterated sequences in eucaryotes. Also I have noted their use in studying the behavior of repeated and deleted sequences in crosses. It is my hope that this experimental system may prove in the near future to be much more than a biological oddity of yeast by providing valuable insights to some basic genetic phenomena that occur in most organisms.

ACKNOWLEDGEMENTS

Original research contained in this paper was supported by Research Grant PHS RO1 GM19607 from the Institute of General Medical Sciences, National Institutes of Health, U.S. Public Health Service. Capable technical assistance was provided by Ms. Mary Townsend. Some of the studies were done by my students, Ms. Shao-hui Kao and Ms. Kris Givler. I wish to thank my colleague C. W. Birky, Jr. for stimulating discussion and critical comments on the manuscript. Professor H. R. Mahler, Department of Chemistry, Indiana University, has been an important contributor to these studies as my research advisor in his laboratory and as a collaborator in more recent and ongoing studies. I also thank him for permitting me to cite his unpublished studies contained in table 8. Yeast strains used in these studies were obtained from Drs. P. P. Slonimski, R. Kleese, D. Wilkie, R. Mortimer, B. Cox, E. Moustacchi, R. S. Criddle, and U. Flury.

LITERATURE CITED

Bastos, R. N., and H. R. Mahler. 1974a. A synthesis of labelled ethidium bromide. Arch. Biochem. Biophys. 160:643–46.

Bastos, R. N., and H. R. Mahler. 1974b. Molecular mechanisms of mitochondrial genetic activity: Effects of ethidium bromide on the DNA and energetics of isolated mitochondria. J. Biol. Chem. 249:6617–24.

Bech-Hansen, N. T., and G. H. Rank. 1973a. Cytoplasmically inherited ethidium bromide resistance in suppressive petites of *Saccharomyces cerevisiae*. Canad. J. Genet. Cytol. 15:381–87.

Bech-Hansen, N. T., and G. H. Rank. 1973b. The bivious suppressiveness of cytoplasmic petites of *S. cerevisiae* lacking in mitochondrial DNA. Molec. Gen. Genet. 120:115–24.

Berk, A. J., and D. A. Clayton. 1974. Mechanism of mitochondrial DNA replication in mouse L-cells: Asynchronous replication of strands, segregation of circular daughter molecules, aspects of topology, and turnover of an initiation sequence. J. Mol. Biol. 86:801–24.

Bernardi, G., M. Faures., G. Piperno, and P. P. Slonimski. 1970. Mitochondrial DNA's from respiratory-sufficient and cytoplasmic respiratory-deficient mutant yeast. J. Mol. Biol. 48:23–42.

Bernardi, G., G. Piperno, and G. Fonty. 1972. The mitochondrial genome of wild-type yeast cells. I. Preparation and heterogeneity of mitochondrial DNA. J. Mol. Biol. 65:173–89.

Bernardi, G., and S. N. Timasheff. 1970. Optical rotary dispersion and circular dichroism properties of yeast mitochondrial DNA. J. Mol. Biol. 48:43–52.

Blamire, J., D. Cryer, D. B. Finkelstein, and J. Marmur. 1972. Sedimentation properties of yeast nuclear and mitochondrial DNA. J. Mol. Biol. 67:11–24.

Bolotin, M., D. Coen, J. Deutsch, B. Dujon, P. Netter, E. Petrochilo, and P. P. Slonimski. 1971. La recombinaison, des mitochondries chez *Saccharomyces cerevisiae.* Bull. Inst. Pasteur 69:215–39.

Borst, P., 1972. Mitochondrial nucleic acids. Ann. Rev. Biochem. 41:333–76.

Brack, C., E. Delain, and G. Riou. 1972. Replicating covalently closed circular DNA from kinetoplasts of *Trypanosoma cruzi.* Proc. Nat. Acad. Sci. USA. 69:1642–46.

Brown, D. D., and K. Sugimoto. 1973. 5sDNA's from *Xenopus laevis* and *Xenopus mulleri:* Evolution of a gene family. J. Mol. Biol. 78:397–415.

Brown, D. D., P. Wensink, and E. Jordan. 1972. A comparison of the ribosomal DNA's of *Xenopus laevis* and *Xenopus mulleri.* J. Mol. Biol. 63:57–73.

Carnevali, F., G. Morpurgo, and G. Tecce. 1969. Cytoplasmic DNA from petite colonies of *Saccharomyces cerevisiae:* A hypothesis on the nature of the mutation. Science 162:1331–33.

Christiansen, C., G. Christiansen, and A. L. Bak. 1974. Heterogeneity of mitochondrial DNA from *Saccharomyces carlsbergensis:* Renaturation and sedimentation studies. J. Mol. Biol. 84:65–82.

Clarkson, S. G., M. C. Birnstiel, and I. F. Purdom. 1973b. Clustering of transfer RNA genes of *Xenopus laevis.* J. Mol. Biol. 79:411–29.

Clarkson, S. G., M. L. Birnstiel, and V. Serra. 1973a. Reiterated transfer RNA genes of *Xenopus laevis.* J. Mol. Biol. 79:391–410.

Coen, D., J. Deutsch, P. Netter, E. Petrochilo, and P. P. Slonimski. 1970. Mitochondrial genetics. I. Methodology and phenomenology. Symp. Soc. Exp. Biol. 24:449–96.

Cohen, M., J. Casey, M. Rabinowitz, and G. S. Getz. 1972. Hybridization of mitochondrial transfer RNA and mitochondrial DNA in petite mutants of yeast. J. Mol. Biol. 63:441–51.

Cohen, M., and M. Rabinowitz. 1972. Analysis of grande and petite yeast mitochondrial DNA by tRNA hybridization. Biochim. Biophys. Acta 281:192–201.

Corneo, G., C. Moore, D. R. Sanadi, L. I. Grossman, and J. Marmur. 1966. Mitochondrial DNA in yeast and some mammalian species. Science 151:687–89.

Cox, B. S., and J. M. Parry. 1968. The isolation, genetics, and survival characteristics of ultraviolet light sensitive mutants in yeast. Mut. Res. 6:37–55.

Deutsch, J., B. Dujon, P. Netter, E. Petrochilo, P. P. Slonimski, M. Bolotin-Fukuhara, and D. Coen. 1974. Mitochondrial genetics. VI. The petite mutation in *Saccharomyces cerevisiae:* Interrelations between the loss of the ρ^+ factor and the loss of the drug resistance mitochondrial genetic markers. Genetics 76:195–219.

Ephrussi, B., and S. Grandchamp. 1965. Etudes sur la suppresivité des mutants à déficience respiratoire de la levure. I. Existence au niveau cellulaire de divers "Degrés de Suppressivité." Heredity 20:1–7.

Ephrussi, B., and H. Hottinguer. 1951. Cytoplasmic constituents of heredity: On an unstable state in yeast. Cold Spring Harbor Symp. on Quan. Biol. 16:75-85.

Ephrussi, B., H. Hottinguer, and A. Chimenes. 1949. Action de l'acriflavine sur les levures. I. La mutation petite colonie. Ann. Inst. Pasteur 76:351-367.

Ephrussi, B., H. Hottinguer, and H. Roman. 1955. Suppressiveness: A new factor in the genetic determinism of the synthesis of respiratory enzymes in yeast. Proc. Nat. Acad. Sci. USA. 41:1065-70.

Ephrussi, B., H. Hottinguer, and J. Tavlitzki. 1949. Action de l'acriflavine sur les levures. II. Etude genètique du mutant "Petite Colonie." Ann. Inst. Pasteur 76:419-42.

Ephrussi, B., H. Jakob, and S. Grandchamp. 1966. Etudes sur la suppressivité des mutants à déficience respiratoire de la levure. II. Etapes de la mutation grande en petite provoquée par le facteur suppressif. Genetics 54:1-29.

Ephrussi, B., P. L'Heritier, and H. Hottinguer. 1949. Action de l'acriflavine sur les levures. VI. Analyse quantitative de la transformation des populations. Ann. Inst. Pasteur 77:64-83.

Fauman, M. A., M. Rabinowitz, and H. Swift. 1973. Comparison of the mitochondrial ribonucleic acid from a wild-type grande and a cytoplasmic petite yeast by ribonucleic acid-deoxyribonucleic acid hybridization. Biochemistry 12:124-28.

Faye, G., H. Fukuhara, C. Grandchamp, J. Lazowska, F. Michel, J. Casey, G. Getz, J. Locker, M. Rabinowitz, M. Bolotin-Fukuhara, D. Coen, J. Deutsch, B. Dujon, P. Netter, and P. P. Slonimski. 1973. Mitochondrial nucleic acids in the petite colonie mutants: Deletions and repetitions of genes. Biochimie 55:779-92.

Flury, U., H. R. Mahler, and F. Feldman. 1974. A novel respiration-deficient mutant of *Saccharomyces cerevisiae.* I. Preliminary characterization of phenotype and mitochondrial inheritance. J. Biol. Chem. 249:6130-37.

Fukuhara, H. 1969. Relative proportions of mitochondrial and nuclear DNA in yeast under various conditions of growth. Europ. J. Biochem. 11:135-39.

Fukuhara, H., M. Faures, and C. Genin. 1969. Comparison of RNA's transcribed from mitochondrial DNA of cytoplasmic and chromosomal respiratory deficient mutants. Molec. Gen. Genet. 104:264-81.

Fukuhara, H., G. Faye, F. Michel, J. Lazowska, J. Deutsch, M. Bolotin-Fukuhara, and P. P. Slonimski. 1974. Physical and genetic organization of petite and grande yeast mitochondrial DNA. I. Studies by RNA-DNA hybridization. Molec. Gen. Genet. 130:215-38.

Gingold, E. B., G. W. Saunders, E. B. Lukins, and A. W. Linnane. 1969. Biogenesis of mitochondria. X. Reassortment of the cytoplasmic genetic determinants for respiratory competence and erythromycin resistance in *Saccharomyces cerevisiae.* Genetics 62:735-44.

Goldring, E. S., L. Grossman, D. Krupnick, D. Cryer, and J. Marmur. 1970. The petite mutation in yeast: Loss of mitochondrial deoxyribonucleic acid during induction of petites with ethidium bromide. J. Mol. Biol. 52:323-35.

Gordon, P., and M. Rabinowitz. 1973. Evidence for deletion and changed sequence in the mitochondrial deoxyribonucleic acid of a spontaneously generated petite mutant of *Saccharomyces cerevisiae*. Biochemistry 12:116-23.

Grimes, G. W., H. R. Mahler, and P. S. Perlman. 1974. Nuclear gene dosage effects on mitochondrial mass and DNA. J. Cell Biol. 61:565-74.

Hollenberg, C. P., P. Borst, R. A. Flavell, C. F. van Kreijl, E. F. van Bruggen, and A. C. Arnberg. 1972. The unusual properties of mtDNA from a low-density petite mutant of yeast. Biochim. Biophys. Acta 277:44–58.

Hollenberg, C. P., P. Borst, R. W. J. Thuring, and E. F. J. van Bruggen. 1969. Size, structure, and genetic complexity of yeast mitochondrial DNA. Biochim. Biophys. Acta 186:417–19.

Hollenberg, C. P., P. Borst, and E. F. J. van Bruggen. 1970. Mitochondrial DNA. V. A 25 micron closed circular duplex DNA molecule in wild-type yeast mitochondria: Structure and genetic complexity. Biochim. Biophys. Acta 209:1–15.

Hollenberg, C. P., P. Borst, and D. F. J. van Bruggen. 1972. Mitochondrial DNA from cytoplasmic petite mutants of yeast. Biochim. Biophys. Acta 277:35–43.

Howell, N., P. L. Molloy, A. W. Linnane, and H. B. Lukins. 1974. Biogenesis of mitochondria. 34. The synergistic interaction of nuclear and mitochondrial mutations to produce resistance to high levels of mikamycin in *Saccharomyces cerevisiae.* Molec. Gen. Genet. 128:43–54.

Hourcade, D., D. Dressler, and J. Wolfson. 1973. The amplification of ribosomal RNA genes involves a rolling circle intermediate. Proc. Nat. Acad. Sci. USA. 70:2926–30.

Inman, R. B. 1967. Denaturation maps of the left and right sides of the Lambda DNA molecule determined by electron microscopy. J. Mol. Biol. 28:103–16.

Kao, S.-h. 1973. Genetic analysis of cytoplasmic mutagenesis in *Saccharomyces cerevisiae.* M.S. thesis, Ohio State University, Columbus, Ohio.

Lazowska, J., F. Michel, G. Faye, H. Fukuhara, and P. P. Slonimski. 1974. Physical and genetic organization of petite and grande yeast mitochondrial DNA. II. DNA-DNA hybridization studies and buoyant density determinations. J. Mol. Biol. 85:393–410.

Leon, S., and H. R. Mahler. 1968. Isolation and properties of mitochondrial RNA from yeast. Arch. Biochem. Biophys., 126:305–19.

Lindegren, C. C., S. Nagai, and H. Nagai. 1958. Induction of respiratory deficiency in yeast by manganese, copper, cobalt, and nickel. Nature. 182:446–48.

Linnane, A. W., G. W. Saunders, E. B. Gingold, and H. B. Lukins. 1968. The biogenesis of mitochondria. V. Cytoplasmic inheritance of erythromycin resistance in *S. cerevisiae.* Proc. Nat. Acad. Sci. USA. 59:903–10.

Locker, J., M. Rabinowitz, and G. S. Getz. 1974. Tandem inverted repeats in mitochondrial DNA of petite mutants of *Saccharomyces cerevisiae.* Proc. Nat. Acad. Sci. USA. 71:1366–70.

Luck, D. J. L., and E. Reich. 1964. DNA in mitochondria of *Neurospora crassa.* Proc. Nat. Acad. Sci. USA. 52:931–38.

Mackler, B., H. C. Douglas, S. Will, D. C. Hawthorne, and H. R. Mahler. 1964. Biochemical correlates of respiratory deficiency. IV. Composition and properties of respiratory particles from mutant yeasts. Biochemistry 4:2016–20.

Mahler, H. R. 1973a. Biogenetic autonomy of mitochondria. CRC Crit. Rev. Biochem. 1:381–460.

Mahler, H. R. 1973b. Structural requirements for mitochondrial mutagenesis. J. Supramolec. Struct. 1:449–60.

Mahler, H. R. 1973c. Genetic autonomy of mitochondrial DNA, pp. 181–208. *In* B.

A. Hamkalo and J. Papaconstantinou (eds.), Molecular cytogenetics. Plenum, New York.

Mahler, H. R., and R. N. Bastos. 1974a. A novel reaction of mitochondrial DNA with ethidium bromide. FEBS Letters 39:27–34.

Mahler, H. R., and R. N. Bastos. 1974b. Coupling between mitochondrial mutation and energy transduction. Proc. Nat. Acad. Sci. USA. 71:2241–45.

Mahler, H. R., B. D. Mehrotra, and P. S. Perlman. 1971. Formation of yeast mitochondria. V. Ethidium bromide as a probe for the functions of mitochondrial DNA. Prog. Molec. Subcell. Biol. 2:274–96.

Mahler, H. R., and P. S. Perlman. 1972a. Mitochondrial membranes and mutagenesis by ethidium bromide. J. Superamolec. Struct. 1:105–24.

Mahler, H. R., and P. S. Perlman. 1972b. Effects of mutagenic treatment by ethidium bromide on cellular and mitochondrial phenotype. Arch. Biochem. Biophys. 148:115–29.

Mahler, H. R., and P. S. Perlman. 1973. Induction of respiration deficient mutants in *Saccharomyces cerevisiae* by berenil. I. Berenil, a novel non-intercalating mutagen. Molec. Gen. Genet. 121:285–94.

Mehrotra, B. D., and H. R. Mahler. 1968. Characterization of some unusual DNA's from the mitochondria from certain "petite" strains of *Saccharomyces cerevisiae.* Arch. Biochem. Biophys. 128:685–703.

Michaelis, G., S. Douglass, M. Tsai, and R. S. Criddle. 1971. Mitochondrial DNA and suppressiveness of petite mutants in *Saccharomyces cerevisiae.* Biochem. Genet. 5:487–95.

Michaelis, G., E. Petrochilo, and P. P. Slonimski. 1973. Mitochondrial genetics III. Recombined molecules of mitochondrial DNA obtained from crosses between cytoplasmic petite mutants of *Saccharomyces cerevisiae:* Physical and genetic characterization. Molec. Gen. Genet. 123:51–65.

Michel, F., J. Lazowska, G. Faye, H. Fukuhara, and P. P. Slonimski. 1974. Physical and genetic organization of petite and grande yeast mitochondrial DNA. III. High resolution melting and reassociation studies. J. Mol. Biol. 85:411–31.

Mounolou, J. C., H. Jakob, and P. P. Slonimski. 1966. Mitochondrial DNA from yeast "petite" mutants: Specific changes in buoyant density corresponding to different cytoplasmic mutations. Biochem. Biophys. Res. Commun. 24:218–24.

Moustacchi, E. 1971. Evidence for nucleus independent steps in control of repair of mitochondrial damage. I. UV-induction of the cytoplasmic petite mutation in UV-sensitive nuclear mutants of *Saccharomyces cerevisiae.* Molec. Gen. Genet. 114:50–58.

Moustacchi, E. 1972. Determination of the degree of suppressivity of *Saccharomyces cerevisiae* strain RD1A. Biochim. Biophys. Acta 277:59–60.

Nagai, S., N. Yanagishima and H. Nagai. 1961. Advances in the study of respiration deficient (RD) mutation in yeast and other micro-organisms. Bact. Rev. 25:404–26.

Nagley, P., E. B. Gingold, H. B. Lukins, and A. W. Linnane. 1973. Biogenesis of mitochondria. 25. Studies on the mitochondrial genomes of petite mutants of yeast using ethidium bromide as a probe. J. Mol. Biol. 78:335–50.

Nagley, P., and A. W. Linnane. 1970. Mitochondrial DNA deficient petite mutants of yeast. Biochem. Biophys. Res. Comm. 39:989–96.

Nagley, P., and A. W. Linnane. 1972. Biogenesis of mitochondria. 21. Studies on the nature of the mitochondrial genome in yeast: The degenerative effect of ethidium bromide on mitochondrial genetic information in a respiratory competent strain. J. Mol. Biol. 66:181-93.

Nass, M. M. K., and S. Nass. 1963a. Intramitochondrial fibers with DNA characteristics. I. Fixation and electron staining reactions. J. Cell Biol. 19:593-611

Nass, S., and M. M. K. Nass. 1963b. Intramitochondrial fibers with DNA characteristics. II. Enzymatic and other hydrolytic treatments. J. Cell. Biol. 19:613-29.

Ogur, M., R. St. John, and S. Nagai. 1957. Tetrazolium overlay technique for population studies of respiration deficiency in yeast. Science 125:928-29.

Perlman, P. S., and C. W. Birky, Jr., 1974. Mitochondrial genetics in bakers' yeast: A molecular mechanism for recombinational polarity and suppressiveness. Proc. Nat. Acad. Sci. USA. 71:4612-16.

Perlman, P. S., and H. R. Mahler. 1970a. Intracellular localization of enzymes in yeast. Arch. Biochem. Biophys. 136:245-59.

Perlman, P. S., and H. R. Mahler. 1970b. Formation of yeast mitochondria. III. Biochemical properties of mitochondria isolated from a cytoplasmic petite mutant. Bioenergetics 1:113-38.

Perlman, P. S., and H. R. Mahler. 1971a. Molecular consequences of ethidium bromide mutagenesis. Nature New Biol. 231:12-16.

Perlman, P. S., and H. R. Mahler. 1971b. A premutational state induced in yeast by ethidium bromide. Biochem. Biophys. Res. Commun. 44:261-67.

Piperno. G., G. Fonty, and G. Bernardi. 1972. The mitochondrial genome of wild-type yeast cells. II. Investigations on the compositional heterogeneity of mitochondrial DNA. J. Mol. Biol. 65:191-205.

Prunell, A., and G. Bernardi. 1974. The mitochondrial genome of wild-type yeast cells. IV. Genes and spacers. J. Mol. Biol. 86:825-41.

Rank, G. H., and C. Person. 1969. Reversion of spontaneously arising respiratory deficiency in *Saccharomyces cerevisiae.* Can. J. Genet. Cytol. 11:716-28.

Rank, G. H. 1970. Genetic evidence for "Darwinian" selection at the molecular level. I. The effect of the suppressive factor on cytoplasmically inherited erythromycin-resistance in *Saccharomyces cerevisiae.* Can. J. Genet. Cytol. 12:129-36.

Raut, C., and W. L. Simpson. 1955. The effects of X-rays and ultraviolet light of different wavelengths on the production of cytochrome-deficient yeasts. Arch. Biochem. Biophys. 56:218-28.

Robberson, D. L., H. Kasamatsu, and J. Vinograd. 1972. Replication of mitochondrial DNA. Circular replicative intermediates in mouse L-Cells. Proc. Natl. Acad. Sci. USA. 69:737-41.

Rodarte-Ramon, U., and R. K. Mortimer. 1972. Radiation-induced recombination in *Saccharomyces:* Isolation and genetic study of recombination-deficient mutants. Radiation Res. 49:133-47.

Sager, R. 1972. Cytoplasmic genes and organelles. Academic Press, New York.

Sanders, P. P. M., R. A. Flavell, P. Borst, and J. N. Mol. 1973. Nature of the base sequence conserved in the mitochondrial DNA of a low density petite. Biochim. Biophys. Acta 312:441-57.

Saunders, G. W., E. B. Gingold, M. K. Trembath, H. B. Lukins, and A. W. Linnane. 1971. Mitochondrial genetics in yeast: Segregation of a cytoplasmic determinant in crosses and its loss or retention in the petite, pp. 185–93. *In* N. K. Boardman, A. W. Linnane, and R. M. Smillie (eds.), Autonomy and biogenesis of mitochondria and chloroplasts. North Holland Publishing Co., Amsterdam.

Schäfer, K., and H. Küntzel. 1972. Mitochondrial genes in *Neurospora:* A single cistron for ribosomal RNA. Biochem. Biophys. Res. Commun. 46:1312–19.

Sena, E. 1971. Mitochondrial DNA replication in yeast. Ph.D. diss., University of Wisconsin, Madison, Wis.

Sena, E., D. Radin, and S. Fogel. 1973. Synchronous mating in yeast. Proc. Nat. Acad. Sci. USA. 70:1373–77.

Sherman, F. 1959. The effects of elevated temperatures on yeast. II. Induction of respiratory deficient mutants. J. Cell. Comp. Physiol. 54:37–52.

Sherman, F., and B. Ephrussi. 1962. The relationship between respiratory deficiency and suppressiveness in yeast as determined with segregational mutants. Genetics 47:695–700.

Slonimski, P. 1949. Action de l'acriflavine sur les levures. IV. Mode d'utilization du glucose par les mutants "Petite Colonie," Ann. Inst. Pasteur 76:510–30.

Slonimski, P. P., and B. Ephrussi. 1949. Action de l'acriflavine sur les levures. V. Le système des cytochromes des mutants "Petite Colonie," Ann. Inst. Pasteur 77:47–63.

Slonimski, P. P., G. Perrodin, and J. H. Croft. 1968. Ethidium bromide induced mutation of yeast mitochondria: Complete transformation of cells into respiratory deficient nonchromosomal "petites." Biochem. Biophys. Res. Comm. 30:232–39.

Smith, D. G., R. Marchant, N. G. Maroudas, and D. Wilkie. 1969. A comparative study of the mitochondrial structure of petite strains of *Saccharomyces cerevisiae.* J. Gen. Micro. 56:47–54.

Stuart, K. O. 1970. Cytoplasmic inheritance of oligomycin and rutamycin resistance in yeast. Biochem. Biophys. Res. Comm. 39:1045–51.

Subik, J., J. Kolarov, and L. Kovac. 1972. Obligatory requirement of intramitochondrial ATP for normal functioning of the eucaryotic cell. Biochem. Biophys. Res. Commun. 49:192–98.

Suda, K., and A. Uchida. 1974. The linkage relationship of the cytoplasmic drug-resistance factors in *Saccharomyces cerevisiae.* Molec. Gen. Genet. 128:331–39.

Tavlitzki, J. 1949. Action de l'acriflavine sur les levures. III. Etude de la croissance des mutants "Petite Colonie." Ann. Inst. Pasteur 76:497–509.

Tewari, K. K., W. Votsch, H. R. Mahler, and B. Mackler. 1966. Biochemical correlates of respiratory deficiency. VI. Mitochondrial DNA. J. Mol. Biol. 20:453–81.

Thomas, D., and D. Wilkie. 1968. Inhibition of mitochondrial synthesis in yeast by erythromycin: Cytoplasmic and nuclear factors controlling resistance. Genet. Res. 11:33–41.

Uchida, A. 1972. Effect of acriflavine and ultraviolet-light irradiation on suppressive petites of *Saccharomyces cerevisiae.* Japan J. Genet. 47:159–63.

Uchida, A., and K. Suda. 1973. Ethidium bromide induced loss and retention of cytoplasmic drug resistance factors in yeast. Mutation Res. 19:57–63.

Upholt, W. B., and P. Borst. 1974. Accumulation of replicative intermediates of mitochondrial DNA in *Tetrahymena pyriformis* grown in ethidium bromide. J. Cell Biol. 61:383–97.

Wakabayshi, K., and N. Gunge. 1970. Extrachromsomal inheritance of oligomycin resistance in yeast. FEBS Letters. 6:302–4.

Yotsuyanagi, Y. 1962. Etudes sur le chondriome de la levure. II. Chondriomes des mutants à déficience respiratoire. J. Ultrastruct. Res. 7:141–58.

C. WILLIAM BIRKY, JR.

Mitochondrial Genetics in Fungi and Ciliates

5

When Mendel began his ambitious attempt to understand the phenomenon of heredity, he had little knowledge and but one tool at his disposal. Of the mechanics of reproduction, he knew only that a single pollen grain suffices to fertilize a single egg cell; chromosomes were decades in the future. His only tool was hybridization: to carefully mate plants of known phenotypes, count the progeny, and from the phenotypic ratios of those progeny to infer the laws governing the behavior of genes. The explanation of those laws in terms of phenomena at the level of the cell had to wait for the discovery of chromosomes and meiosis. The extension of that explanation to the molecular level was again delayed until phage geneticists and molecular biologists provided the basic tools and information about gene structure and function nearly a century after Mendel's work.

Today, we who study the genetics of mitochondria and chloroplasts are luckier. Our subject matter dates back to 1908, when Correns and Baur discovered the first cases of chloroplast inheritance. Nevertheless, like Mendel, the Drosophila Group, and the Phage Group in the past, we are fortunate enough to know the excitement of having for our subject entire genetic systems for which the most basic aspects of transmission genetics are scarcely understood. But unlike the pioneers of genetics, we have in addition a vast and detailed body of knowledge of cells, organelles, genes, and DNA molecules and their behavior, together with an impressive arsenal of tailor-made experimental organisms and precision techniques to aid us in our task.

Department of Genetics, Ohio State University, Columbus, Ohio 43210

That task, for the geneticist, includes the elucidation of both the phenogenetics and the transmission genetics of organelles. By phenogenetics I mean the study of gene structure and of gene action and its control. By transmission genetics I mean the study of how genes are transmitted from generation to generation, how they recombine and segregate, and how one explains the numerical proportions of different kinds of offspring produced in a cross. Our knowledge of organelle phenogenetics has advanced more rapidly than that of organelle transmission genetics, partly because of the current popularity of molecular genetics in general and partly because of the complexity of transmission genetic phenomena and the scarcity of suitable experimental systems. But an understanding of transmission genetics is essential before we can evaluate the role of organelle genes in evolutionary processes; it is of importance in practical applications, especially in plant breeding; and besides, it is a fascinating challenge!

In this paper I will concentrate on that challenge. The paper will deal exclusively with mitochondrial genetics of fungi and ciliates, but, as will become apparent, this constitutes almost *all* of mitochondrial transmission genetics at the present time. I will begin by describing some of the basic phenomena encountered in organelle genetics in general, and in mitochondrial genetics in particular, and point out the important differences in phenomena and methods used in the study of organelle and nuclear genes. Following a brief description of the kinds of mitochondrial genes available for study, I will discuss a series of genetic phenomena at the organelle and molecular levels: complementation, intracellular selection, recombination, input ratios, and vegetative segregation. Finally, these same phenomena will be invoked in attempts to understand the numerical results of crosses involving mitochondrial genes, including especially the phenomena of zygote heterogeneity, bias, polarity, and suppressiveness.

ORGANELLE AND NUCLEAR GENETICS COMPARED AND CONTRASTED

Figure 1 summarizes some basic features of nuclear genetics, using a hypothetical plant and a gene controlling pigment synthesis. The geneticist is confronted with *output ratios* or proportions of offspring of different phenotypes. He attempts to explain these in terms of *meiotic segregation*, recombination, and *random fertilization* which result in the production of homozygous or heterozygous zygotes. During vegetative growth of the zygotes to produce adult organisms (or colonies of microorganisms) there is generally no further segrega-

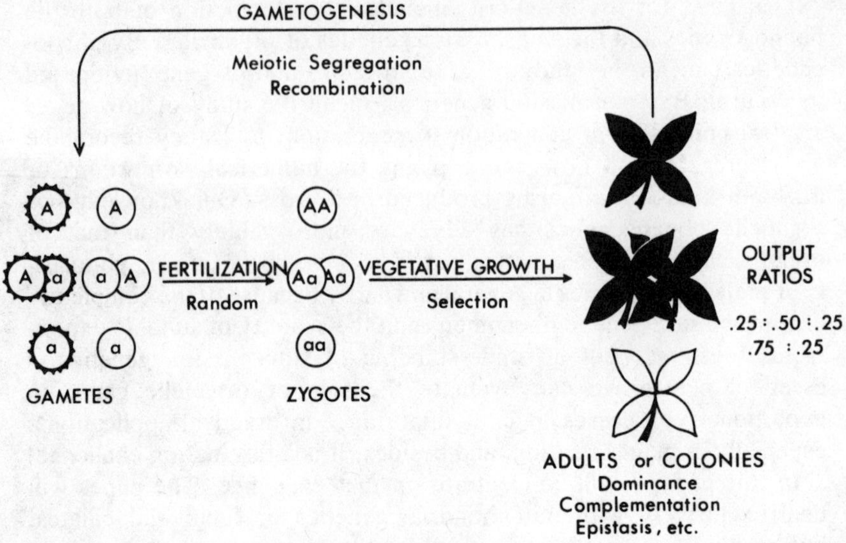

Fig. 1. Some basic features of the transmission genetics of nuclear genes in a hypothetical plant, for comparison with figure 2. See text for explanation.

tion. *Selection* intervenes here and at other points in the life cycle to help determine the observed genotypic ratios; these genotypic ratios are related to the phenotypic ratios by various forms of interaction between alleles or between genes, such as *dominance, complementation,* and *epistasis.* In this figure, meiosis is shown taking place after an interval of vegetative growth, but of course in some organisms it occurs shortly after zygote formation; in that case phenotypic ratios are observed in haploid cells where dominance and complementation are not involved.

Figure 2 shows a cross involving green and white (mutant) chloroplasts. It illustrates that mitochondria and chloroplasts, and their genomes, may be present in multiple copies in gametes and zygotes. We must therefore take into account the *input ratio* of organelle genomes in the zygote when explaining the results of the cross. The fusion of gametes with genetically different organelles results in a zygote referred to as heteroplasmic, by analogy to cells which are heterozygous for nuclear genes. *Recombination* and *complementation* may be observed in such cells. Since each organelle contains its own genome and machinery for protein synthesis, and is self-replicating, the organelles in a cell enjoy a certain degree of autonomy which

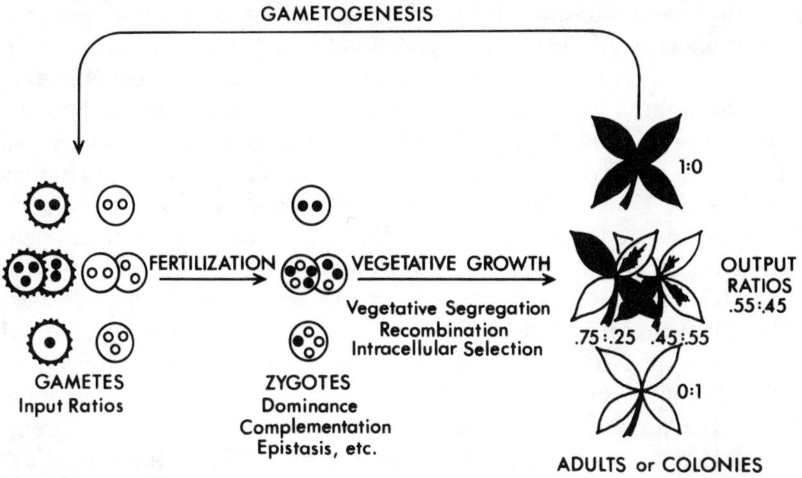

Fig. 2. Some basic features of the transmission genetics of organelle genes in a hypothetical plant, for comparison with figure 1. See text for explanation.

can result in *intracellular selection* for organelles of particular genotypes. The heteroplasmic state is generally short-lived, since organelles rapidly segregate during vegetative reproduction to produce homoplasmic cells pure for one organelle genotype or another. Output ratios may be determined for the progeny cells of a single zygote as well as for the total progeny of the cross, and each zygote may have a unique output ratio (*zygote heterogeneity*). In such homoplasmic cells, genotypic and phenotypic ratios are identical. (The ratios shown in figure 2 are possible examples, and have no special significance.) At present very little information is available on dominance or on epistasis and other interactions between different genes in mitochondria, so these phenomena will not be discussed further.

TYPES OF MITOCHONDRIAL MUTANTS AVAILABLE FOR ANALYSIS

Cytoplasmic mutants were first reported in baker's yeast (*Saccharomyces cerevisiae*)[1] by Ephrussi and his collaborators in 1949 (see Ephrussi et al., 1955 for references). These *vegetative petite* (ρ^-) mutants are respiratory-deficient; they have since been shown to result

1. The laboratory stocks called *Saccharomyces cerevisiae* are actually derived originally from hybrids between *S. cerevisiae* and closely related species, such as *S. carlsbergensis* (Sherman and Lawrence, 1974).

from the loss of varying portions of the mitochondrial genome, together with duplication of the remaining portion (if any). A detailed discussion of ρ^- mutants and references are given by Perlman in this volume.

Mitochondrial genetics in yeast was given a strong impetus by the isolation and characterization of mitochondrial mutants resistant to the antibiotics erythromycin, spiramycin, or paromomycin (Thomas and Wilkie, 1968a; Linnane et al., 1968). Other mutants have been found for resistance to chloramphenicol (Coen et al., 1970), mikamycin (see Howell et al., 1974), and oligomycin and other inhibitors of oxidative phosphorylation (Wakabayashi and Gunge, 1970; Stuart, 1970; Griffiths, this volume).[2] A mitochondrial mutant with modified mutation rate from ρ^+ to ρ^- has been isolated by Handwerker et al. (1973) but is likely to be difficult to use in transmission genetics, as are the respiratory-deficient mutants isolated by Mahler's group (Mahler, this volume) and by Storm et al. (1974). The *omega* (ω) locus (Bolotin et al., 1971) is described below. That each of these genes actually resides on mitochondrial DNA has been shown by demonstrating that they are lost in ρ^- mutants that lack mitochondrial DNA. The biochemical basis of antibiotic resistance is reviewed by Linnane et al. (1972).

In *Neurospora crassa* Mitchell and Mitchell (1952) obtained a slow growing mutant deficient in cytochrome oxidase and called it *poky;* it was later named *mi-1* (maternally-inherited). A number of similar mutants have since been isolated and are variously named *abn* (abnormal), *stp* (stopper), *exn* (extranuclear), and *SG* (slow growth) (see Bertrand and Pittenger, 1972a,b, for references; also Kuriyama and Luck, 1974, *re* biochemical studies). I will refer to them collectively as poky mutants. Microinjection experiments by Diacumakos et al. (1956) demonstrated that these cytoplasmically inherited mutations are located in the mitochondria.

In *Aspergillus nidulans* at least two probable mitochondrial mutants have been discovered: one confers resistance to oligomycin (Rowlands and Turner, 1973), and one confers cold sensitivity (Waldron and Roberts, 1973). Both are inherited as cytoplasmic genes, and are linked (Rowlands and Turner, 1974a). Since the oligomycin-resistant mutant

2. For simplicity, mitochondrial alleles conferring resistance or sensitivity to antibiotics will be designated as C^R/C^S, E^R/E^S, M^R/M^S, O^R/O^S, P^R/P^S, and S^R/S^S for chloramphenicol, erythromycin, mikamycin, oligomycin, paromomycin, and spiramycin, respectively. Other conventions are in use for yeast but are less suitable for a non-technical review.

shows several physiological alterations of the mitochondria (Rowlands and Turner, 1974b), it appears likely that both are mitochondrial mutations.

Among the ciliates the only extensive studies of mitochondrial genetics have utilized *Paramecium aurelia.* Eight phenotypic classes of mutants resistant to various combinations of erythromycin, chloramphenicol, spiramycin, and mikamycin are available (Beale, 1969; Adoutte and Beisson, 1970; Adoutte, 1974). They have been identified as mitochondrial mutants on the basis of their cytoplasmic inheritance, phenotypes, and transmission by microinjection of mitochondria (Beale et al., 1972; Adoutte et al., 1972).

An unfortunate aspect of this list is its brevity; this is even more distressing when one considers that there are no verified mitochondrial mutants available in any other organism (although there are probable cases in *Tetrahymena:* Roberts and Orias, 1973; in chlamydomonas: Alexander et al., 1974; in cultured mouse cells: Bunn et al., 1974; and elsewhere: Sager, 1972). Under these circumstances, the comparative method that has been so very fruitful in other areas of biology cannot be applied. Even in those species that are being studied intensively, the number of different kinds of mutants and of loci available is small. Presumably both of these problems will be overcome in time; the intensive study of mitochondrial genetics is, after all, less than a decade old.

COMPLEMENTATION BETWEEN MITOCHONDRIAL GENES

The complementation test has proved to be a powerful tool in genetics. It is used to determine if two recessive mutations are in the same gene, where "gene" is used in the classical sense and is synonymous with "cistron." The two mutations are introduced by appropriate crosses into the same cell or organism: if the cell phenotype is mutant, i.e., if the mutations do not complement, then they belong to the same gene (they are alleles). If on the other hand the cell phenotype is wild-type, the mutations are said to complement each other because each one entered the cell along with the dominant wild-type allele of the other, and they belong to different genes.

The complementation test has been applied with apparent success to mitochondrial genes in *Neurospora* by Bertrand and Pittenger (1972b). In a series of carefully designed crosses, with all appropriate genetic controls, they made heteroplasmons for most pairwise combinations of different poky-type mutations. Complementation was seen in some combinations as initially normal growth rates. The poky mutants could be divided into three groups, based on their phenotypes and interactions with nuclear genes. It was found that mutants in groups I and II fall into one complementation group, and those in group III constitute a second complementation group. A complication in the interpretation of these data is that complementation was never complete: cytochrome spectra were always abnormal, and growth rates declined in older hyphae of *every* combination. This could conceivably be analogous to the partial complementation that is sometimes seen between two mutations in the *same* gene in studies of nuclear genes.

The negative results obtained in paramecia are more difficult to interpret, and illustrate an important point about complementation tests with organelle genes. Adoutte and Beisson (1972; Adoutte et al., 1973) have performed the cross $C^R E^S \times C^S E^R$ with cytoplasmic exchange, using several different combinations of C^R and E^R alleles. When the exconjugants were placed immediately into culture medium containing both chloramphenicol and erythromycin, all failed to grow and divide and eventually died. There are two possible interpretations of this result: (1) All C^R and E^R mutations are in the same gene, i.e., all result in the modification of the same ribosome component (likely to be a ribosomal protein: Beale et al., 1972). The heteroplasmic cells thus are unable to form ribosomes containing subunits resistant to both antibiotics. (2) In the heteroplasmic cell neither type of mitochondrion supplies ribosome components to the other.

The second possibility points up a general feature of complementation tests in the case of organelles. If the mitochondria in a particular species never share the products of their gene action, then no complementation can be expected under any circumstances *unless* the two different mutant genomes can be incorporated inside the same organelle, e.g., by fusion of mitochondria. Complementation tests between mutations that are known in advance to affect different genes (a likely example would be O^R and E^R mutations) would thus provide information about the extent of mitochondrial interactions such as fusion or sharing of gene products. In the specific case of paramecia, the failure to find recombination between C^R and E^R (see below) also indicates that either organelle fusion is rare or C^R and E^R are mutations in the same gene.

In yeast several crosses have been done between $C^R E^S$ and $C^S E^R$ strains and the zygotes plated on medium containing both chloramphenicol and erythromycin (Coen et al., 1970; Rank and Bech-Hansen,

1972; Birky, unpublished data). In no case did any zygotes begin colony formation immediately. Colonies formed later are attributable to intracellular selection for C^R E^R recombinant genophores. This result is intriguing: we know that mitochondria exchange DNA in yeast; the C^R and E^R loci are separable by recombination; and the E^R mutation appears to reside in the gene coding for the 23 S ribosomal RNA of the mitochondrion, while the C^R mutational site lies near to, but outside of, this gene (Faye et al., 1974). This suggests that, although mitochondria exchange DNA, they do not efficiently share the products of the C^R and E^R genes.

INTRACELLULAR SELECTION

A number of observations point to the existence of a fair degree of autonomy of individual organelles in a cell. Each mitochondrion, for example, has its own protein-synthesizing machinery; in many plants, phenotypically mutant and wild-type chloroplasts can exist side by side in the same cell (reviewed by Kirk and Tilney-Bassett, 1967). Is it possible that in a heteroplasmic cell each mitochondrion could replicate at a rate determined by its *own* genotype and environment? Differences in replication rates of plastids with different genotypes has been invoked to explain genetic ratios in higher plants (Kirk and Tilney-Bassett, 1967), and there is evidence for such a phenomenon in fungi and ciliates.

In paramecia, heteroplasmons are made by conjugation with cytoplasmic exchange or by complete fusion of two cells to produce "doublets." These cells are then placed in media with or without antibiotics and observed over a number of generations (Adoutte and Beisson, 1972; Adoutte et al., 1973; Perasso and Adoutte, 1974). When a cell with a minority of E^R and a majority of E^S mitochondria is placed in erythromycin, it does not reproduce for several days. The cell then begins to grow and divide, and morphologically normal mitochondria begin to appear in a population that was initially uniformly abnormal. Such a cell, if placed in medium without antibiotic, will produce a mixture of homoplasmic E^R and E^S cells. After as few as three generations, all mitochondria in the cells are morphologically normal and all cells are homoplasmic for E^R. The rapidity with which the mitochondrial population changes from mainly E^S to completely E^R indicates that (1) E^R mitochondria do replicate and (2) E^S mitochondria do not replicate and either degenerate or are "transformed" in some manner by mitochondrial DNA carrying the E^R gene.

Adoutte and Beisson also grew paramecia having various proportions of antibiotic-resistant and -sensitive mitochondria in media without antibiotics (Adoutte and Beisson, 1972; Adoutte et al., 1973). In this case, it was found that the antibiotic-resistant organelles replicate more slowly than the antibiotic-sensitive organelles and are replaced by them in time. A number of different C^R and E^R mutants were tested, and it was found that each mitochondrial genotype has its own characteristic rate of replication.

The occurrence of intracellular selection suggests a solution to the problem of the origin of mitochondrial mutant cells. The problem is that each cell contains a large number of mitochondrial genomes (about fifty in the case of haploid yeast; several thousands in a paramecium). A mutation, whether spontaneous or induced, is unlikely to occur in more than one or a few of these mitochondria in any one cell. How is it that a resistant cell, which must contain a majority of mutant mitochondria in paramecia and presumably in other organisms as well, can arise in the presence of the antibiotic? In the initial studies of paramecia, Beale (1969) and Adoutte and Beisson (1970) observed that sensitive cells placed in erythromycin or chloramphenicol stop dividing, and none grow for a period of one or a few weeks. Then a few cells start to grow after varying periods of time and produce resistant clones. It seems possible that these cells initially have one or a few antibiotic-resistant mitochondria that continue to divide even though the cells are not growing. When a cell acquires enough resistant mitochondria to support cellular metabolism, the cell itself begins to grow and forms a resistant clone.

Support for this suggestion was obtained in yeast (Birky, 1973). In these experiments it was shown, by use of the classical Newcombe experiment, that antibiotic-resistant *cells* are not present in significant numbers in a wild-type population of yeast until *after* exposure to the antibiotic. The Newcombe experiment permits two interpretations: (1) resistant cells arise by intracellular selection of a few resistant mitochondria, or (2) the antibiotic induces resistant mutations. The latter possibility was ruled out by showing that erythromycin-resistant cells do not arise in the presence of *both* chloramphenicol and erythromycin. Under these conditions, the hypothetical mutation induction should continue to occur, but E^R genomes would be prohibited from expression by the chloramphenicol and hence would not be selected.

Comparable experiments have not been done with mitochondria in other organisms. They have been done with chloroplast mutants in *Chlamydomonas,* where it was shown that streptomycin-resistant cells appear only after exposure to the antibiotic (Gillham and Levine, 1962; Sager, 1962). In this case, however, mutation induction by streptomycin has not been ruled out. But from the viewpoint of evolutionary biology, the most important result of these studies is that mutant cells appear in response to the selective agent, rather than being selected from a preexisting pool of mutant cells. This has potential implications for the population genetics and evolution of organelles that should be thoroughly explored.

RECOMBINATION AND MAPPING OF MITOCHONDRIAL GENES

Recombination of mitochondrial genes was first described in yeast in 1968. Two-factor crosses involving antibiotic-resistance mutations were reported by Thomas and Wilkie (1968b): the mating $E^S P^R \times E^R P^S$ produced diploid cells of the two parental genotypes $E^S P^R$ and $E^R P^S$, and in addition the two recombinant types $E^R P^R$ and $E^S P^S$ were found. (The "three-factor" crosses $E^R S^R P^S \times E^S S^S P^R$ reported in that paper are actually two-factor crosses involving a mutation that confers simultaneous resistance to erythromycin and spiramycin. The rare $E^R S^S$ and $E^S S^R$ "recombinants" obtained could have been $E^R S^R$ cells, which in my experience are easily misclassified.) At the same time, Linnane et al. (1968; Gingold et al., 1969) reported recombination of the E^R and ρ^- phenotypes.

The question immediately arose as to whether the observed recombination was due to independent segregation of genetic markers located on different molecules. This question is answered by physical studies of mitochondrial DNA that have shown in a number of organisms (including yeast, *Neurospora,* and *Tetrahymena*) that there is but a single molecular species of mitochondrial DNA (reviewed by Linnane et al., 1972). In yeast, for example, the mitochondrial DNA molecule has a molecular weight of about 50×10^6 daltons; this molecule contains all of the mitochondrial genetic information, and is present in about fifty copies per haploid cell. (The number of copies may vary with the physiological condition of the cells in some stocks: Goldthwaite et al., 1974).

There is some confusion in the terminology applied to DNA molecules and genomes in prokaryotes, eukaryotes, and organelles. The term "chromosome" was originally applied to the "colored bodies" of the eukaryotic nucleus, which contain histones and other proteins and RNA, and which undergo mitosis and meiosis. The application of this term to the naked DNA molecules of prokaryotes and organelles is clearly inappropriate. Ris (1961) has suggested the use of the term "genophore" for the physical entity corresponding to a linkage group; following Stanier (1970), I will use this term for mitochondrial and chloroplast DNA molecules.

A set of ingenious experiments by Shannon et al. (1972) and Michaelis et al., (1973) utilized petite mutants with mitochondrial DNA of altered buoyant density to demonstrate that recombination in yeast involves the breakage and reunion of DNA molecules. The large alterations in buoyant density of the recombinant molecules found in these experiments argues that recombination of mitochondrial genes can involve the exchange of genophore segments (crossing-over). But this does not exclude the possibility that much, or even most, of the recombination of single markers is non-reciprocal (gene conversion), as has been found in many other systems.

The experiments of Michaelis et al. (1973) involved $\rho^- \times \rho^-$ matings; since there is little or no protein synthesis in ρ^- mitochondria, there could have been little or no expression of mitochondrial genes in the zygotes. The fact that recombination still occurred indicates that nuclear genes code for all of the enzymes that are essential for mitochondrial gene recombination. In my laboratory, Alice Fraenkel (1974) has tested twelve different nuclear gene mutations, representing at least ten loci, that are believed to reduce or eliminate recombination of genes in the yeast nucleus, and has found that none of these has a large effect on mitochondrial gene recombination. It thus appears that recombination in the nucleus and in the mitochondria involves different sets of genes and thus different enzymes, at least in part.

In *Neurospora* no careful attempts have been made to detect recombination of mitochondrial markers. Gowdridge (1956) looked for, and failed to find, recombination between *mi-1* and *mi-3*, but her experimental design could only have detected recombinants produced at very high frequency. We now know that this is not to be expected, since these two mutations belong to the same complementation group (Bertrand and Pittenger, 1972b).

Recombination between the mitochondrial genes conferring oligomycin-resistance and cold sensitivity has been detected in *Aspergillus* (Rowlands and Turner, 1974a). If additional types of mutants can be obtained and technical problems with quantitative measurements of recombination can be overcome, this system may prove the most useful in the study of recombination in mycelial fungi.

It is only in yeast, then, that recombination furnishes a useful tool for the study of mitochondrial genetics at present. In this organism extensive mapping experiments have been performed independently in a number of different laboratories, using a variety of nuclear gene backgrounds and independently isolated mutants. It is therefore a pleasant surprise to find a fair degree of agreement in the results. The data, summarized in figure 3, indicate that there is a region of the yeast mitochondrial genophore spanning about 10 map units that contains four or five loci: the ω locus, which affects output ratios (see "Yeast: Polarity" below); the R_I locus, usually characterized by C^R mutants; the R_{II} and R_{III} loci, represented by E^R or S^R mutants; and (assuming the *cap1* locus of Linnane's group corresponds to the R_I locus of Slonimski's group) a *spi4* locus (S^R). This region will be referred to as the R region of the genophore or the R linkage group; the use of the letter "R" to designate these loci is based on their control of antibiotic resistance in mitochondrial ribosomes. Additional linkage groups have been found by Lancashire and Griffiths (1975: O_I and O_{III}) and by Howell et al. (1974: *mik1* and *oli1*). These linkage groups, and the P and O_{II} loci, are separated from each other in all possible combinations by about 20% recombination. It is thus not possible to establish an unambiguous map order for the various linkage groups at the present time, and they are considered to be formally unlinked; the situation is precisely analogous to that in the early period of phage genetics. The interpretation of the data in terms of a random mating pool of genophores, and genetic evidence that all of the above mitochondrial mutants are on the same genophore, will be discussed in detail below (see "Yeast: The Phage-Analogy Model").

There has been an intensive search for recombination between various C^R and E^R mutations in *Paramecium:* heteroplasmic exconjugants from the cross $C^R E^S \times C^S E^R$ are reared either in chloramphenicol plus erythromycin to select for $C^R E^R$ recombinants or in drug-free medium to select for $C^S E^S$ recombinants (Adoutte and Beisson, 1972; Adoutte, 1974). The frequency of recombinant types obtained in these experiments is so low as to be not clearly distinguishable from the frequency of back-mutations, suggesting that recombination between mitochondria is rare or non-existent. This is an organism in which cytoplasmic exchange between conjugants, needed for the formation of heteroplasmons, is probably quite rare in nature. With the opportunities for recombination thus being limited, the evolutionary loss of the recombination mechanism might follow.

This discussion of mapping has considered only the classical tech-

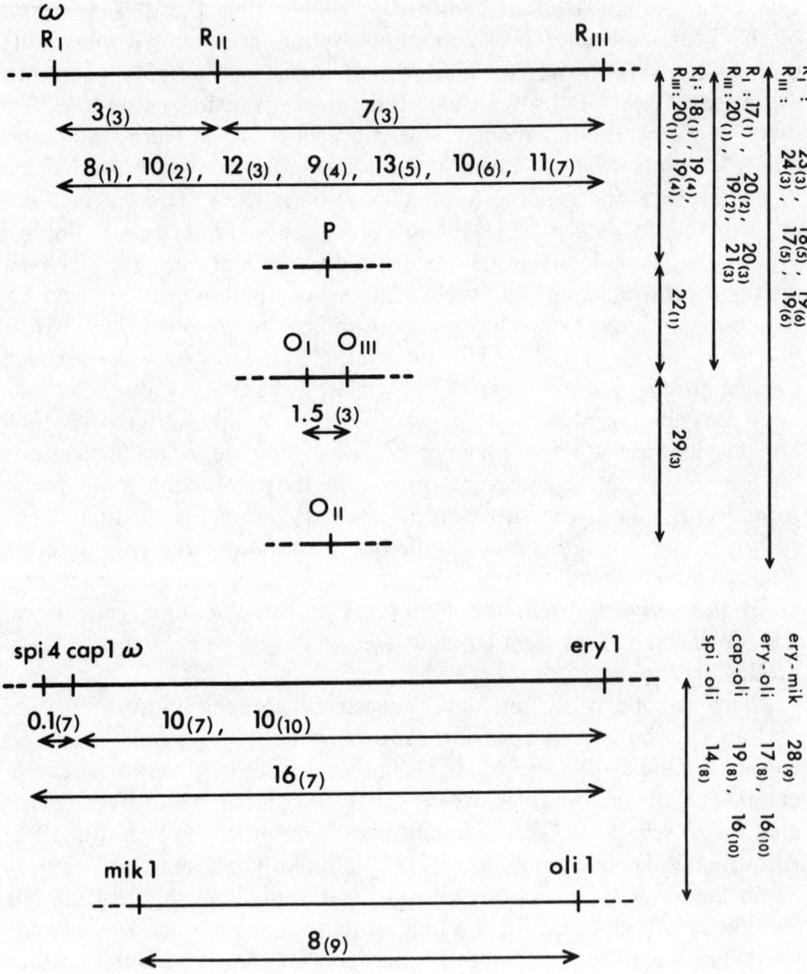

ω
R I
R II
R III
3(3)
7(3)
8(1), 10(2), 12(3), 9(4), 13(5), 10(6), 11(7)
P
O I
O III
1.5 (3)
O II
R I : 23(3) ; 18(5) ; 19(6)
R III : 24(3) ; 17(5) ; 19(6)
R I : 17(1) , 20(2) ; 20(3)
R III : 20(1) , 19(2) , 21(3)
R I : 18(1) , 19(4) ,
R III : 20(1) , 19(4) ,
22(1)
29(3)
spi 4 cap1 ω
ery 1
0.1(7)
10(7), 10(10)
16(7)
mik 1
oli 1
8(9)
ery-mik 28(9)
ery-oli 17(8) , 16(10)
cap-oli 19(8) , 16(10)
spi - oli 14(8)

niques and their application to so-called homopolar crosses. In the heteropolar crosses (defined later) markers can be mapped relative to the ω locus by the efficiency of their transmission to daughter cells or by the effect of prior UV irradiation on that transmission (Coen et al., 1970; Bolotin et al., 1971; Avner et al., 1973; Wolf et al., 1973; Dujon et al., 1974) as well as by classical methods (Linnane et al, 1974). Finally, the close linkage of the R_I and R_{III} loci, and their greater separation from O_I, has been confirmed by studies of the frequency of coincidental deletion of these markers in ρ^- mutants (Deutsch et al., 1974; Linnane et al., 1974; Suda and Uchida, 1974).

Recombination of plastid genes has been looked for, and found, only in *Chlamydomonas*. The reader is referred to Sager (1972) for a review of her pioneering work in mapping chloroplast genes.

INPUT EFFECTS

In studies of Mendelian genes the geneticist is normally certain that his gametes contain precisely one copy of each gene and his zygotes contain precisely two copies. The organelle geneticist, however, must deal with the likelihood that there are multiple copies of the organelle genome not only in the zygotes but also in the gametes of his experimental organism; moreover, the two gametes that fuse

---

Fig. 3. Map data for mitochondrial genes of *Saccharomyces cerevisiae*, obtained from homopolar crosses. Horizontal solid lines represent linkage groups. Numbers on the arrows represent map distances (references) between loci (horizontal arrows) or linkage groups (vertical arrows). The upper diagram is based principally on data from Slonimski's laboratory (1, 2) or from other laboratories using mutants and stocks derived in part from the Slonimski collection (3, 7), but the distances agree well with data obtained for independently-isolated C^R (R_I), E^R (R_{II}), O^R, and P^R mutants (4, 5, 6). The lower diagram is based on data from Linnane's laboratory (8, 9, 10, 11) and includes an *spi4* (S^R) and *mik1* (M^R) locus with no obvious homologies to loci from other labs. Most distances are simple averages of the results of several crosses using the same mitochondrial genes but different nuclear genotypes. The data for references (5) and (6) are arbitrarily shown to involve the O_{II} locus, but in fact the allelic relationships of their O mutants to O_I, O_{II}, and O_{III} are unknown. Not shown are the data of Callen (1974a,b) dealing with two different S^R mutants and two different O^R mutants, which show about 18% recombination with each other in all possible combinations. Callen (personal communication) has a C^R locus that lies about 10 map units from his S_2^R locus, so these may correspond to the R_I and R_{III} loci of Slonimski, while his other S^R locus appears to be on the other side of C^R. References: (1) Wolf et al., 1973; (2) Avner et al., 1973; (3) Griffiths et al., 1975; (4) Kleese et al., 1972; (5) Rank, 1973; (6) Suda and Uchida, 1973; (7) Birky, unpublished data; (8) Trembath et al., 1973; (9) Howell et al., 1973; (10) Howell et al., 1974.

to form a zygote are known to contain very different numbers of organelle genophores in many cases. Thus arises the possibility of input effects: providing the organelle genomes replicate at similar rates, the output ratios of genotypes in the homoplasmic progeny will reflect the input ratios of genomes in the gametes.

Evidence that such effects do exist comes from studies on *Paramecium*. For example, following the cross $C^R \times C^S$ with cytoplasmic exchange, the originally C^S exconjugant will produce C^S homoplasmic progeny after about thirty generations, and by the eightieth generation the clone will consist entirely of C^S homoplasmons. In contrast, the originally C^R exconjugant will have produced few or no homoplasmic C^S cells by the eightieth generation. These results reflect the preponderance of C^S and C^R mitochondria, respectively, in the two exconjugants.

These results are backed up by experimental evidence obtained with yeast. Perlman and Demko (1974) and Young (personal communication) have performed crosses between haploid and diploid yeast: diploids in general contain twice as many mitochondrial DNA molecules as do haploids. In the homoplasmic progeny of the triploid zygotes, mitochondrial genes contributed by the diploid parent were always present in excess over their alleles contributed by the haploid parent (table 1). Input effects have also been observed in crosses in which the amount of mitochondrial DNA in one parent was increased by treatment with cycloheximide (Chou, 1973; Perlman and Demko, 1974), or by catabolite derepression (Goldthwaite et al., 1974).

It is important to note that the "input" ratio that affects output ratios in crosses is not necessarily synonymous with the ratio of

TABLE 1

INPUT EFFECTS SEEN IN HOMOPOLAR CROSSES INVOLVING MITOCHONDRIAL GENES IN *SACCHAROMYCES*

$C^SE^S \times C^RE^R$	% Diploid or Triploid Cells					
	C^SE^S	C^RE^R	C^SE^R	C^RE^S	C^S	E^S
Haploid × Haploid	33.9	58.0	4.7	3.5	38.6	37.4
Diploid × Haploid	72.3	22.6	3.5	1.6	75.8	73.9

Haploid cells of stock 2-36 (mitochondrial genotype $\omega^+C^RE^R$) were mated either with haploid stock 4810 ($\omega^+C^SE^S$) or with a diploid stock obtained by diploidizing 4810 and thus isogenic with it. The mating mixtures were plated on minimal medium for prototrophic selection of zygotes. After about twenty generations of growth of the zygotes and their diploid progeny, the cells produced by at least 10^3 zygotes were washed off the plate and the genotypes of a random sample of diploids or triploids were determined by standard replica-plating techniques. The results of this total progeny analysis are expressed as percentage of diploid (or triploid) cells having a particular genotype. Data of Robert Young.

organelle genes in the gametes. One can imagine that two gametes might have equal numbers of organelle genes but that one gamete might not contribute all of its cytoplasmic contents to the zygote. An example of this is seen in the tunicate, in which all of the sperm mitochondria remain outside the egg at fertilization (Ursprung and Schabtach, 1965). This, of course, would not be a factor in organisms such as yeast where the zygote is formed by the complete fusion of two cells, or in doublet cells formed by the complete fusion of conjugating paramecia.

SELECTIVE REPLICATION AND DESTRUCTION

It is also necessary to bear in mind that the input ratio as defined above refers to the initial number of organelle genophores in the zygote. But this is not necessarily the same as the number of *genetically functional* genophores. The possibility exists that mitochondria contributed by one parent and/or their DNA molecules can be selectively destroyed in the zygote. There is no direct evidence for this in ciliates or fungi, but sperm mitochondria are seen to degenerate in the eggs of some animal species (e.g., Anderson, 1968). The destruction of chloroplasts from one parent is known to occur in the zygotes of some algae (e.g., Bråten, 1973), and Sager and her collaborators (Sager, 1972) have presented evidence for the enzymatic degradation of chloroplast DNA from the mt^- parent in *Chlamydomonas* zygotes. Another possibility, for which there is no direct evidence in mitochondria or chloroplasts, is the selective replication of genophores from one parent. It has been suggested, for instance, that bacterial cells contain a limited number of sites to which plasmid (Jacob et al., 1963) or phage (Sinsheimer et al., 1968) DNA molecules can bind for replication. Gillham et al. (1974) have postulated that the same may be true in chloroplasts, and that the plastid DNA molecules that gain access to these sites are the only ones to be replicated and hence to be recovered in the progeny.

VEGETATIVE SEGREGATION

Vegetative segregation of mitochondrial genes in *Paramecium* has been studied by Adoutte and Beisson (1972). We have already seen that, following matings of the form $C^R \times C^S$ or $E^R \times E^S$ with cytoplasmic exchange, the sensitive exconjugant contains a mixture of a few antibiotic-resistant mitochondria and a majority of sensitive mitochondria. When such cells are grown in medium without antibiotic, homoplasmic sensitive cells appear in the clone after as few as ten fissions; such cells are identified by their inability to produce clones when transferred to medium containing antibiotic. A similar experiment can be done in which resistant and sensitive cells are fused completely during conjugation, producing double animals with approximately equal numbers of resistant and sensitive mitochondria. A few homoplasmic cells can be found after ten generations of growth of such double animals, with substantial numbers being seen only after twenty or more generations.

Vegetative segregation is seen in yeast as the appearance of homoplasmic diploid cells among the progeny of heteroplasmic zygotes. The process is extremely rapid. Even the first bud produced by a zygote is frequently homoplasmic (Wilkie and Thomas, 1973; Waxman et al., 1973; Callen, 1974c; Strausberg and Perlman, 1974; Lukins et al., 1973; Dujon et al., 1974), and segregation is complete (all cells are homoplasmic) after about ten generations of reproduction by mitosis and budding. Vegetative segregation of poky and wild-type mitochondria has also been demonstrated in mycelia of *Neurospora* (e.g., Gowdridge, 1956). Unfortunately, the interpretation of such studies is complicated by the variable phenotype of the poky mutants. The oligomycin-resistant and cold-sensitive mutants of *Aspergillus* are more useful in this respect; both show clear-cut segregation from wild-type in heteroplasmic mycelia (Rowlands and Turner, 1974a).

In contrast to chromosomes, mitochondria and chloroplasts are usually not closely associated with the mitotic or meiotic spindles (for some classical exceptions, see Wilson, 1925). They give the impression of being distributed passively, and therefore randomly, between daughter cells during cell division. If in fact the distribution of organelle genophores is random with respect to their genotype, then eventually all cells will inevitably become homoplasmic (the process is frequently called ''sorting-out''). The number of generations required to complete sorting-out, i.e., to produce a population in which all cells are homoplasmic, depends upon the number of segregating units, and can be calculated using hypergeometric probabilities (Michaelis, 1955). To a first approximation, if an organism has N segregating units per cell immediately after cell division, it will take $10N$ generations for complete sorting-out. This figure is independent of the intitial ratio of different genotypes in the heteroplasmic parent cell, although the ratio will affect the rate at which homoplasmic cells of a particular genotype will be produced. In a diploid yeast cell, there are approximately 100 mitochondrial DNA molecules, so that complete sorting-out should require 1,000 generations. In the case of *Paramecium,* the probability of obtaining a homoplasmic cell after twenty generations in double animals having several thousand mitochondria is extremely small.

Random segregation is clearly not sufficient to explain the high rates of vegetative segregation seen in these organisms. A number of hypotheses can be invoked to explain this rapid production of homoplasmic cells, but all fall into two basic categories. In the first category, it is supposed that the number of genophores actually segregating is smaller than the total population in the cell. Specifically: (1) In the case of yeast it has been suggested that only a small number, not half, of the mitochondrial DNA molecules in a cell enter the bud (Dujon et al., 1974). This would speed up the segregation process, but would additionally require some mechanism to ensure that the molecules which entered the bud replicated faster than those which remained behind in the mother cell. It is possible that the nucleus occludes the narrow neck between mother and bud during mitosis, and so prevents communication between these cells. Mitochondria in the smaller bud might then replicate faster than those in the mother. (2) Another device for reduction of the number of segregating units would be the enzymatic destruction of most of the genophores in the zygote. This would be particularly effective if genophores from one parent or the other were destroyed preferentially.

Elissa Sena (Sena et al., 1973, and personal communication) has studied directly the mitochondrial DNA of isolated zygotes and their first buds in yeast. Her data indicate that, for the particular cross studied, there is no loss of mitochondrial DNA from either parent in the zygote prior to the production of the first bud. However, the zygote contains more than the "diploid" level of mitochondrial DNA, while the first bud contains the "diploid" amount. These data argue against destruction of genophores, selective or otherwise, but are compatible with the hypothesis that only a portion of those genophores replicate or enter the first bud. No genetic studies have been done on this particular cross, and it may be dangerous to generalize from these results in view of the great variability, between crosses, of such factors as extent of premating DNA synthesis, budding pattern, and output ratios. The same caveat holds for the experiments of Williamson and Fennell (1974), which showed that all mitochondrial DNA molecules are conserved and replicated in a diploid stock of yeast.

An argument against the first of these hypotheses can be developed from the results of studies on the progeny of buds produced by zygotes and isolated by micromanipulation (Wilkie and Thomas, 1973; Callen, 1974c). The colonies produced by such buds often include a very small minority of cells of one genotype, e.g., 1% of the total. This is most easily interpreted as meaning that about one hundred genophores entered that bud, one of which was of the minority genotype (Callen, 1974c). Suppose, at the other extreme, that two genophores entered the bud, one of which was of the "minority" genotype. But the "minority" type would have to be *excluded* from the first six buds produced by this cell, since they will produce 63/64 of the final colony. Since the "minority" genotype represents about half the genophores in the mother cell, the probability of this sequence of events is approximately 1/64.

The second category of hypotheses supposes that segregation is actually *non*-random, with a tendency for molecules of like genotype to segregate together. Specifically: (1) the distribution of organelles in the zygote might not be random. In the case of yeast, cell lineage studies frequently show that the genotypes of the progeny produced by a zygote bud are correlated with the position of the bud on the zygote (Waxman et al., 1973; Callen, 1974c). Strausberg and Perlman (1974) have examined zygotes in which the mitochondria from one parent were stained, and have found that mitochondria from the two parents are not completely mixed by the time the first bud is produced. They further observed that first buds formed in the neck of the dumbbell-shaped zygote, where mixing should occur most rapidly, had a higher frequency of recombinant genotypes than did buds formed at the end of the zygote, where most of the mitochondria are derived from one parent or the other. There is thus substantial evidence in favor of a non-random distribution of mitochondria in yeast zygotes. However, it is difficult to see how incomplete mixing could be a factor in the case of double-animal paramecia, where the cells contributing the mitochondria have fused side-by-side, but cell division separates anterior and posterior halves.

(2) In a number of organisms there are many mitochondrial DNA molecules in each mitochondrion. In yeast, for example, some stocks have their mitochondrial DNA molecules packaged four or more to the organelle, depending upon the physiological state of the cells (Grimes et al., 1974; Perlman, personal communication). This raises the possibility that when a mitochondrial DNA molecule replicates, the daughter molecules, having identical genotypes, might tend to remain in the same mitochondrion and hence enter the same cell

during cell division. In algae that have but one chloroplast, or in yeast stocks having a single giant mitochondrion (Hoffman and Avers, 1973), one could postulate that daughter molecules tend to remain in the same region of the organelle, possibly because of their attachment to a membrane. When the organelle divided during cell division, they would tend to remain in the same fragment and thus enter the same cell (Callen, 1974c).

(3) Another device for increasing the genetic homogeneity of genophores in a mitochondrion is gene conversion. In this process of non-reciprocal recombination, two DNA molecules of differing genotype (e.g., E^R and E^S) pair and form heteroduplex (hybrid) molecules containing mismatched base pairs. Repair of these mismatched pairs can result in both molecules becoming E^R or both becoming E^S. Russell Skavaril and I are doing computer simulations of this stochastic process; the results to date indicate that an organelle with four genomes will become homoplasmic after an average of six rounds of random pairing and gene conversion. Since there is evidence that multiple rounds do occur, gene conversion might play a significant role in producing homoplasmic cells. A mechanism of this sort is especially attractive to explain the segregation of mutant and wild-type organelles in the mycelia of *Aspergillus*, where mixing of mitochondria is not prevented by cell walls.

(4) Finally, it is conceivable that there actually exists a mechanism that ensures that genophore segregation is non-random. Segregation of daughter molecules after replication might be controlled by the segregation of special sites on the organelle membrane to which they are attached, for example. Mechanisms of this sort have been proposed to account for the segregation of chloroplast genes in *Chlamydomonas* (Sager, 1972) and of mitochondria in *Saccharomyces* (Callen, 1974b). A consequence of both models is that the rate of segregation should be different for different loci on the organelle genophore. The data of Callen (1974b) suggest that this variation does in fact exist. But these data can be understood on the basis of other hypotheses as well. For example, in line with hypothesis (3) above, if different loci have different conversion frequencies, they will appear to segregate at different rates.

A number of these hypotheses are directly testable, and some are being tested in yeast in various laboratories at the present time. It is of course possible, and in fact seems likely *a priori*, that several factors are involved and operate simultaneously to move a population of cells toward the homoplasmic condition.

QUANTITATIVE ASPECTS OF MITOCHONDRIAL GENETICS: ACCOUNTING FOR PHENOMENA SEEN IN OUTPUT RATIOS

In the preceding sections I presented evidence for the occurrence of several phenomena at the organelle and molecular level that must be taken into account in any attempt to understand mitochondrial transmission genetics. These phenomena include intracellular selection, recombination, input effects, and non-random segregation. The possibility of selective destruction or replication of organelles in the zygote must also be considered, although there is no direct evidence for these phenomena at present. We can now turn to an examination of the quantitative aspects of mitochondrial genetics. Here the results of crosses will be described in terms of the output ratios of cell genotypes produced in the cross. Some new phenomena appear: uniparental inheritance, seen most strikingly in *Neurospora,* and zygote heterogeneity, bias, polarity, and suppressiveness in *Saccharomyces.* I will suggest how these phenomena might be explained in terms of the organelle and molecular phenomena already discussed.

The Ciliates: Uniparental Inheritance, Input Effect, and Intracellular Selection

Cytoplasmic exchange of organelles as large as mitochondria has never been observed in *Tetrahymena,* and may never occur in nature. The inheritance of organelle genes is thus strictly uniparental, with all of the progeny of an exconjugant being genetically identical to the parent cell. In *Paramecium* cytoplasmic exchange occurs in the laboratory only rarely, and is likely to be equally rare in nature. When it does occur, the proportions of mitochondrial genotypes (input ratio) in the exconjugants will be heavily biased toward the genotype of the parent cell. In the presence of antibiotic, both exconjugants produce exclusively resistant progeny: intracellular selection seems to play an overriding role. In the absence of antibiotic both exconjugants can produce a mixture of resistant and sensitive progeny. The proportions of each appear to be explained adequately by the interaction of two factors: the biased input, and selection, both intracellular and intercellular.

Neurospora: Uniparental Inheritance

Heteroplasmons can be formed in *Neurospora* in two ways: by hyphal fusion, or by fertilization of protoperithecia by conidia. The results of hyphal fusion are variable and difficult to generalize. In poky + wild-type heteroplasmons, both mutant and wild-type mycelia are produced in some instances (e.g., *stp*, McDougall and Pittenger, 1966; *mi-3*, Gowdridge, 1956). In other cases all mycelia are wild-type (*mi-1*, Gowdridge, 1956) or mutant (*mi-1* and *mi-4*, Pittenger, 1956). The differences in results obtained with different mutants or in different laboratories with the same mutant (*mi-1*) make it difficult and unprofitable to discuss mechanisms at the present time.

During sexual reproduction in *Neurospora*, small spores or conidia from one ("paternal") parent fuse with a specialized segment of hypha, the protoperithecium, of the other ("maternal") parent. Both "gametes" contain mitochondria, so that the "zygote," which is a segment of hypha, presumably contains mitochondria as well as nuclei from both parents. (No electron micrographs have been made of the fertilization process, so it remains possible but unlikely that only nuclei are donated by the conidium to the zygote.) The heteroplasmic state is only transient, however, for nuclear fusion and meiosis quickly take place and the "zygote" differentiates into a perithecium with the haploid nuclear products of meiosis packaged into ascospores. When these ascospores are plated, it is found that the vast majority produce mycelia having the mitochondrial genotype of the protoperithecial parent (see Sager, 1972, for references and St. Lawrence and Chalmers, 1973, for exceptions to maternal inheritance). This is a striking case of uniparental inheritance. The mechanism is unknown: possibilities include an input highly biased in favor of the protoperithecium (which does include a larger cytoplasmic volume), selective destruction of genophores donated by the conidium, or selective replication of genophores donated by the protoperithecium. No relevant experimental observations have been made.

Yeast: The Phage-Analogy Model

The more extensive and detailed studies of mitochondrial genetics in yeast have produced a wealth of information and unearthed new phenomena, so that the picture is now quite complicated. I will begin by describing a cross in which some of these complications are absent, and in which the output ratios can be accounted for by a simple model of events that take place in an "average" zygote. I will then deal with the complications introduced when one looks at the genetic behavior of individual zygotes. Finally, the poorly understood phenomena of bias and polarity ("mitochondrial sex") will be considered.

The results of a simple cross are summarized in table 2. This is a "total progeny" analysis, in which we have determined the percentage

TABLE 2

OUTPUT RATIOS CHARACTERISTIC OF AN UNBIASED HOMOPOLAR CROSS INVOLVING MITOCHONDRIAL GENES IN *SACCHAROMYCES*

% Diploid Cells					
C^SE^S	C^RE^R	C^SE^R	C^RE^S	C^S	E^S
48.1	44.8	4.2	2.8	52.3	49.0

Haploid parent cells of stocks D243-4 (mitochondrial genotype $\omega^+C^SE^S$) and 2-3/6 ($\omega^+C^RE^R$) were mated and a total progeny analysis was performed as described in table 1 (note) and the text.

of homoplasmic diploid cells of each genotype present in the pooled progeny of a large number (usually a few thousand) zygotes, without regard to their origin from specific zygotes. Here we are asking questions about the average behavior of zygotes and can see phenomena that are obscured by the variability in behavior of individual zygotes. The output ratios can be used to draw inferences about the mitochondrial genotypes produced in the "average" zygote or its progeny and about their transmission and replication.

The following features of the total progeny output ratios should be noted: (1) the parental genotypes are recovered in approximately equal frequencies; (2) reciprocal recombinants (e.g., C^SE^R and C^RE^S) are also recovered in approximate equality; (3) for each of the markers in the cross, the two alleles, one derived from each parent, are transmitted about equally well to the progeny. These results are compatible with a simple model (fig. 4; Wolf et al., 1973; Dujon et al., 1974; Linnane et al., 1974) that is based on an analogy with the phage-infected cell, in which phage genomes of two different genotypes are replicating side-by-side and simultaneously pairing and recombining (Visconti and Delbrück, 1958). It can be assumed that the zygote in a yeast cross contains a pool of at least one hundred mitochondrial DNA molecules, initially of two parental genotypes. These molecules are permitted to undergo repeated rounds of random mating (pairing and recombination), presumably by repeated fusions and fissions of the mitochondria in which they are packaged. At the same time they are replicating and being segregated between daughter cells (mother and bud). If all mitochondrial genophores replicate at approximately the same rate, regardless of genotype, and if there is no selection for cells of a particular genotype, then one

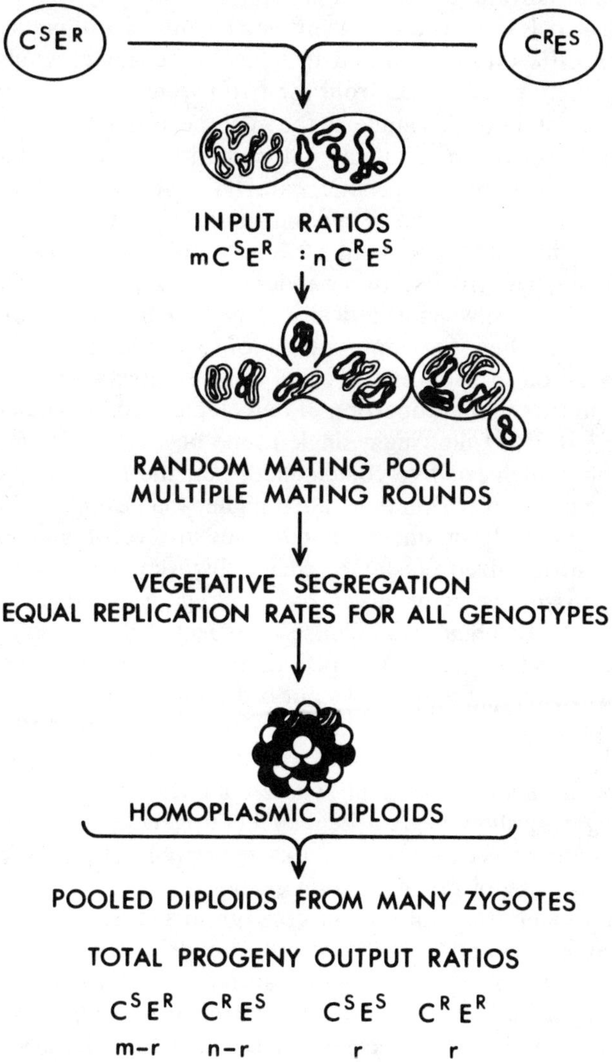

Fig. 4. A simple model of events taking place in a mitochondrial genetics cross in yeast, based on analogy with the phage-infected cell. The diagram shows the effect of input ratios on output ratios, but does not illustrate the consequences of multiple rounds or random mating. For further explanation, see the text.

can predict that the ratio of parental genotypes in the progeny will equal the input ratio of parental genophores in the zygotes. The input ratio will also determine the transmission frequencies of pairs of alleles. In the specific cross illustrated in table 2, the model would suggest approximately equal inputs from the two parents.

A basic feature of this model is the hypothesis that the mitochondrial genomes in a cross constitute a random mating pool. This is supported by two features of the genetic data (Wolf et al., 1973; Dujon et al., 1974). First, with the exception of the unusual V^RT^R mutants, there is a maximum of about 20% recombination between several different linkage groups. In a random mating pool a portion of all matings will be between genetically identical genophores; this portion will be 1/2 when the input ratios of parental molecules is 1:1. Recombination in such matings will not be detectable, and this will reduce the maximum proportion of detectable recombinants from 50% to 50%/2 = 25% following a single round of mating. Second, genetic data show a high positive coincidence, i.e., the frequency of double and triple cross-over events is much higher than expected by chance. This is probably not due to a tendency for recombination events to occur in localized clusters, because high coincidence is seen for double crossovers in widely separated regions such as $R_{III} - O_I$ and $O_I - P$. But high coincidence is expected if only some matings produce detectable recombinants, as in a random mating pool. An alternative explanation that has not been adequately ruled out is that only a part of the yeast population participates in recombination. Studies described in the following section ("Zygote heterogeneity") on output ratios of individual zygote colonies, or of individual buds from zygotes, show a great deal of heterogeneity in the proportion of recombinant genotypes, and this may well reflect heterogeneity in the frequency of recombination events.

Even though the various linkage groups described above (see "Recombination and mapping of mitochondrial genes") are formally unlinked, the genetic data suggest that they are on a single genophore. This is indicated by their coordinate transmission frequencies (feature [3] above). It is also indicated by the fact that the assortment of unlinked markers is not random in a cross (Lancashire and Griffiths, 1975). Neither of these two features is shown by the V^RT^R mutants, which also show 50 percent recombination with all loci (Lancashire and Griffiths, 1975). These mutants thus behave as if they were on different, independently segregating DNA molecules.

Yeast: Zygote Heterogeneity

Let us now turn to table 3 and examine data on output ratios of individual zygotes. The most striking aspect of zygote colony analyses in yeast is their variabiity. All zygotes were produced in the same mass mating, and yet each behaves in a unique fashion. Many zygotes produce progeny of only one or a few genotypes, even though all initially contained mitochondria of at least two different genotypes. Very few zygotes produce output ratios resembling the total progeny: the "average zygote" is thus virtually nonexistent except as a statistical construct. These phenomena have been seen in every study of zygote colonies (Coen et al., 1970; Rank and Bech-Hansen, 1972; Lukins *et al.*, 1973; Wilkie and Thomas, 1973; Waxman, et al., 1973; Birky, 1974, and unpublished data).

The data are of course subject to a variety of types of random error, including, for example, experimental error and cell death, but these can scarcely explain the extreme variations seen here. Random genetic drift and non-reciprocal recombination operating in a population of 100 genophores would rarely produce the more extreme output ratios seen in some zygotes. These stochastic phenomena *could* produce such great variability if the actual number of segregating genetic units were very small. This possibility, suggested by Dujon et al. (1974), has already been discussed above (see "Vegetative segregation"). Another source of variability in yeast zygotes is the combination

TABLE 3

MITOCHONDRIAL GENETICS IN *SACCHAROMYCES:* ZYGOTE HETEROGENEITY ILLUSTRATED BY OUTPUT RATIOS OF SIX ZYGOTES FROM A HOMOPOLAR CROSS

Zygote Number	% DIPLOID CELLS HAVING A PARTICULAR GENOTYPE AT THE C E O LOCI							
	RSR	SSR	RRR	RSS	SRR	SSS	RRS	SRS
12	100	...	...	...	...	...	...	...
34	67	...	...	26	...	7	...	...
5	42	...	...	...	...	1	...	57
9	21	...	...	53	...	1	24	1
21	3	...	...	...	...	...	...	97
11	0.1	...	0.2	2	...	0.1	17	80
Total progeny	38	3	2	4	2	4	4	47

The three-factor cross D6 $\omega^+C^SE^RO^S \times$ 6-2/7 $\omega^+C^RE^SO^R$ was performed as in Table 1, except that in addition to the total progeny analysis, the diploid progeny of 50 individual zygotes were collected and the genotypes of a random sample of cells from each zygote were determined. The results for six representative zygotes are shown here. Glycerol was used as the carbon source.

of incomplete mixing of mitochondria, previously mentioned, and variable budding patterns. Strausberg and Perlman (1974) have shown that central buds contain a higher proportion of recombinant genophores. Since the first bud from a zygote will produce about one-half of the cells in the zygote colony, a zygote whose first bud is central will produce colonies with a higher percent of recombinant cells than will a zygote whose first bud arises from one end or the other. Strausberg and Perlman also note that crosses in which the majority of zygotes produce first end buds show a lower recombination frequency, and a shorter map distance, between two loci than do crosses in which the majority of zygotes produce first central buds. This factor needs to be taken into account in future mapping studies.

The data for zygote colony analyses can also be examined in another fashion that brings to light a new phenomenon. Frequency distributions (figs. 5-7) of genetic parameters, e.g., the transmission of individual alleles, show that the zygotes of a particular cross often fall into two or more distinct classes with respect to the transmission of mitochondrial genes from the two parents. For example, in the cross D6 × 6-2/7 three classes can be recognized: some zygotes preferentially replicate and transmit genes from the D6 parent, some transmit preferentially genes from the 6-2/7 parent, and some transmit genes about equally well from both. The phenomenon is reminiscent of the classes of zygotes found in Tilney-Bassett's studies of plastid gene inheritance in the geranium (this volume). Three sorts of hypotheses can be framed to explain the existence of these multiple classes. (1) They could represent classes of zygotes in which the mitochondrial DNA of one parent or the other was preferentially destroyed by enzymes. Classes showing efficient transmission of markers from both parents could represent zygotes in which destruction was incomplete, or in which markers were rescued by recombination from molecules that were to be destroyed and incorporated into those that were to be transmitted. (2) They could represent classes of zygotes in which the mitochondrial genomes of one parent or the other were preferentially replicated, e.g., because they had access to early buds or to specific sites of replication present in limited numbers on mitochondrial membranes, as suggested by Wilkie (1972). (3) They could represent classes of matings with different input ratios of mitochondrial genes. It has been shown that cells of at least some yeast stocks undergo a round of mitochondrial DNA replication without cell division immediately prior to zygote formation (Sena et al., 1973).

This replication is not essential for mating, because mating takes

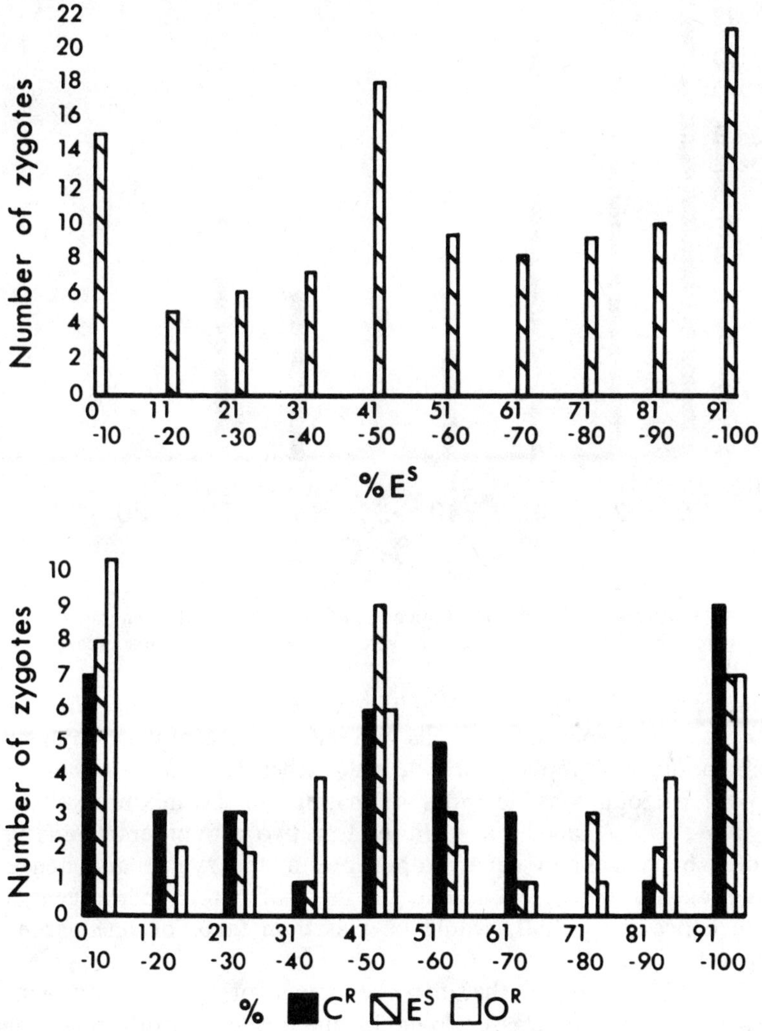

Fig. 5. Multiple classes of zygotes with respect to transmission of mitochondrial genes in yeast. A frequency distribution was made of percent transmission (in intervals of 10%) of the E^S allele, from the 6-2/7 parent, versus number of zygotes; data were taken from the cross described in table 2. The lower graph shows data for the E^S allele from a smaller number of zygotes, compared with data for the C^R and O^R alleles in the same zygotes.

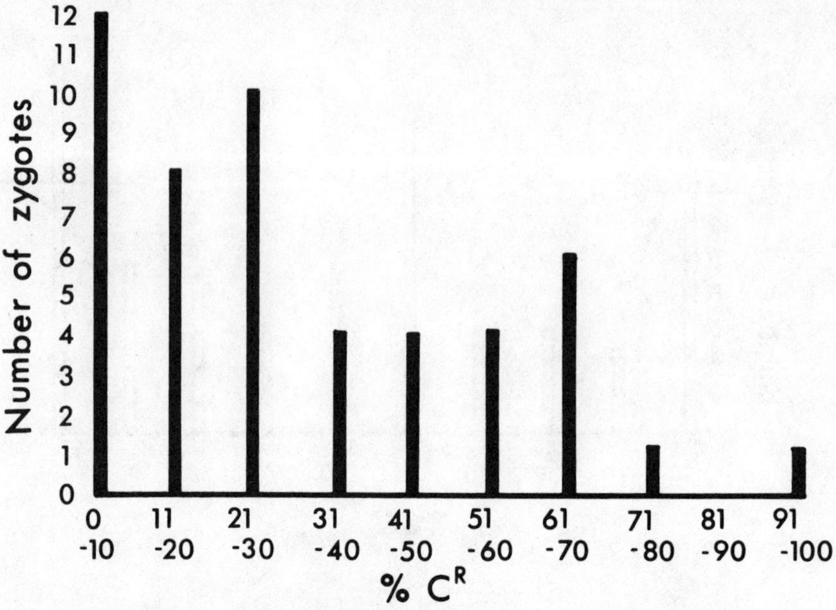

Fig. 6. Frequency distribution illustrating zygote heterogeneity; the same cross as shown in figure 5, but with glucose rather than glycerol used as the carbon source.

place when it has been inhibited by hydroxyurea (Perlman and Demko, 1974). If only some of the cells in a mating mixture were to undergo premating DNA replication (+), while others did not (−), then there would be four possible types of matings in the mixture: + × +, − × −, + × −, and − × +. In the first two both parental genotypes could be present in equal proportions in the zygote and hence in its progeny. In the second two the input of mitochondrial genomes, and hence the output, would be biased in favor of one parent or the other.

It should be noted that the phenomena of zygote heterogeneity and multiple zygote classes do not invalidate the use of the phage-analogy model of yeast mitochondrial genetics introduced above. Rather, they indicate that the model in its present state (Dujon et al., 1974) is not sufficiently complex to predict the results of zygote clone analysis and will need modification. Even in its present state the model provides a useful framework for discussion and theorizing, and it remains adequate to explain a number of phenomena seen in total progeny analyses, to which we now return.

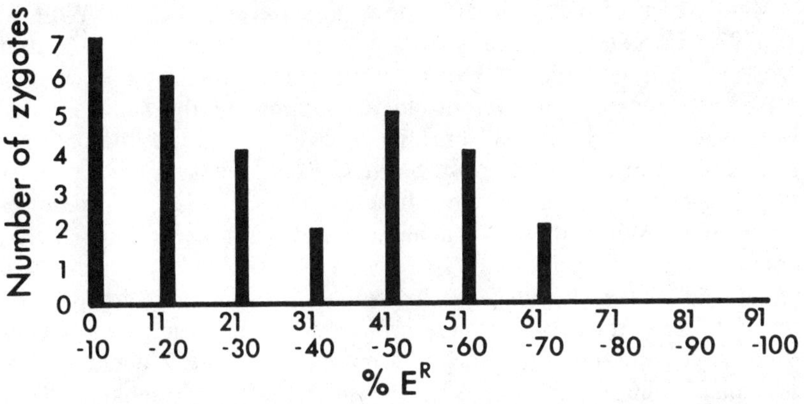

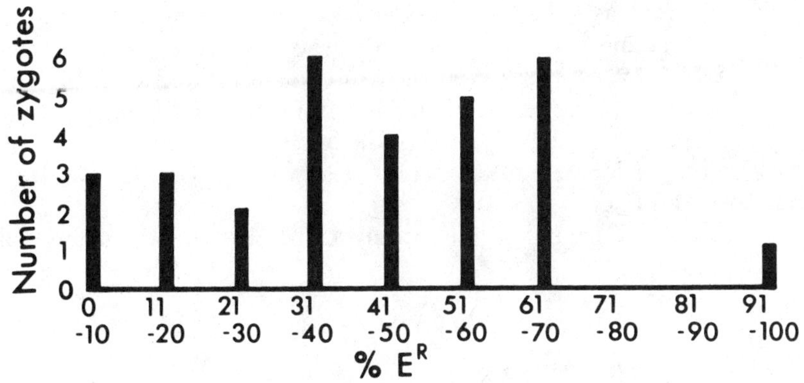

Fig. 7. Frequency distributions illustrating zygote heterogeneity. The matings are IL16 $\omega^- E^S \times$ 5L3-12 $\omega^- E^R$ (upper) and IL16 $\omega^- E^S \times$ 5L1-1 $\omega^- E^R$ (lower). The two different E^R stocks have identical mitochondrial genomes but differ in nuclear genotypes. The number of zygotes studied is small, but the differences in pattern are verified by significant differences in total progeny output ratios and by differences seen in a previous experiment following a different protocol.

Yeast: Bias

Many pairs of yeast stocks show *bias* (terminology of Wolf et al., 1973) in their output ratios when mated; an example of a biased cross is seen in table 4. The transmission of alleles at all loci is biased in favor of one parent, to approximately the same extent. Bias is also seen in the preferential recovery of one of the parental genotypes at the expense of the other. Recombinaton, however, is still reciprocal. The degree and direction of bias in a cross depends upon the physiological state of the cells at the time of mating (Birky, unpublished); this effect can be seen in tables 2 and 4, which show results of crosses with glucose-repressed (table 4) and derepressed (table 2) cells. It is also strongly influenced by nuclear genes (Avner et al., 1973), incuding specifically the mating type locus (Callen, 1974a), the *rad12* locus, and a gene in the *rho5* mutant (Fraenkel, 1974).

In the model of Dujon et al. (1974) bias is interpreted in terms of biased inputs of mitochondrial genomes from the two parents. The two parental stocks might, for example, have different contents of mitochondrial DNA prior to mating. The experiments of Perlman and Demko (1974), Chou (1974), Young (unpublished data), and Goldthwaite et al. (1974) described previously (see "Input effects") fit this interpretation. Perlman and Demko (1974) have obtained evidence that bias in some crosses is due to differences in the amount of pre-mating mitochondrial DNA synthesis that takes place immediately before zygote formation in a mixture of *a* and α cells. The model predicts that the recombination frequency between two loci, e.g., C and O, should be greatest in unbiased crosses and should decline proportionately to the extent of bias in a cross; this has been verified (Dujon et al., 1974).

However, bias cannot be due entirely to differences in mitochondrial

TABLE 4

Output Ratios Characteristic of a Biased Homopolar Cross Involving Mitochondrial Genes in *Saccharomyces*

% Diploid Cells					
C^SE^R	C^RE^S	C^SE^S	C^RE^R	C^S	E^R
63.0	26.1	4.4	6.5	67.4	69.5

Total progeny analysis performed as described in table 1 (note). This is the same cross shown in table 3 (D6 × 6-2/7), but with glucose used as the carbon source. For simplicity, the data for only two of the three loci involved are shown here.

gene dosage in the parents prior to zygote formation. The output ratios of the haploid × diploid matings are biased by much more than the factor of two predicted on that basis. And haploid × haploid crosses often give output ratios as extreme as 0.85:0.15, which would require a sixfold difference in mitochondrial DNA contents of the parents. When the mutant N123 *rho5* is mated to a wild-type stock, the transmission of mitochondrial genes from the *rho5* parents is 80%. This extreme bias is largely a reflection of the fact that 44% of the zygotes are uniparental, transmitting *only rho5* mitochondrial genes (Birky, unpublished). Some instances of bias may be due in part, or entirely, to selection, intracellular or intercellular, for mitochondria of certain genotypes (see "Yeast: Other forms of asymmetry in output ratios" below). The former is difficult to study, and controls for the latter have not been done in most instances. But the *rho5* case, the haploid × diploid matings, and Callen's study of effects of the mating type locus indicate that selection cannot be a general mechanism for bias.

Yeast: Polarity

In 1970 Slonimski's group (Coen et al., 1970) reported the discovery of the polarity phenomenon. Polarity was subsequently shown to depend upon a mitochondrial gene locus designated ω (omega) and existing in two allelic forms: ω^+ and ω^-. A hypothetical mechanism for polarity was proposed by Bolotin et al. (1971) based upon an analogy with bacterial conjugation; the phenomenon was then referred to as mitochondrial sex, and the relevant types of crosses were designated homosexual ($\omega^+ \times \omega^+$ or $\omega^- \times \omega^-$) and heterosexual ($\omega^+ \times \omega^-$). The original mechanism has now been disproved (Dujon et al., 1974) and the reference to polarity as a form of mitochondrial sexuality appears to me to be unjustified; accordingly, I and my colleagues refer to it only as polarity and call the two types of crosses homopolar and heteropolar.

The crosses discussed hitherto in this review are homopolar ($\omega^+ \times \omega^+$ or $\omega^- \times \omega^-$). The phenomenon of polarity is detectable only in heteropolar crosses ($\omega^+ \times \omega^-$), an example of which is shown in table 5. There are three aspects of polarity: (1) preferential recovery of the ω^+ parental genotype (not seen in every heteropolar cross); (2) non-reciprocal recombinants, with preferential recovery of those recombinants carrying the + allele for the locus closest to ω on the map; and (3) a locus-specific effect on transmission, favoring alleles

TABLE 5

OUTPUT RATIOS CHARACTERISTIC OF AN UNBIASED HETEROPOLAR CROSS INVOLVING MITOCHONDRIAL GENES IN SACCHAROMYCES

% DIPLOID CELLS							
$C^+E^+O^+$	$C^-E^-O^-$	$C^+E^-O^-$	$C^-E^+O^+$	$C^+E^+O^-$	$C^-E^-O^+$	$C^+E^-O^+$	$C^-E^+O^-$
46.0	2.8	18.0	0.6	26.2	0.4	5.2	0
C^+E^+	C^-E^-	C^+E^-	C^-E^+	C^+	E^+	O^+	
72.8	3.2	23.2	0.6	96.0	73.4	52.8	

Total progeny analysis was performed as described in table 1 (note), for the cross 6-2/5 $\omega^+C^RE^SO^R \times$ 2-19 $\omega^-C^SE^RO^S$. On the top line the data are shown for the three-factor cross. On the bottom line are given the output ratios for the C and E loci only, as well as the individual transmission values of the C, E, and O alleles contributed by the ω^+ parent. Data of Philip Perlman.

donated by the ω^+ parent, and decreasing in extent with increasing distance of the locus from ω. It is important to distinguish this phenomenon from bias, which can result in unequal parentals (1) but not in unequal recombinants in a two-factor cross (2), and which has the same effect on the transmission of all loci regardless to their linkage relationship to ω (cf. 3).

That ω is a mitochondrial gene locus is shown by its vegetative segregation, the fact that it segregates 4:0 during meiosis, and its apparent linkage to the R region of the mitochondrial map (Bolotin et al., 1971; Linnane et al., 1974). According to Slonimski's group, it is located immediately to the left of R_I, but no data have been published to substantiate this; the data of Linnane et al. (1974) suggest a location to the right of R_I, between R_I and R_{III}. The important point is that, in either event, it is quite close to R_I.

Polarity can then be viewed as the loss, in the zygote or its early progeny, of genes that are linked to ω^- and the preferential recovery of genes linked to ω^+. The preferential recovery of ω^+ and its linked loci could be accounted for by any of three basic mechanisms: (1) preferential replication of ω^+; (2) preferential destruction of ω^-; or (3) preferential conversion of ω^- to ω^+. The first hypothesis cannot readily explain the nearly complete loss of the R_I allele from the ω^- parent; there would have to be *no* replication of ω^- for at least four generations. The second hypothesis is also *a priori* unlikely, according to the following argument. The mechanism for degradation must operate only in heteroplasmic zygotes, and never in zygotes from the cross $\omega^- \times \omega^-$. This can only be insured if the enzymes required for degradation are products of the ω allele or of a closely linked locus. But polarity is still seen in $\rho^-\omega^+ \times \rho^+\omega^-$ crosses (Coen et al., 1970), in which protein synthesis in the $\rho^-\ \omega^+$ mitochondria is limited or nonexistent because of the ρ^- lesion.

Dujon et al. (1974) adopted the third alternative as their working hypothesis. They assume that, during matings of ω^+ and ω^- mitochondrial DNA molecules in the zygote, gene conversion begins preferentially at the ω locus and proceeds for a variable distance through the R segment. Conversion is always in the same direction, with genes on the ω^- molecule being converted to their alleles on the

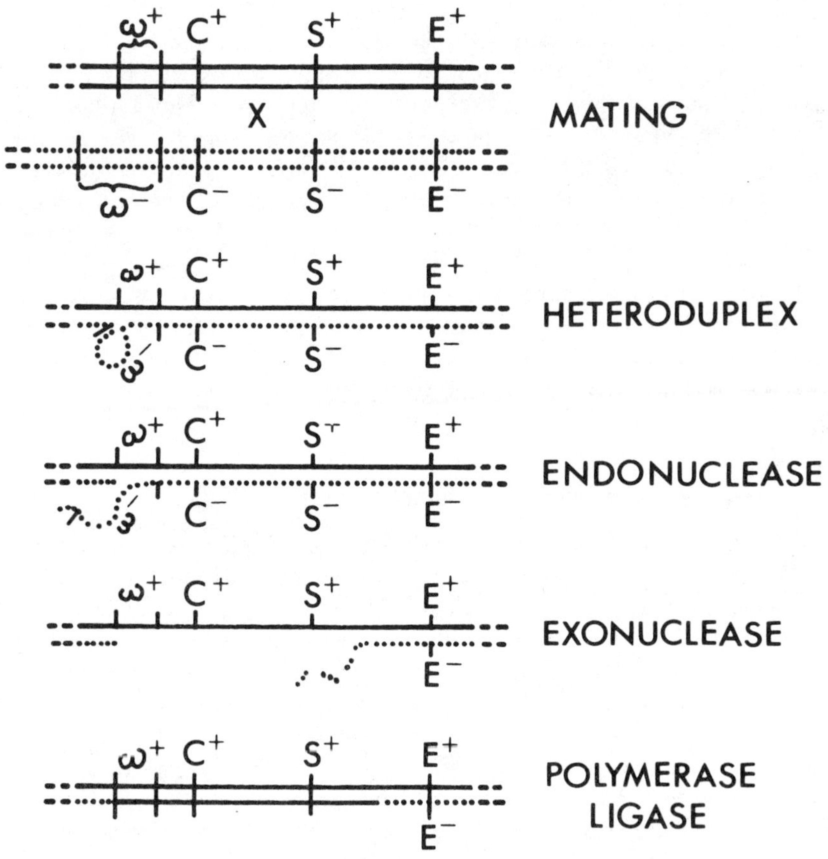

Fig. 8. A hypothetical mechanism for polarity in yeast mitochondrial genetics, according to Perlman and Birky (1974). See text for further explanation.

ω^+ molecule. The insertion of this process into their mathematical model led to quantitative predictions, about relationships between variables in outputs from heteropolar crosses, that were consistent with their experimental data.

Philip Perlman and I (Perlman and Birky, 1974) developed a slightly modified version of this hypothesis that included a plausible molecular mechanism for the preferential conversion of ω^- to ω^+ (fig. 8). We assume that the ω^+ and ω^- loci differ primarily in length, with ω^- being longer (it could, for instance, be a duplication of ω^+). During the course of mating and recombination, conversion is assumed to begin with the formation of a heteroduplex DNA region containing paired strands from genetically different molecules, as appears to be the case in other systems. If a heteroduplex is formed that includes the ω locus, the ω^- strand will form a single-stranded loop of DNA that has no homologous region with which to pair. This loop could be recognized by an appropriate nuclease and excised, along with varying portions of the rest of the ω^- strand. Repair of the gap would then have to use the ω^+ strand as a template; the result would be the net loss of one ω^- strand and the gain of one ω^+ strand. Repeated rounds of mating would cause the progressive loss of ω^- and linked loci. This model is particularly attractive because the preferential conversion of longer to shorter genes has been observed in phage (references in Perlman and Birky, 1974), and can also be used to explain the phenomenon of suppressiveness seen in $\rho^+ \times \rho^-$ crosses (discussed by Perlman in this volume).

Yeast: Other Forms of Asymmetry in Output Ratios

Callen (1974b) has done a careful study of the behavior of four unlinked loci (O_1^R, O_2^R, S_1^R, and S_2^R) in a series of two-factor crosses in which the nuclear background differed only at the mating type locus. All four loci showed a bias in output ratios, which could be corrected for. When this correction was made, it was found that the O_1^R gene was consistently recovered more frequently than its O_1^S allele. This was interpreted as intracellular selection for O_1^R genomes. However, the possibility of *inter*cellular selection was not completely ruled out. Moreover, I am unable to find any reason in his data for ruling out the possibility of a modest degree of preferential conversion of O_1^S to O_1^R; such conversion would affect the frequency of these two alleles in the parental as well as in the recombinant genotypes.

Preferential gene conversion was, in fact, invoked by Callen to explain the preferential recovery of O_1^R and of S_2^S over their respective alleles in *recombinant* genotypes. This was expressed as non-reciprocal recombinants in the output; e.g., the cross $O_1^R S_1^S \times O_1^S S_1^R$ produced 12.3% $O^R S^R$ and 4.5% $O^S S^S$ recombinants.

Such cases of non-reciprocal recombination in homopolar crosses have also been seen by Linnane et al., (1974; Howell et al., 1973) and in my laboratory as well. Linnane and his collaborators refer to this phenomenon as "low polarity"; their data indicate that the degree and direction of this polarity is under the control of cytoplasmic (mitochondrial) genetic factors (Linnane et al., 1974). Low polar crosses show differential transmission levels of different loci, but, unlike ω-polarity, these transmission differences are not clearly related to the map position of the loci. Also, the effects on parental genotypes do not show a consistent pattern. More data are needed before mechanisms can be fruitfully discussed. The data of Suda and Uchida (1972; 1974) show that, for a set of ten crosses involving an O^R mutation, an E^R mutation, and in some cases a C^R mutation, reciprocal recombinant genotypes as well as parentals were often unequal. Moreover, when the transmission frequencies of individual markers were calculated, the deviations from 50% were often unequal. These deviations were always greatest for the O locus, less for E, and least for C. A further analysis of their results shows a consistent preferential recovery of the O^S allele; this suggests that the simplest explanation of their results would be selection (intra- or intercellular or both) favoring the O^S phenotype, with the E and C loci being affected to lesser extents due to their physical linkage to O, or due to weaker selection for C^S and E^S. The data also indicate an over-all bias in favor of mitochondrial genes from the α parent, such as was found by Callen (1974a).

CLOSING THOUGHTS

The Utility of the Phage-Analogy

In the hands of Slonimski and his collaborators, the simple phage-analogy model of mitochondrial genetics has been successful in accounting for some aspects of the yeast studies, especially those involving recombination. It is clearly *too* simple to give a complete explanation of the phenomena seen in the study of total progenies, and does not even begin to cope with the individualistic behavior of zygotes. Faced with this latter problem, some workers (e.g., Linnane et al., 1974; Wilkie and Thomas, 1973) have preferred to focus attention on zygote analyses in the apparent hope of finding a unitary explanation for polarity, bias, *and* zygote heterogeneity. So far, my colleagues and I have found the phage-analogy model more useful as a starting point for generating explicit, testable hypotheses about the molecular and cellular mechanisms of polarity and bias. This may be because the model ignores a number of complexities and thus isolates two specific phenomena (polarity and bias) and two mechanisms (gene conversion and input effects) for careful study. Separate (but not independent) efforts are being made in our laboratory to understand zygote heterogeneity in terms of these and other mechanisms. To divide is often to conquer when dealing with biological complexity.

How Much Progress Have we Made?

It should be clear from this review that two accomplishments mark our progress in understanding mitochondrial genetics. First, enough crosses have been made and enough output ratios obtained to enable us to identify the major features of transmission genetics that require explanation (at least in three microorganisms). Second, a number of phenomena have been identified at the cell and molecular levels that have enabled us to explain some of those features of output ratios and to frame useful hypotheses about most of the others. Complete and final explanations and understanding still elude us, and some surprises may await us, but I believe that a strong conceptual basis has been laid. This, coupled with the other advantages mentioned in the introduction, should permit rapid progress to be made.

General Theories and Laws of Organelle Genetics

Our ultimate goal is presumably to rationalize as much of the subject as possible by developing general laws (theories) of organelle transmission genetics, analogous to Mendel's laws, the chromosome theory of heredity, and so on. My review dealt only with mitochondria, but it will be clear to the reader of this volume that some likely candidates for such laws are already available. For instance, uniparental inheritance and rapid vegetative segregation are, to varying degrees, characteristics of plastid inheritance also. A real danger exists, however, that the very small number of experimental organisms being used may mislead us into promoting, as general laws, phenomena that are actually typical only of microorganisms, or only of fungi and ciliates. In this connection, polarity is not a good candidate for a law of mitochondrial genetics (cf. Bolotin et al., 1971). The nearly complete loss of the ω^- allele in heteropolar crosses puts it at a marked selective disadvantage whenever sexual reproduction occurs. Unless it enjoys some compensating advantage during asexual reproduction (none has yet been noticed), natural populations should be largely ω^+ and most matings would be homopolar. Polarity is best viewed as a tool for the study of mitochondrial genetics, and possibly as a model system for the investigation of one-way gene conversion.

What Needs to Be Done?

The answer is simple: further work on the well-developed systems described here should be accompanied by the study of mitochondrial transmission genetics in new experimental organisms from diverse taxa. Only when this has been done will the goal of general laws be in sight.

ACKNOWLEDGMENTS

I gratefully acknowledge the many helpful discussions and contributions of ideas provided by my graduate students and colleagues, in particular Dr. Philip S. Perlman. I am grateful to D. Callen, B. Dujon, C. Goldthwaite, P. Perlman, and R. Young for sharing their unpublished manuscripts and data with me. Suggestions and criticisms on the manuscript were made by Drs. T. J. Byers, P. S. Perlman, and R. L. Scholl. The experimental work described here is to a great extent the results of skillful technical assistance from Mary Townsend and, in the early studies of zygote heterogeneity, Jeffrey Knight. Supported by NIH research grant 5 R01 GM-19607.

LITERATURE CITED

Adoutte, A. 1974. Mitochondrial mutations in *Paramecium:* Phenotypical characterization and recombination, pp. 263–71. *In* A. M. Kroon and C. Saccone (eds.), The biogenesis of mitochondria. Academic Press, New York.

Adoutte, A., M. Balmefrézol, J. Beisson, and J. Andre. 1972. The effects of erythromycin and choloramphenicol on the ultrastructure of mitochondria in sensitive and resistant strains of *Paramecium.* J. Cell Biol. 54:8–19.

Adoutte, A., and J. Beisson. 1970. Cytoplasmic inheritance of erythromycin resistant mutations in *Paramecium aurelia.* Molec. Gen. Genet. 108:70–77.

Adoutte, A., and J. Beisson. 1972. Evolution of mixed populations of genetically different mitochondria in *Paramecium aurelia.* Nature 235:393–96.

Adoutte, A., A. Sainsard, M. Rossignol, and J. Beisson. 1973. Aspects génétiques de la biogenèse des mitochondries chez la *Paramécia.* Biochimie 55:793–99.

Alexander, N. J., N. W. Gillham, and J. E. Boynton. 1974. The mitochondrial genome of *Chlamydomonas.* Induction of minute colony mutations by acriflavin and their inheritance. Molec. Gen. Genet. 130:275–90.

Anderson, W. A. 1968. Structure and fate of the paternal mitochondrion during early embryogenesis of *Paracentrotus lividus.* J. Ultrast. Res. 24:311–21.

Avner, P. R., D. Coen, B. Dujon, and P. P. Slonimski. 1973. Mitochondrial genetics. IV. Allelism and mapping studies of oligomycin resistant mutants in *S. cerevisiae.* Molec. Gen. Genet. 125:9–52.

Beale, G. H. 1969. A note on the inheritance of erythromycin-resistance in *Paramecium aurelia.* Genet. Res. 14:341–42.

Beale, G. H., J. K. Knowles, and A. Tait. 1972. Mitochondrial genetics in *Paramecium.* Nature 235:396–97.

Bertrand, H., and T. H. Pittenger. 1972a. Isolation and classification of extranuclear mutants of *Neurospora crassa.* Genetics 71:521–33.

Bertrand, H., and T. H. Pittenger. 1972b. Complementation among cytoplasmic mutants of *Neurospora crassa.* Molec. Gen. Genet. 117:82–90.

Birky, C. W., Jr. 1973. On the origin of mitochondrial mutants: Evidence for intracellular selection of mitochondria in the origin of antibiotic-resistant cells in yeast. Genetics 74:421–32.

Birky, C. W., Jr. 1974. Uniparental inheritance of organelle genes in yeast and in general. Genetics 77:s55 (Abstr.)

Bolotin, M., D. Coen, J. Deutsch, B. Dujon, P. Netter, E. Petrochilo, and P. P. Slonimski. 1971. La recombinaison des mitochondries chez la levure. Bull. Inst. Pasteur 69:215–39.

Bråten, T. 1973. Autoradiographic evidence for the rapid disintegration of one chloroplast in the zygote of the green alga *Ulva mutabilis.* J. Cell Sci. 12:385–89.

Bunn, C. L., D. C. Wallace, and J. M. Eisenstadt. 1974. Cytoplasmic inheritance of chloramphenicol resistance in mouse tissue culture cells. Proc. Nat. Acad. Sci. USA. 71:1681–85.

Callen, D. F. 1974a. The effect of mating type on the polarity of mitochondrial gene transmission in *Saccharomyces cerevisiae.* Molec. Gen. Genet. 128:321–29.

Callen, D. F. 1974b. Recombination and segregation of mitochondrial genes in *Saccharomyces cerevisiae.* Molec. Gen. Genet. 134:49–63.

Callen, D. F. 1974c. Segregation of mitochondrially inherited antibiotic resistance genes in zygote cell lineages of *Saccharomyces cerevisiae.* Molec. Gen. Genet. 143:65–76.

Chou, Chyau-bin. 1973. The effect of mitochondrial DNA content on mitochondrial marker transmission in baker's yeast, *Saccharomyces cerevisiae.* M.S. thesis, Ohio State University.

Coen, D., J. Deutsch, P. Netter, E. Petrochilo, and P. P. Slonimski. 1970. Mitochondrial genetics. I. Methodology and phenomenology. Symp. Soc. Exp. Biol. 24:449–96.

Deutsch, J., B. Dujon, P. Netter, E. Petrochilo, P. P. Slonimski, M. Bolotin-Fukuhara, and D. Coen. 1974. Mitochondrial genetics. VI. The petite mutation in *Saccharomyces cerevisiae:* Interrelations between the loss of the ρ^+ factor and the loss of the drug resistance mitochondrial markers. Genetics 76:195–219.

Diacumakos, E. G., L. Garnjobst, and E. L. Tatum. 1965. A cytoplasmic character in *Neurospora crassa.* J. Cell Biol. 26:427–43.

Dujon, B., P. P. Slonimski, and L. Weill. 1974. Mitochondrial genetics. IX. A model for recombination and segregation of mitochondrial genomes in *Saccharomyces cerevisiae.* Genetics 78:415–37.

Ephrussi, B., H. de Margerie-Hottinguer, and H. Roman. 1955. Suppressiveness: A new factor in the genetic determinism of the sythesis of respiratory enzymes in yeast. Proc. Nat. Acad. Sci. USA. 41:1065–71.

Faye, G., K. Chantal, and H. Fukuhara. 1974. Physical and genetic organization of *petite* and *grande* yeast mitochondrial DNA. IV. *In vivo* transcription products of mitochondrial DNA and localization of 23 S ribosomal RNA in *petite* mutants of *Saccharomyces cerevisiae.* J. Mol. Biol. 88:185–203.

Fraenkel, A. M. 1974. The effects of nuclear mutations for recombination and repair functions and of caffeine on mitochondrial recombination. Ph.D. thesis, Ohio State University.

Gillham, N. W., J. E. Boynton, and R. W. Lee. 1974. Segregation and recombination of non-Mendelian genes in *Chlamydomonas.* Genetics 78:439–57.

Gillham, N. W., and R. P. Levine. 1962. Studies on the origin of streptomycin resistant mutants in *Chlamydomonas reinhardi.* Genetics 47:1463–74.

Gingold, E. B., G. W. Saunders, H. B. Lukins, and A. W. Linnane. 1969. Biogenesis of mitochondria. X. Reassortment of the cytoplasmic genetic determinants for respiratory competence and erythromycin resistance in *Saccharomyces cerevisiae.* Genetics 62:735–44.

Goldthwaite, C. D., D. R. Cryer, and J. Marmur. 1974. Effect of carbon source on the replication and transmission of yeast mitochondrial genomes. Molec. Gen. Genet. 133:87–104.

Gowdridge, B. M. 1956. Heterocaryons between strains of *Neurospora crassa* with different cytoplasms. Genetics 41:780–89.

Griffiths, D. E., R. L. Houghton, W. E. Lancashire, and P. M. Meadows. 1975. Isolation and properties of mitochondrial venturicidin-resistant mutants of *Saccharomyces cerevisiae.* Eur. J. Biochem. 51 (in press).

Grimes, G. W., H. R. Mahler, and P. S. Perlman. 1974. Nuclear gene dosage effects on mitochondrial mass and DNA. J. Cell Biol. 61:565–74.

Handwerker, A., R. J. Schweyen, K. Wolf, and F. Kaudewitz. 1973. Evidence for an extrakaryotic mutation affecting the maintenance of the rho factor in yeast. J. Bact. 113:1307–10.

Hoffman, H. P., and C. J. Avers. 1973. Mitochondrion of yeast: Ultrastructural evidence for one giant, branched organelle per cell. Science 181:749–51.

Howell, N., M. K. Trembath, A. W. Linnane, and H. B. Lukins. 1973. Biogenesis of mitochondria. 30. An analysis of polarity of mitochondrial gene recombination and transmission. Molec. Gen. Genet. 122:37–51.

Howell, N., P. L. Molloy, A. W. Linnane, and H. B. Lukins. 1974. Biogenesis of mitochondria. 34. The synergistic interaction of nuclear and mitochondrial mutations to produce resistance to high levels of mikamycin in *Saccharomyces cerevisiae.* Molec. Gen. Genet. 128:43–54.

Jacob, F., S. Brenner, and F. Cuzin. 1963. On the regulation of DNA replication in bacteria. Cold Spring Harbor Symp. Quant. Biol. 28:329-48.

Kirk, J. T. O., and R. A. E. Tilney-Bassett. 1967. The plastids. W. H. Freeman, London.

Kleese, R. A., R. C. Grotbeck, and J. R. Snyder. 1972. Recombination among three mitochondrial genes in yeast (*Saccharomyces cerevisiae*). J. Bact. 112:1023-25.

Kuriyama, Y., and D. J. L. Luck. 1974. Synthesis and processing of mitochondrial ribosomal RNA in wild-type and poky strain of *Neurospora crassa*, pp. 117-33. *In* A. M. Kroon and C. Saccone (eds.), The biogenesis of mitochondria. Academic Press, New York.

Lancashire, W. E., and D. E. Griffiths. 1975. Isolation, characterisation and genetic analysis of trialkyltin resistant mutants of *Saccharomyces cerevisiae*. Eur. J. Biochem. 51 (in press).

Linnane, A. W., N. Howell, and H. B. Lukins. 1974. Mitochondrial genetics, pp. 193-213. *In* A. M. Kroon and C. Saccone (eds.), The biogenesis of mitochondria. Academic Press, New York.

Linnane, A. W., G. W. Saunders, E. B. Gingold, and H. B. Lukins. 1968. The biogenesis of mitochondria. V. Cytoplasmic inheritance of erythromycin resistance in *Saccharomyces cerevisiae*. Proc. Nat. Acad. Sci. USA. 59:903-10.

Lukins, H. B., W. R. Tate, G. W. Saunders, and A. W. Linnane. 1973. The biogenesis of mitochondria. 26. Mitochondrial recombination: The segregation of parental and recombinant mitochondrial genotypes during vegetative division of yeast. Molec. Gen. Genetics 120:17-25.

McDougall, K. J., and T. H. Pittenger, 1966. A cytoplasmic variant of *Neurospora crassa*. Genetics 54:551-65.

Michaelis, P. 1955. Über Gesetzmässigkeiten der Plasmon—Umkombination und über eine Methode zur Trennung einer Plastiden-, Chondriosomen-, resp. Sphaerosomen-, (Mikrosomen)—und einer Zytoplasmavererbung. Cytologia 20:315-38.

Michaelis, G., E. Petrochilo, and P. P. Slonimski. 1973. Mitochondrial genetics. III. Recombined molecules of mitochondrial DNA obtained from crosses between cytoplasmic petite mutants of *Saccharomyces cerevisiae:* Physical and genetic characterization. Molec. Gen. Genet. 123:51-65.

Mitchell, M. B., and H. K. Mitchell. 1952. A case of "maternal" inheritance in *Neurospora crassa*. Proc. Nat. Acad. Sci. USA. 38: 442-49.

Perasso, R., and A. Adoutte. 1974. The process of selection of erthromycin-resistant mitochondria by erythromycin in *Paramecium*. J. Cell Sci. 14:475-97.

Perlman, P. S., and C. W. Birky, Jr. 1974. Mitochondrial genetics in baker's yeast: A molecular mechanism for recombinational polarity and suppressiveness. Proc. Nat. Acad. Sci. USA. 71:4612-16.

Perlman, P. S., and C. A. Demko. 1974. Effects of gene dosage on mitochondrial marker transmission in *Saccharomyces cerevisiae*. Genetics 77:s50-51 (Abstr).

Pittenger, T. H. 1956. Synergism of two cytoplasmically inherited mutants in *Neurospora crassa*. Proc. Nat. Acad. Sci. USA. 42:747-52.

Rank, G. H. 1973. Recombination in 3-factor crosses of cytoplasmically inherited antibiotic-resistance mitochondrial markers in *S. cerevisiae*. Heredity 30:265-71.

Rank, G. H., and N. T. Bech-Hansen. 1972. Somatic segregation, recombination, asymmetrical distribution, and complementation tests of cytoplasmically-inherited antibiotic-resistance mitochondrial markers in *S. cerevisiae.* Genetics 72:1–15.

Ris, H. 1961. Ultrastructure and molecular organization of genetic systems. Can. J. Genet. Cytol. 3:95–120.

Roberts, C. T., Jr., and E. Orias. 1973. Cytoplasmic inheritance of choloramphenicol resistance in *Tetrahymena.* Genetics 73:259–72.

Rowlands, R. T., and G. Turner. 1973. Nuclear and extranuclear inheritance of oligomycin resistance in *Aspergillus nidulans.* Molec. Gen. Genet. 126:201–16.

Rowlands, R. T., and G. Turner. 1974a. Recombination between the extranuclear genes conferring oligomycin resistance and cold sensitivity in *Aspergillus nidulans.* Molec. Gen. Genet. 133:151–61.

Rowlands, R. T., and G. Turner. 1974b. Physiological and biochemical studies of nuclear and extranuclear oligomycin-resistant mutants of *Aspergillus nidulans.* Molec. Gen. Genet. 132–73–88.

Sager, R. 1962. Streptomycin as a mutagen for nonchromosomal genes. Proc. Nat. Acad. Sci. USA. 48:2018–26.

Sager, R. 1972. Cytoplasmic genes and organelles. Academic Press, New York.

St. Lawrence, P., and J. H. Chalmers, Jr. 1973. Stability of poky strains of *N. crassa.* Genetics 74:s237 (Abstr.)

Sena, E., J. Welch, D. Radin, and S. Fogel. 1973. DNA replication during mating in yeast. Genetics 74:s248 (Abstr.)

Shannon, C., A. Rao, S. Douglass, and R. S. Criddle. 1972. Recombination in yeast mitochondrial DNA. J. Supramolec. Struct. 1:145–52.

Sherman, F., and C. W. Lawrence. 1974. *Saccharomyces*, pp. 359–93. *In* R. C. King (ed.), Handbook of genetics. Plenum Press, New York.

Sinsheimer, R. L., R. Knippers, and T. Komano. 1968. Stages in the replication of bacteriophage φX174 DNA *in vivo.* Cold Spring Harbor Symp. Quant. Biol. 33:443–47.

Stanier, R. Y. 1970. Some aspects of the biology of cells and their possible evolutionary significance, pp. 1–38. *In* H. P. Charles and B. C. J. G. Knight (eds.), Organization and control in prokaryotic and eukaryotic cells (20th Symp. Soc. Gen. Microbiol.). At the University Press, Cambridge.

Storm, E. M., K-B Lam, and J. Marmur. 1974. A non-Mendelian temperature sensitive respiratory deficient mutant of *Saccharomyces cerevisiae.* Genetics 77:s62 (Abstr.).

Strausberg, R. L., and P. S. Perlman. 1974. Cellular analysis of mitochondrial inheritance in *Saccharomyces cerevisiae.* Genetics 77:s62–63 (Abstr.).

Stuart, K. D. 1970. Cytoplasmic inheritance of oligomycin and rutamycin resistance in yeast. Biochem. Biophys. Res. Comm. 39:1045–51.

Suda, K., and A. Uchida. 1972. Segregation and recombination of cytoplasmic drug-resistance factors in *Saccharomyces cerevisiae.* Jap. J. Genet. 47:441–44.

Suda, K., and A. Uchida. 1974. The linkage relationship of the cytoplasmic drug-resistance factors in *Saccharomyces cerevisiae.* Molec. Gen. Genet. 128:331–39.

Thomas, D. Y., and D. Wilkie. 1968a. Inhibition of mitochondrial synthesis in yeast by erythromycin: Cytoplasmic and nuclear factors controlling resistance. Genet. Res. 11:33–41.

Thomas, D. Y., and D. Wilkie. 1968b. Recombination of mitochondrial drug-resistance factors in *Saccharomyces cerevisiae*. Biochem. Biophys. Res. Comm. 30:368-72.

Trembath, M. K., C. L. Bunn, H. B. Lukins, and A. W. Linnane. 1973. Biogenesis of mitochondria. 27. Genetic and biochemical characterization of cytoplasmic and nuclear mutations to spiramycin resistance in *Saccharomyces cerevisiae*. Molec. Gen. Genet. 121:35-48.

Ursprung, H., and E. Schabtach. 1965. Fertilization in tunicates: Loss of the paternal mitochondria prior to sperm entry. J. Exp. Zool. 159:379-81.

Wakabayashi, K., and N. Gunge. 1970. Extrachromosomal inheritance of oligomycin resistance in yeast. FEBS Letters 6:302-4.

Waldron, C., and C. F. Roberts. 1973. Cytoplasmic inheritance of a cold-sensitive mutant in *Aspergillus nidulans*. J. Gen. Microbiol. 78:379-81.

Waxman, M. F., N. Eaton, and D. Wilkie. 1973. Effect of antibiotics on the transmission of mitochondrial factors in *Saccharomyces cerevisiae*. Molec. Gen. Genet. 127:277-84.

Wilkie, D. 1972. Genetic aspects of mitochondria. Proc. Eighth Meeting Fed. Europ. Biochem. Soc. 28:85-94.

Wilkie, D., and D. Y. Thomas. 1973. Mitochondrial genetic analysis by zygote cell lineages in *Saccharomyces cerevisiae*. Genetics 73:368-77.

Williamson, D. H., and D. J. Fennell. 1974. Apparent dispersive replication of yeast mitochondrial DNA as revealed by density labelling experiments. Molec. Gen. Genet. 131:193-207.

Wilson, E. B. 1925. The cell. (3d ed.) MacMillan Company, New York.

Wolf, K., B. Dujon, and P. P. Slonimski. 1973. Mitochondrial genetics. V. Multifactorial mitochondrial crosses involving a mutation conferring paromomycin-resistance in *Saccharomyces cerevisiae*. Molec. Gen. Genet. 125:53-90.

Visconti, N., and M. Delbrück. 1953. The mechanism of genetic recombination in phage. Genetics 38:5-33.

J. KENNETH HOOBER
W. J. STEGEMAN

Regulation of Chloroplast Membrane Synthesis

6

Chloroplast membranes have played an indispensable role in the scheme of life. These membranes are capable of capturing light energy released by the sun and converting it into a form that can be used to drive replenishment of the carbon lost as CO_2 during oxidative metabolism. In addition, these membranes provide the oxygen needed for oxidative reactions in aerobic organisms. Considering the importance of this membrane system, it is imperative that we learn something about its function, structure, and mode of assembly. A large and growing body of literature is already available on the photosynthetic functions of the thylakoid membranes in chloroplasts (see, for example, reviews by Boardman, 1968; Cheniae, 1970; Bishop, 1971; Ke, 1973; Borisov and Godik, 1973). Less is known about the structure of these membranes, but investigations employing the recent developments in electron microscopy (Branton and Park, 1967; Goodenough and Staehelin, 1971; Wang and Packer, 1973), x-ray diffraction (Sadler et al., 1973), spin-labeling (Torres-Pereira et al., 1974), and chemical fractionation (Boardman, 1970; Levine et al., 1972; Jennings and Eytan, 1973; Anderson and Levine, 1974; Klein and Vernon, 1974) of chloroplast membranes should soon enable an understanding of this aspect. Least known about thylakoid, or any other, membranes are the mechanisms of assembly and the influence of environmental factors on the functional formation of these structures.

Greening of the y-1 strain of *Chlamydomonas reinhardtii*, a single-celled green alga, provides a remarkable opportunity to study specifi-

Department of Biochemistry, School of Medicine, Temple University, Philadelphia, Pennsylvania 19140 cally the assembly of thylakoid membranes. Within cells grown in continuous light, the single, large chloroplast is nearly filled with these chlorophyll-containing membranes (Ohad et al., 1967a). In the y-1 strain formation of these membranes, but not cell growth, requires light. Following transfer of cultures to the dark, cell growth and division eventually yield cells whose chloroplasts are nearly depleted of membranous structures (Ohad et al., 1967a,b). These etiolated cells, which contain 1 to 2 percent of the amount of chlorophyll in green cells, respond rapidly when returned to the light and achieve a level of chlorophyll comparable to that of fully green cells (30 to 40 μg of chlorophyll per 10^7 cells) in eight to ten hours (Ohad et al., 1967b; Hoober et al., 1969). The amount of thylakoid membranes and the photosynthetic activities characteristic of these membranes increase in parallel with the accumulation of chlorophyll. Since cell division is temporarily suspended during the greening process (Ohad et al., 1967b), formation of thylakoid membranes can be studied without significant interference by increases in other types of cellular structures.

Assembly of functional thylakoid membranes in *C. reinhardtii* y-1 requires the production of chlorophyll, a process dependent upon light (Ohad et al., 1967b); the synthesis of lipids and in particular glycolipids in the chloroplast (Goldberg and Ohad, 1970); and proteins synthesized both on cytoplasmic and chloroplast ribosomes (Hoober et al., 1969; Eytan and Ohad, 1970). The system can be manipulated easily by a variety of means, including (1) the intensity and quality of light provided and (2) the use of selective inhibitors of the different ribosomes. Of the latter, the most widely used have been chloramphenicol, an inhibitor of 70 S ribosomes within chloroplasts and mitochondria, and cycloheximide, an inhibitor of cytoplasmic 82 S ribosomes (Hoober and Blobel, 1969). The following discussion will describe experimental results obtained by these two approaches. Included in the discussion are data with regard to (1) some characteristics of the thylakoid membrane polypeptides of *C. reinhardtii*, (2) sites of synthesis of the membrane polypeptides, and (3) some factors that appear to be involved in regulating the synthesis of the major membrane polypeptides.

CHARACTERISTICS OF THYLAKOID MEMBRANE POLYPEPTIDES

Thylakoid membranes of *C. reinhardtii* y-1 were purified from green cells by a combination of sedimentation and flotation procedures

TABLE 1

PROTEIN TO CHLOROPHYLL RATIO OF THYLAKOID MEMBRANES OF *C. REINHARDTII* Y-1

Treatment	Protein/chlorophyll (mg/mg)
H_2O	5.7 ± 0.2
8 M urea	4.7 ± 0.1
1 M KCl in 8 M urea	4.6 ± 0.1

Membranes were purified from green cells as described previously (Hoober, 1970). Samples of the membrane preparations were suspended in either H_2O, 8 M urea or 8 M urea containing 1 M KCl. After standing at 0 C for 1 hour, the samples were centrifuged for 2 hours at 204,000 × g_{av}. The pellets were extracted with 80% acetone, and chlorophyll in the extracts was determined by Vernon's (1960) procedure. The residue was washed two times with acetone, dried under reduced pressure, and then dissolved in 2% sodium dodecyl sulfate. The amount of protein was determined by the biuret procedure (Gornall et al., 1949) with bovine serum albumin, in 2% sodium dodecyl sulfate, as the standard. The values shown are averages of measurements on 3 different preparations.

(Hoober, 1970). The membrane preparations contained protein and chlorophyll in a ratio of 5.7 mg per mg (table 1). The use of 8 M urea or of 8 M urea containing 1 M KCl to wash the membranes reduced the ratio to 4.7 and 4.6, respectively. The addition of KCl to the wash medium did not remove significantly more protein than did 8 M urea alone, but such drastic procedures should remove all but the most tightly bound, integral components of the membrane (Singer, 1974). An analysis by gel electrophoresis in the presence of sodium dodecyl sulfate of the proteins remaining with the washed membranes revealed the pattern shown in figure 1. Conspicuous were three main polypeptide fractions whose mobilities on the gel corresponded to molecular weights of about 31,000; 28,000; and 24,000. These three fractions are designated for reference purposes as thylakoid main fractions (TMF) *a*, *b*, and *c*, respectively. TMF*a* was resolved into at least two polypeptide species on gels of higher acrylamide concentration, but TMF*b* and TMF*c* behaved, in the main, as single polypeptide species. The latter two polypeptides were present in nearly equal amounts, with each accounting for about 20% of the total protein stain on the gel.

TMF*b* and TMF*c* were purified by preparative gel electrophoresis and subjected to analytical gel electrophoresis. Figure 2 shows that each fraction contained principally polypeptides of a single molecular size. Slight cross contamination of the two preparations was observed, but these results attest to the separate identity of these two polypeptide fractions. An assumption made in these studies, which has not been

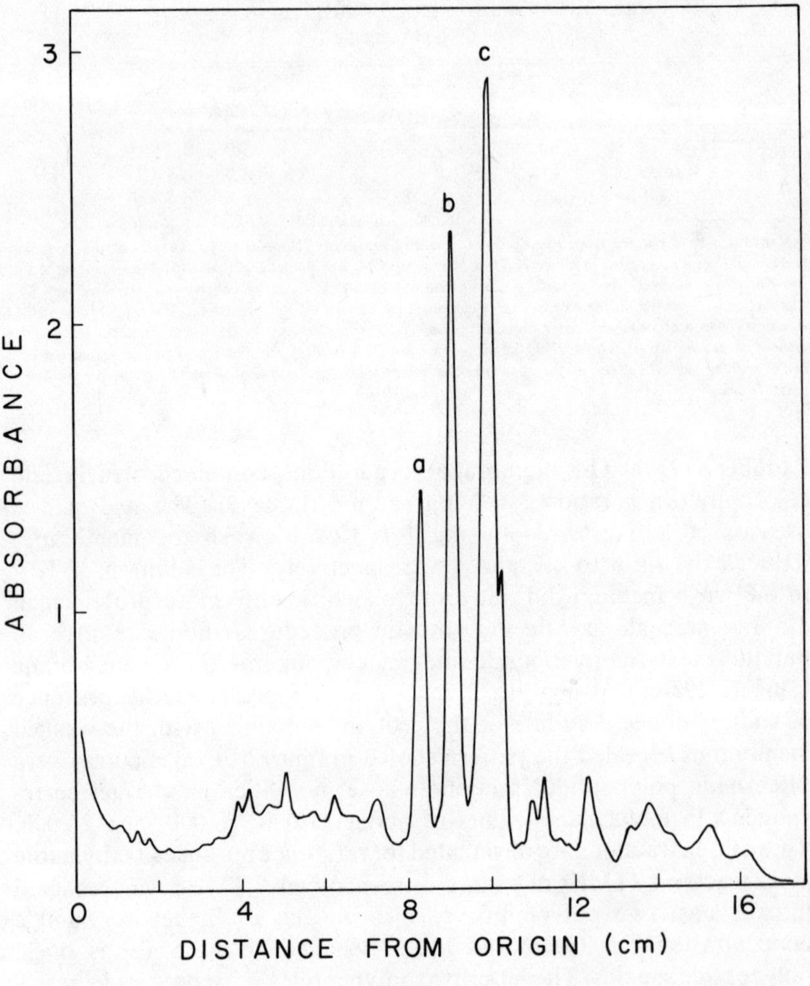

Fig. 1. Electrophoretic analysis of the polypeptides of thylakoid membranes of *C. reinhardtii*. Membranes were purified from green cells and washed with 8 M urea containing 1 M KC1 (see table 1). After extracting lipids with 90% acetone, the proteins were dissolved in buffer containing sodium dodecyl sulfate and subjected to electrophoresis as described previously (Hoober, 1970, 1972). Gels were stained with Coomassie blue and after destaining were scanned at 563 nm.

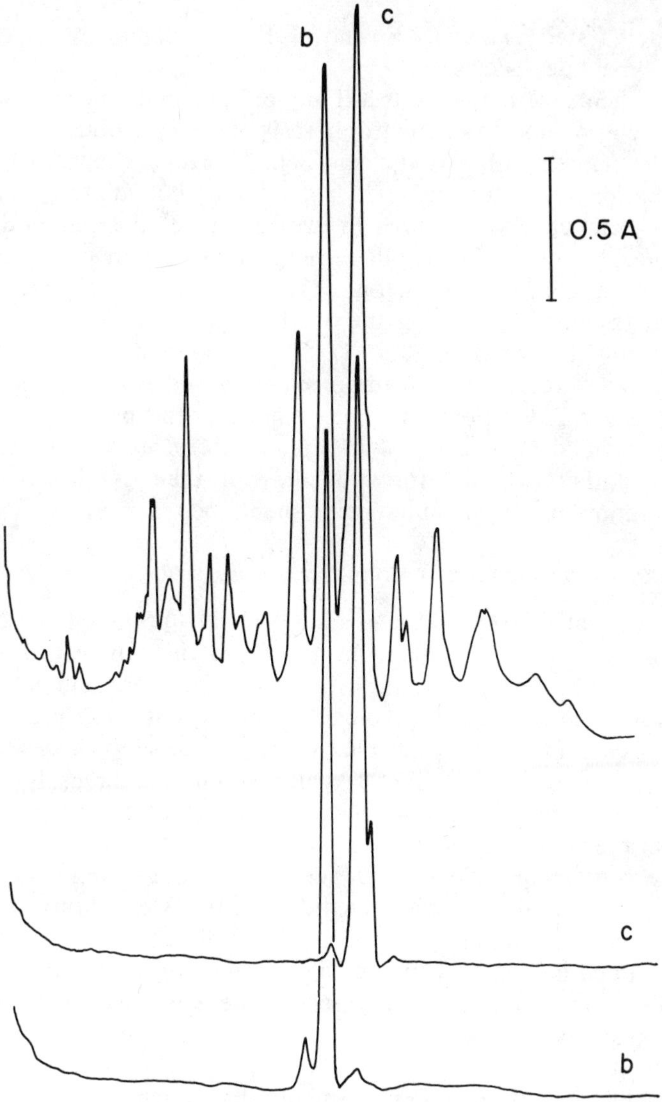

Fig. 2. Purification of TMF*b* and TMF*c* by preparative gel electrophoresis. Thylakoid membranes were dissolved in sodium dodecyl sulfate and subjected to electrophoresis through an 8% polyacrylamide gel, 1.3 × 10 cm. The gel was mounted over a 0 to 0.9 M linear sucrose gradient, 1.3 × 20 cm. Electrophoresis was continued until the polypeptides had migrated into the gradient. The gradient was pumped through a UV analyser and portions containing TMF*b* and TMF*c* were collected. The upper trace shows the pattern for total membrane polypeptides. The lower traces show the patterns for the fractions of the gradient containing either TMF*b* or TMF*c*.

adequately tested, is that TMF*b* and TMF*c* each contain one predominant polypeptide species.

Experiments were done to test if any of the membrane components were glycoproteins. After electrophoresis of total membrane proteins, gels were treated with periodic acid-Schiff reagents (Zacharius et al., 1969; Fairbanks et al., 1971) to stain for carbohydrate groups. A positive reaction was observed in two regions of the gels, a diffuse band near the top of the gel where no peak of protein stain occurred (about 3 cm from the origin, fig. 1), and two peaks coincident with the two fractions containing the smallest polypeptides (e.g., 15 to 17 cm from the origin in fig. 1). Thus, the only clearly defined glycoproteins present in the membrane are the smallest of the polypeptides. Neither TMF*b* nor TMF*c* was stained by this procedure. These results provided assurance that TMF*b* and TMF*c* are simple polypeptides, and that a straightforward study of their synthesis by the incorporation of labeled amino acids should be feasible.

SITES OF SYNTHESIS OF THYLAKOID MEMBRANE POLYPEPTIDES

Hoober et al. (1969) and Eytan and Ohad (1970) found that greening of etiolated cells was dependent upon protein synthesis both on chloroplast and on cytoplasmic ribosomes. Chloramphenicol-treated cells were able to form thylakoid membranes, but those membranes were deficient in photochemical activities. Conversely, in cells treated with cycloheximide, chlorophyll synthesis and membrane formation were strongly inhibited. But if cells containing incomplete membranes, made in the presence of chloramphenicol, were transferred to fresh medium containing cycloheximide, development of the photochemical functions resumed until the specific activities were comparable to those found with control cells (Hoober et al., 1969). These results are illustrated in figure 3 with the Hill reaction, an assay for photosystem II, as an example. Similar results were obtained with assays for photosystem I and photophosphorylation (Bar-Nun et al., 1972; Wallach et al., 1972).

The inhibition of membrane formation by cycloheximide indicated that proteins synthesized on cytoplasmic ribosomes are required for the assembly process. Moreover, the formation of incomplete membranes in the presence of chloramphenicol suggested that proteins synthesized on chloroplast ribosomes are necessary for full functional activity. Figure 3 shows, as indicated by the development of full activity after the removal of chloramphenicol, that proteins newly

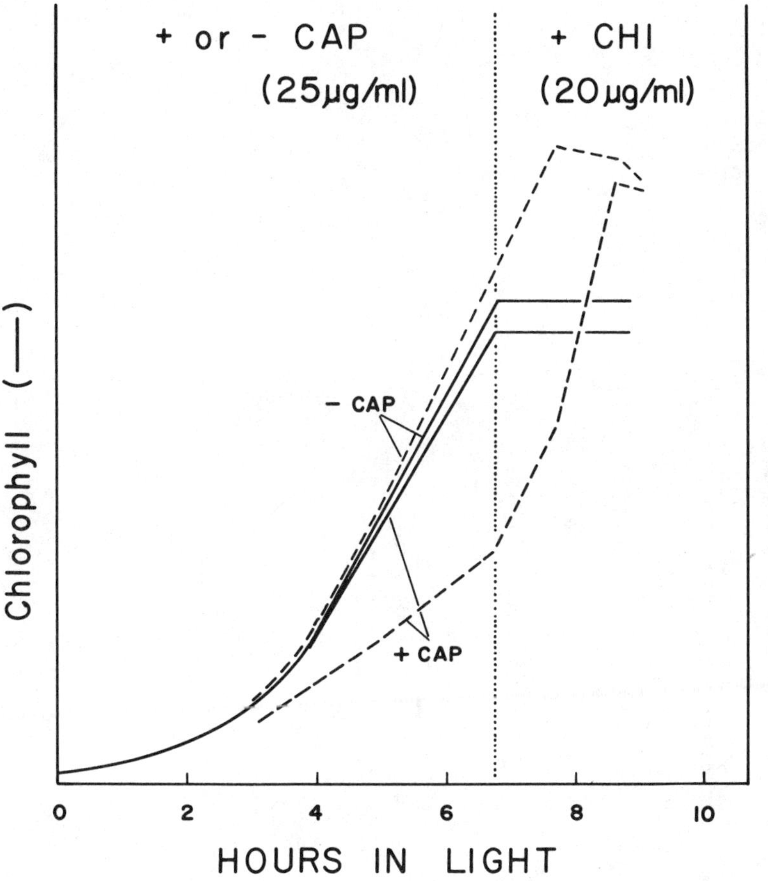

Fig. 3. Diagram illustrating the effects of chloramphenicol and cycloheximide during greening of etiolated cells of *C. reinhardtii* y-1. Dark-grown cells were suspended in fresh medium to 4×10^6 cells per ml. Chloramphenicol was added to one sample, while another served as a control. After 6.5 hours in the light, the cells from both samples were collected, suspended in medium containing cycloheximide, and returned to the light. Analyses for the amount of chlorophyll and activity of the Hill reaction were performed periodically during greening. (From data presented in Hoober et al., 1969).

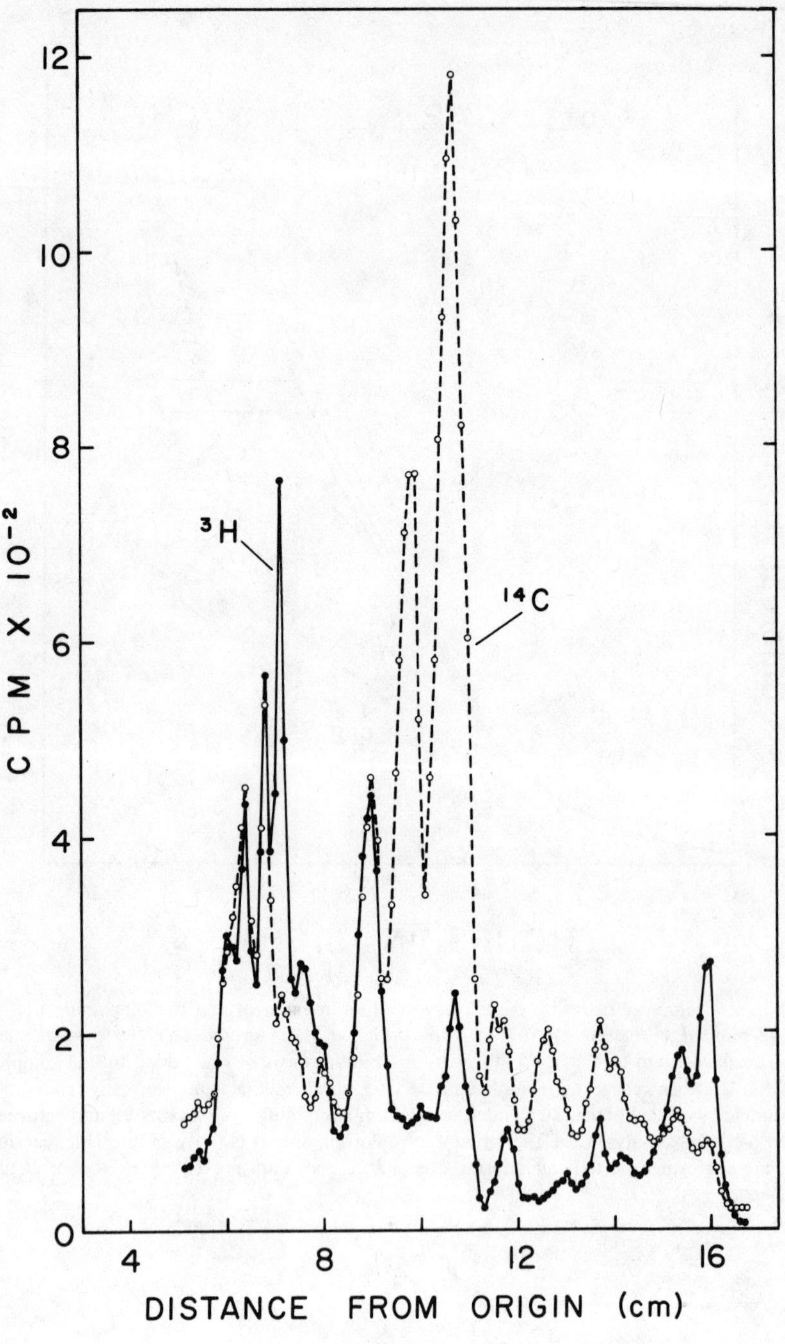

CPM X 10⁻²
12
10
8
6
4
2
0
4
8
12
16
DISTANCE FROM ORIGIN (cm)
³H
¹⁴C

synthesized on chloroplast ribosomes can be incorporated into existing, incomplete membranes. The attainment of full activity is not prevented by cycloheximide. These results provided an opportunity to determine the sites of synthesis of the membrane polypetides by alternating the presence of these two drugs (Hoober, 1970). In an experiment similar to that described in figure 3, cells were allowed to make membranes in the presence of chloramphenicol. [^{14}C]Arginine was provided to label proteins made on cytoplasmic ribosomes with ^{14}C. But since the concentration of chloramphenicol used (25 μg/ml) was insufficient to inhibit chloroplast protein synthesis completely, a small amount of protein of chloroplast origin also became labeled with ^{14}C and incorporated into the growing membranes. After a certain amount of membranes had accumulated, cells were transferred to medium containing cycloheximide, at a concentration (20 μg/ml) sufficiently high to nearly completely inhibit cytoplasmic protein synthesis. [^{3}H]Arginine then was added to label proteins made in the chloroplast with ^{3}H. The doubly labeled thylakoid membranes were purified, and the proteins were subjected to electrophoresis. The resultant patterns of ^{14}C and ^{3}H are shown in figure 4. The site of synthesis of any polypeptide that was sufficiently resolved by the electrophoretic procedure was indicated by its enrichment in either ^{14}C or ^{3}H. Most obvious of those polypeptides enriched in ^{14}C, a characteristic expected for a cytoplasmic site of synthesis, were the major polypeptides, TMF*b* and TMF*c*. Of the minor polypeptides, some were clearly made in the chloroplast, while others were marked as cytoplasmic products (Hoober, 1970).

Synthesis of TMF*b* and TMF*c* on cytoplasmic ribosomes was confirmed without the use of inhibitors by a study of the incorporation of [^{3}H]leucine into proteins *in vivo* (Stegeman and Hoober, 1974).

Fig. 4. Analysis of the sites of synthesis of thylakoid membrane polypeptides in *C. reinhardtii* y-1. Etiolated cells were allowed to green in the presence of chloramphenicol as described under fig. 3. [^{14}C]Arginine (237 mCi/mmole) was added, after 4 hours in the light, to a concentration of 0.04 μCi/ml. After 8.5 hours of greening, cells were removed from medium containing chloramphenicol and [^{14}C]arginine, and suspended in medium containing cycloheximide. [^{3}H]Arginine (7.3 Ci/mmole) then was added, to 2 μCi/ml. After an additional 2 hours in light, membranes were purified from the cells, and the polypeptides were subjected to electrophoresis in the presence of sodium dodecyl sulfate. The gel was sliced into 1-mm sections, and the patterns of ^{14}C and ^{3}H were determined. An enrichment in ^{14}C indicates synthesis on cytoplasmic ribosomes, and an enrichment in ^{3}H indicates synthesis on chloroplast ribosomes (Data taken from Hoober, 1970).

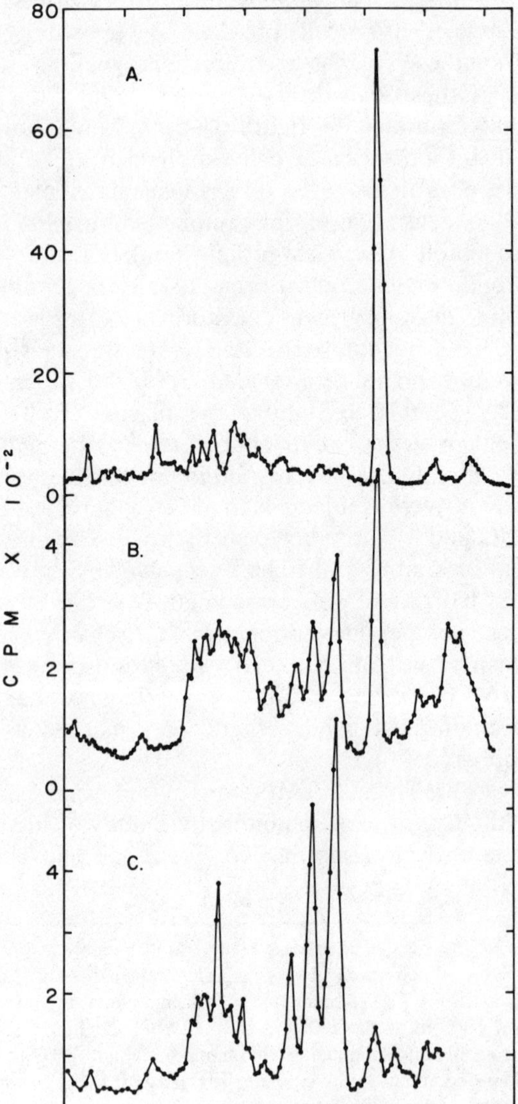

80
60
40
20
0
4
2
0
4
2
0
A.
B.
C.
C P M X 10⁻²
0
4
8
12
DISTANCE FROM ORIGIN (cm)

Figure 5 shows electrophoretic patterns of radioactivity obtained when leucine was present in the medium at two different concentrations. Electrophoresis of total cellular protein, labeled by the addition of [^{3}H]leucine (specific activity 30 to 40 Ci/mmole) to provide a concentration of about 1×10^{-7} M in the medium, yielded the pattern of radioactivity shown in figure 5A. This pattern had no resemblance to the pattern of protein stain or to the pattern of radioactivity obtained with labeled arginine (Hoober, 1972; Hoober and Stegeman, 1973). We have found that under these conditions the [^{3}H]leucine was selectively incorporated into proteins within the mitochondrial compartment of the cell (Stegeman and Hoober, 1974). This incorporation was entirely sensitive to inhibitors of 70 S-type ribosomes but insensitive to cycloheximide, indicating that the [^{3}H]leucine had not entered the cytoplasmic pool of amino acids to an extent sufficient to detectably label proteins synthesized on cytoplasmic ribosomes. Moreover, products of chloroplast 70 S ribosomes, most predominant of which is the large subunit of ribulose 1,5-diphosphate carboxylase (Hoober, 1972; Boynton et al., 1972), also were not detectably labeled at this concentration of leucine. A component that was prominently labeled (at 10.5 cm from the origin in figure 5A) is a polypeptide of about 16,500 daltons in size. This polypeptide is apparently a product of transcription on mitochondrial DNA as well as translation on mitochondrial ribosomes. Its synthesis requires light, but it is made at a high rate only during the early stages of greening.

On the other hand, when the concentration of leucine in the medium was increased several hundredfold to 2 to 3×10^{-5} M by the addition of unlabeled L-leucine, incorporation of ^{3}H into additional polypeptides was observed (fig. 5B). After electrophoresis of total cellular protein labeled at the higher concentration, the most significant finding was the presence of prominent peaks in the pattern of radioactivity at

Fig. 5. Electrophoretic analysis of the incorporation of [^{3}H]leucine into polypeptides of *C. reinhardtii* y-1 at two concentrations of the amino acid. *A.* Etiolated cells were exposed to light for 2 hours. L-[^{3}H]Leucine (30 Ci/mmole) then was added to 2 μCi/ml, providing an extracellular concentration of 5×10^{-8} M, and the cells were incubated for an additional hour. Total cellular protein was subjected to electrophoresis. *B.* Etiolated cells were exposed to light for 4 hours, and then L-[^{1}H]leucine and [^{3}H]leucine (40 Ci/mmole) were added to 3×10^{-5} M and 20 μCi/ml, respectively. The final specific activity was about 1 Ci/mmole. Incubation was continued for an additional hour. Total cellular protein was subjected to electrophoresis. *C.* Cells were labeled for 2 hours as described under *B.* After breaking the cells by sonication, membranes were collected by centrifugation for 20 min at 100,000 × g. Membrane polypeptides then were subjected to electrophoresis.

8.5 and 9.2 cm from the origin. As shown by a comparison with figure 5C, which is the pattern of radioactivity for the membrane fraction under these latter conditions, these peaks correspond to TMF*b* and TMF*c*. Therefore, when provided at the higher concentration, leucine entered the cytoplasmic pool and was incorporated into proteins synthesized on cytoplasmic ribosomes. Since at the higher concentration [^{3}H]leucine was still not significantly incorporated into the large subunit of ribulose 1.5-diphosphate carboxylase (positioned at 5.3 cm from the origin in figure 5B), these results establish that the two major polypeptides of thylakoid membranes are synthesized on cytoplasmic ribosomes. As expected, the incorporation of [^{3}H]leucine into TMF*b* and TMF*c* was insensitive to chloramphenicol (Stegeman and Hoober, 1974). In contrast to previous results with [^{3}H]arginine (Hoober, 1972), these [^{3}H]leucine patterns suggest that TMF*b* and TMF*c* are major products of cytoplasmic protein synthesis during greening of the cells.

A question that these results pose is how polypeptides are transferred into the chloroplast upon completion of synthesis on cytoplasmic ribosomes. Although the mechanism is not known, the transfer of TMF*b* and TMF*c* into the chloroplast occurs rapidly and does not involve a detectable pool of a soluble form of these polypeptides. After a two-minute pulse with [^{3}H]arginine, an amino acid taken up by the cells more readily than leucine, all the TMF*b* and TMF*c* that could be detected by electrophoretically analyzing the soluble and membrane fractions of the cells (Hoober, 1972) was already associated with membranes (Stegeman and Hoober, unpublished results). The patterns for the membrane and soluble fractions obtained after a subsequent thirty-minute chase period were unchanged from those found at the end of two minutes of labeling. If the task of synthesizing membrane proteins were distributed nondiscriminately over the cytoplasmic ribosomes, a pool of soluble in-transit polypeptides would be expected since the bulk of cytoplasmic ribosomes are not attached to membranes (Ohad et al., 1967a; Hoober and Blobel, 1969). The rapidity with which the polypeptides become associated with membranes suggests that their synthesis may occur on ribosomes attached to the chloroplast envelope, analogous to the association in yeast cells of cytoplasmic ribosomes with mitochondria (Kellems et al., 1974; Kellems and Butow, 1974). Thus, the soluble form of TMF*c*, found previously in *C. reinhardtii* y-1 under conditions in which membrane assembly was inhibited (Hoober, 1972), may have accumulated within the chloroplast as the result of vectorial discharge into the organelle during synthesis.

REGULATION OF THE SYNTHESIS OF THE MAJOR MEMBRANE POLYPEPTIDES IN *C. REINHARDTII* Y-1

Although TMF*b* and TMF*c* are made on cytoplasmic ribosomes, their synthesis is regulated by events that apparently occur within the chloroplast. Two factors involved in control of the synthesis of these polypeptides are (1) protochlorophyll,[1] or a more distant precursor of chlorophyll, and (2) a protein synthesized on chloramphenicol-sensitive ribosomes. We have proposed (Hoober and Stegeman, 1973) that these two components are related in the manner outlined on figure 6. The hypothesis presented in this scheme suggests that protochlorophyll, the end-product of chlorophyll synthesis in the dark, is an inhibitor of the synthesis of TMF*b* and TMF*c*. However, for inhibition of the synthesis of these polypeptides to occur, a protein synthesized presumably on chloroplast ribosomes is required. Upon exposure of cells to light, protochlorophyll is reduced to chlorophyll and synthesis of the polypeptides is released from the inhibitory mechanism. Since the mRNA's for TMF*b* and TMF*c* are translated on cytoplasmic ribosomes, whereas the controlling factors appear to be located within the chloroplast, this scheme suggests that regulation of the synthesis of these polypeptides occurs at the level of transcription on chloroplast DNA. Furthermore, the hypothesis suggests that protochlorophyll and the regulatory protein are analogous to a corepressor-repressor combination.

This scheme is difficult to prove, but the following discussion will present some of the evidence upon which it is based. Most of the

Note added in proof: We have recently found that *Chlamydomonas reinhardtii* does not incorporate [^{3}H]leucine when added at the low concentrations used in the experiment described under figure 5A. The incorporation that was observed has been traced to a bacterial contaminant in the cultures. The alga secretes into the medium a substance that induces within the contaminating cells synthesis of the polypeptide illustrated in figure 5 (see W. G. Stegeman and J. K. Hoober, 1975, Nature [in press], for details). The results on the synthesis of chloroplast membrane polypeptides have not been affected by this situation.

1. The terms protochlorophyll and chlorophyll are used in the generic sense to denote, in the absence of specific information, both the esterfied and unesterified forms of the porphyrins.

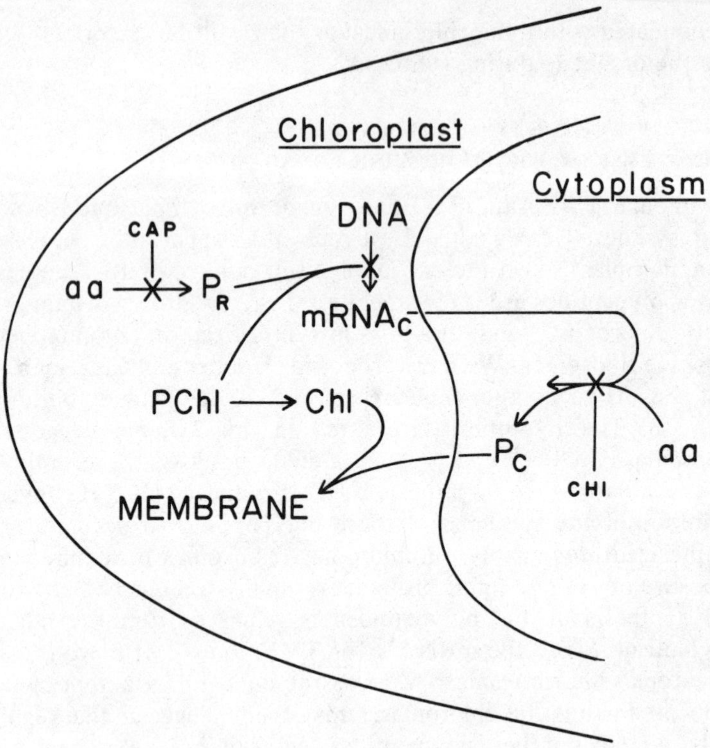

Fig. 6. Scheme depicting the proposed mechanism for control of the synthesis of the major polypeptides of thylakoid membranes in *C. reinhardtii* y-1. P_c, polypeptide TMF*c;* $mRNA_c$, messenger RNA for polypeptide TMF*c;* PChl, protochlorophyll(ide); Chl, chlorophyll(ide); P_R, regulatory protein; aa, amino acids; CHI cycloheximide; CAP, chloramphenicol.

pathway for chlorophyll synthesis, at least the terminal portion, is located within the chloroplast (Rebeiz and Castelfranco, 1973). Moreover, the solubility properties of the substrates and enzymes involved in the final steps of chlorophyll biosynthesis suggest that these reactions occur in association with the growing thylakoid membranes (Rebeiz and Castelfranco, 1973). In the dark the final product of the pathway is protochlorophyll, which accumulates to a small extent in the chloroplast in association with a protein (Kahn et al., 1970; Nielsen and Kahn, 1973). Absorption of light by this complex, the protochlorophyll holochrome, results in the photoreduction of the porphyrin to chlorophyll. Two photoconvertible forms of protochlorophyll are generally found in higher plants, a major form that absorbs maximally in the red region of the spectrum at 650 nm and a minor form that absorbs maximally at about 635 nm (Kahn and Nielsen, 1974). The action spectrum for the synthesis of chlorophyll in higher plants reflects this situation, with a maximum in red light at 650 nm (Koski et al., 1951). A third form of protochlorophyll, which is sometimes present in a small amount in dark-grown cells and which absorbs red light maximally at 628 nm, is considered to exist in a free, nontransformable state (Kahn et al., 1970).

Various values have been reported for the maximum in the red portion of the action spectrum for chlorophyll synthesis in algae. Studies on etiolated mutant strains of *Chlorella vulgaris* (Bryan, 1962, cited by Bogorad, 1966) and *Chlamydomonas reinhardtii* (McLeod et al., 1963) indicated that the action spectrum for greening of these algae is identical to the spectrum in higher plants (Koski et al., 1951), with a maximum at 650 nm. However, Wolken and Mellon (1956) reported a maximum of about 630 nm for the action spectrum in etiolated *Euglena,* a value recently confirmed by Egan et al., (1974). Ohad (1975) also found a maximum at about 630 nm for the greening of *Chlamydomonas reinhardtii* y-1. The reason for these different action spectra is not known, but they may reflect differences in the *in vivo* organization of the phototransformable species of protochlorophyll. Nevertheless, the fact remains that in these etiolated plants the absorption of light by protochlorophyll is required for the synthesis of chlorophyll.

Mutant strains of algae, including *C. reinhardtii,* in the dark accumulate small amounts of protochlorophyll that subsequently is converted to chlorophyll in the presence of light (Matsuda et al., 1971). Greening of these cells requires continuous light. Cells placed in the dark after a period of greening abruptly stop producing chlorophyll, but upon return to the light chlorophyll synthesis resumes at the rate established during the previous light period (Ohad et al., 1967b). As indicated by the action spectrum, the effect of light is to facilitate the photoreduction of protochlorophyll. Therefore, rather than placing cells in complete darkness, it should be possible to manipulate chlorophyll synthesis by varying the quality of light provided. Figure 7 shows the results of an experiment in which cut-off filters were used to eliminate portions of the visible light impinging on greening cells. Two cultures of cells were permitted to green normally for five hours in white light, and then the filters were inserted in the light beams to provide light above 600 nm to one culture and above 650 nm to the other. In the first case, light above 600 nm continued to support

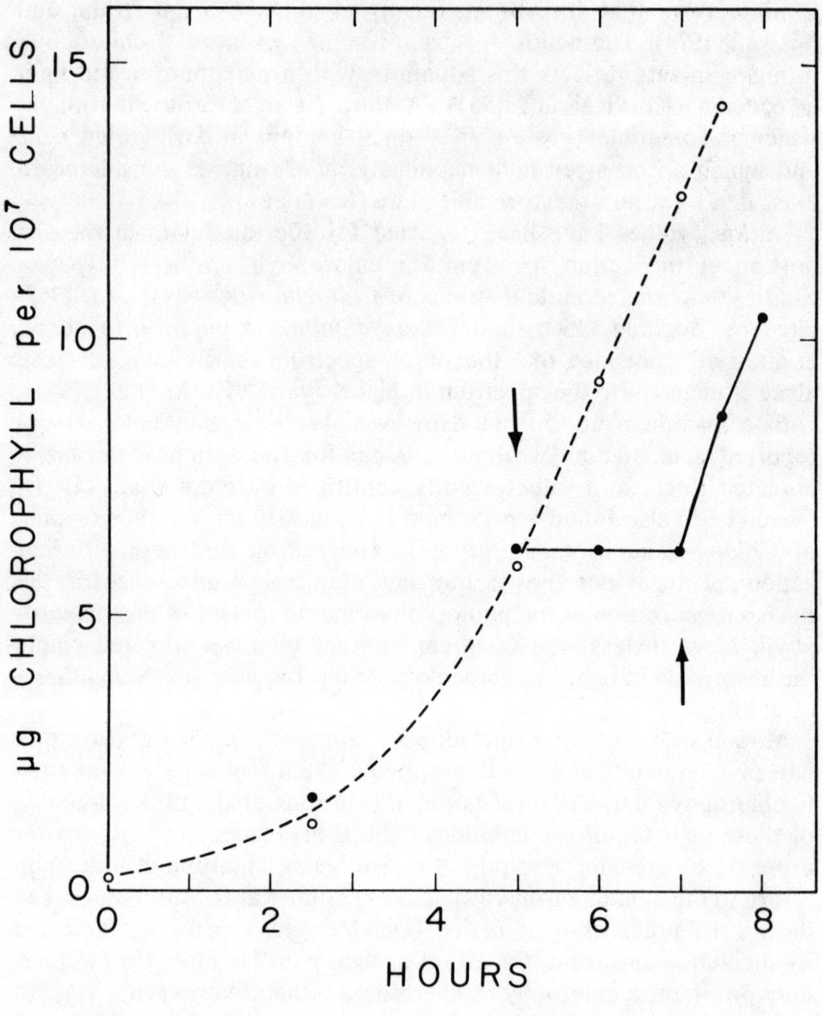

Fig. 7. Effects of light filters on the increase of chlorophyll during greening of *C. reinhardtii* y-1. Etiolated cells were exposed to white light from incandescent lamps. At 5 hours light filters were inserted into the light beams to provide light above 650 nm to one culture (●) and above 600 nm to the control culture (O). The wavelength given for the cutoff is that for 1% transmission through the filter. The filters were removed at 7 hours (from Hoober and Stegeman, 1973. Grateful acknowledgment is made to the *Journal of Cell Biology* to reprint this and other figures in this paper).

chlorophyll synthesis at a rate equal to that supported by white light. However, in cells illuminated with light above 650 nm, chlorophyll synthesis stopped as abruptly as if the cells had been transferred to the dark, even though light was still available to drive photosynthesis (Hoober and Stegeman, 1973).

Experiments of the type described in figure 7 then were done to determine if shifting the cut-off of light from 600 nm to 650 nm also affected synthesis of TMF*b* and TMF*c*. It was already known that TMF*b* and TMF*c* could not be detected after electrophoresis of etiolated cell proteins and that synthesis of these polypeptides does not occur in the dark (Hoober, 1972). Upon exposure of cells to white light, the synthesis and accumulation of the major thylakoid polypeptides paralleled the production of chlorophyll (Eytan and Ohad, 1972; Hoober, unpublished results). Figure 8 shows the effect on the synthesis of these polypeptides when cells were transferred from white light to light above 650 nm. At various times after the transfer, samples of the cell culture were labeled with [^{3}H]leucine, since the pattern of radioactivity for total protein provided by this amino acid after electrophoresis (see fig. 5B) revealed the state of synthesis of TMF*b* and TMF*c* more clearly than did that provided by [^{3}H]arginine (Hoober and Stegeman, 1973). Synthesis of these polypeptides in cells illuminated with light above 650 nm declined over a period of two to three hours, until the pattern obtained after three hours became the same, with respect to the labeling of these polypeptides, as was found with cells labeled in the dark without exposure to light. But synthesis of other proteins, as indicated by the incorporation of [^{3}H]leucine, was not affected by the change in the quality of light.

Within the region of 600 to 650 nm lies one of the two strong absorption maxima for protochlorophyll, the other occurring in the blue region of the spectrum (Koski et al., 1951). The absorption of light at either maximum is sufficient to achieve conversion of protochlorophyll to chlorophyll (Koski et al., 1951). Therefore, the inhibition of chlorophyll synthesis, as shown in figure 7, when the cut-off was shifted from 600 to 650 nm, was most likely the result of eliminating the wavelengths of light effective in the photoreduction of protochlorophyll. The converse experiment was done, in which dark-grown cells were illuminated only with light transmitted through an interference filter (half-band width, 15 nm) with maximal transmission at 628 nm. This filter provided light near the maximum in the red region of the action spectrum for chlorophyll synthesis in *C. reinhardtii* (Ohad, 1975). Variables such as the intensity of light and effect of

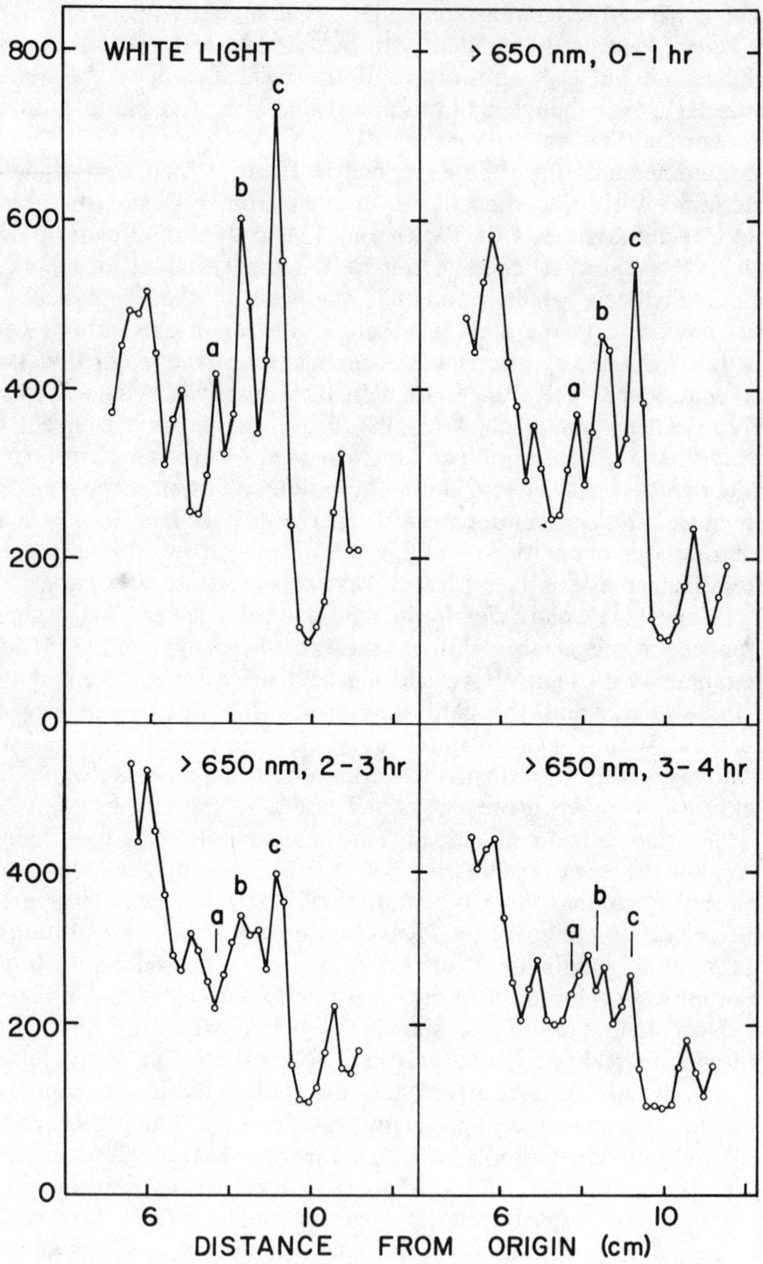

800
600
400
200
0
WHITE LIGHT
> 650 nm, 0-1 hr
> 650 nm, 2-3 hr
> 650 nm, 3-4 hr
a
b
c
6
10
DISTANCE FROM ORIGIN (cm)

illuminating cells first with white light followed by 628 nm light have not been investigated, but in preliminary experiments cells given only light transmitted by the interference filter greened at 50 to 70% of the rate measured for cells illuminated with white light.

The amount of synthesis of TMF*b* and TMF*c* was determined in cells illuminated with light transmitted by the interference filter (maximum transmission, 628 nm) and compared to that in other cells either kept in the dark or exposed to white light. Figure 9 shows that light transmitted by this filter was nearly as effective as white light in eliciting synthesis of the membrane polypeptides. This result was confirmed by Ohad (1975), who found that the action spectrum for synthesis of the major membrane polypeptides coincided with that for chlorophyll synthesis, with a maximum at about 630 nm. Therefore, light between 600 and 650 nm is required for the synthesis of TMF*b* and TMF*c* as well as for the synthesis of chlorophyll. These results suggest that the photoreduction of protochlorophyll, an event that occurs within the chloroplast, does indeed control synthesis of the major membrane polypeptides.

Synthesis of TMF*b* and TMF*c* was not inhibited when cells were treated with chloramphenicol (Hoober, 1970; Stegeman and Hoober, 1974). Moreover, after several hours in the presence of this antibiotic, the synthesis of the membrane polypeptides was released from control by protochlorophyll. In experiments similar to that described in figure 8, repression of synthesis of these polypeptides failed to occur when chloramphenicol-treated cells were transferred from white light to light above 650 nm (Hoober and Stegeman, 1973). The results of such an experiment, in which cells were labeled with [^{3}H]arginine are shown in figure 10. In control cells illuminated with light above 600 nm, synthesis of TMF*b* and TMF*c* proceeded as usual during greening (fig. 10A). If such cells were transferred to light above 650 nm, an inhibition of the synthesis of these polypeptides resulted (Hoober and Stegeman, 1973; also fig. 8). But when chloramphenicol-

---

Fig. 8. Extent of synthesis of TMF*b* and TMF*c* after illumination was changed from white light to light above 650 nm. Etiolated cells of *C. reinhardtii* y-1 were exposed to light from an incandescent lamp for 3 hours, and then a light filter, which transmitted light above 650 nm (described under fig. 7), was inserted into the light beam. Portions of the culture were labeled with [^{3}H]leucine (1 Ci/mmole) as described under fig. 5B during the last hour in white light, and the first, third, and fourth hours in light above 650 nm. Total cellular protein was subjected to electrophoresis. The region of the gels containing TMF*b* and TMF*c* were sliced into 2-mm sections, and the patterns of radioactivity were determined.

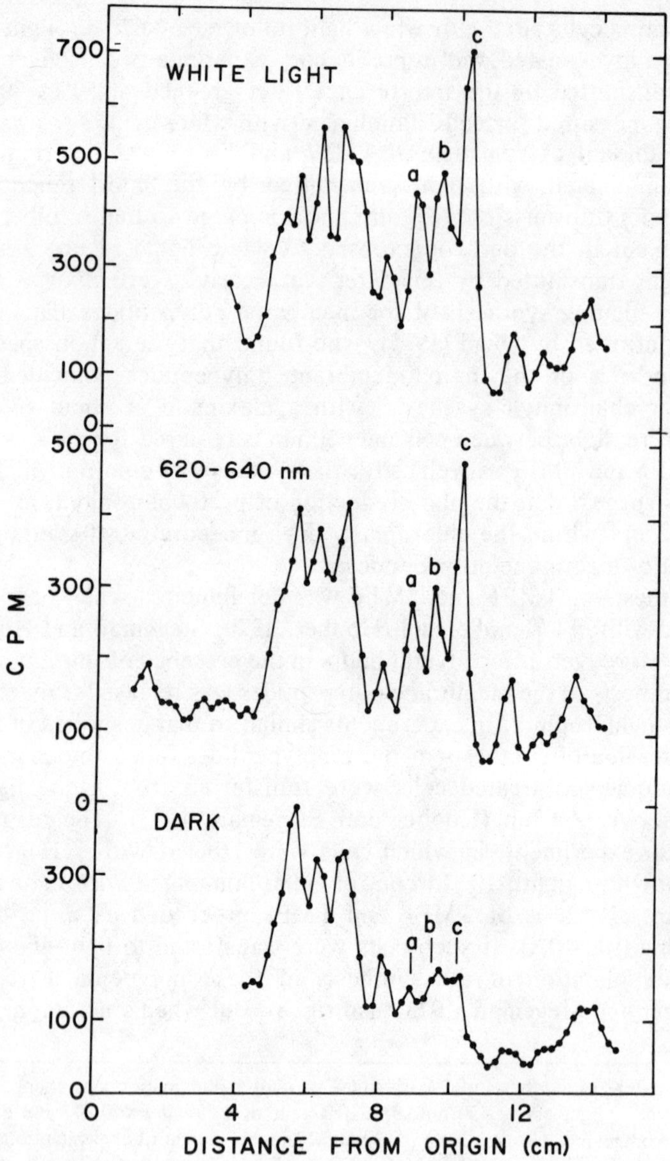

Fig. 9. Effect of light of 620 to 640 nm on the synthesis of TMF*b* and TMF*c*. Etiolated cells of *C. reinhardtii* y-1 were either maintained in the dark, exposed to white light (15 J $m^{-2}s^{-1}$) from a fluorescent lamp, or exposed to light (6 J $m^{-2}s^{-1}$) transmitted by an interference filter (maximum transmission at 628 nm, half-band width 15 nm). After 4.5 hours the cells were labeled with [^{3}H]leucine (1 Ci/mmole) as described under fig. 5B for 80 min. Total protein was subjected to electrophoresis. Portions of the gels were sliced into 2-mm sections and the pattern of radioactivity was determined.

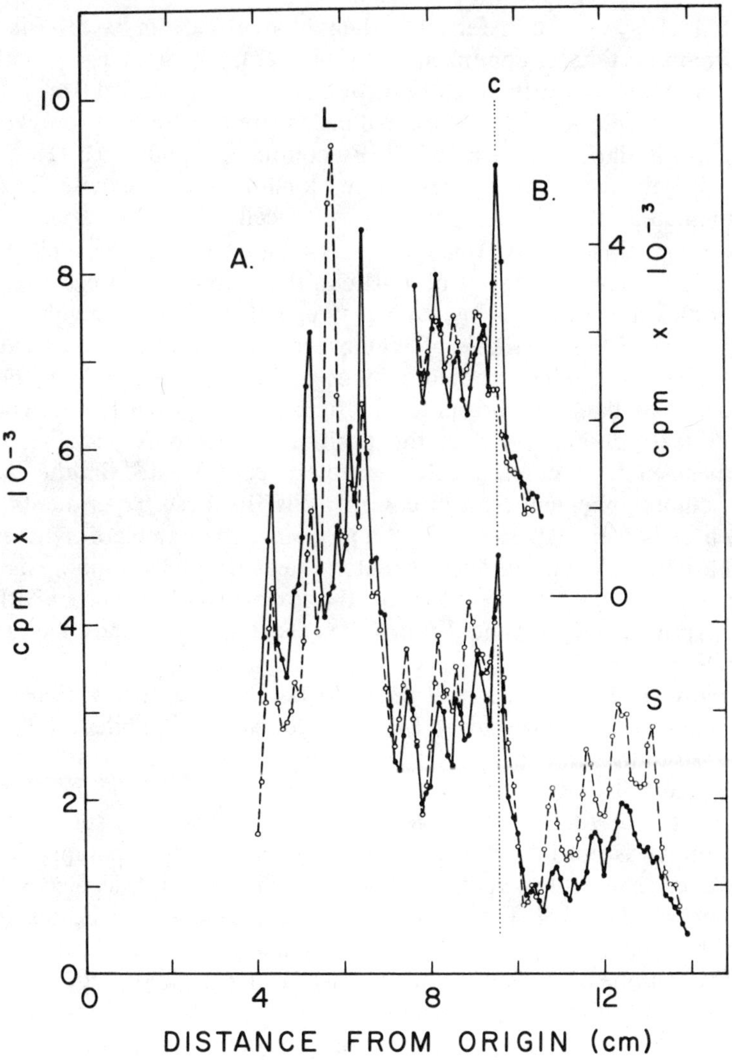

Fig. 10. Effect of chloramphenicol (200 μg/ml) on the synthesis of TMF*c*. One sample of etiolated cells of *C. reinhardtii* y-1 was treated with chloramphenicol, while a second served as a control. After 4.5 hours of exposure to light from incandescent lamps, chloramphenicol-treated and control cells were provided light above 650 nm and above 600 nm, respectively, as described under fig. 7. [^{3}H]Arginine (26 Ci/mmole) was added 30 min later to 1 μCi/ml, and the cells were labeled for 1 hour. *A.* Total cellular protein was subjected to electrophoresis. *B.* Cells were broken by sonication and centrifuged at 204,000 × g_{av} for 2 hours at 2°. Soluble proteins were subjected to electrophoresis. The patterns of radioactivity were determined for the region of the gels containing TMF*c*. O, control cells; ●, chloramphenicol-treated cells. (From Hoober and Stegeman, 1973).

treated cells were transferred to light above 650 nm, synthesis of TMF*c*, in particular, continued at a rate at least as great as in control cells, although no further synthesis of chlorophyll occurred (fig. 10A). Even when chloramphenicol-treated cells were transferred from white light to the dark, synthesis of TMF*c* continued uninhibited (Hoober and Stegeman, 1973). Under these conditions the excess TMF*c* accumulated in the soluble fraction of the cells (fig. 10B).

These results suggest that protein synthesis in the chloroplast in some way affects synthesis of TMF*c* in the cytoplasm. If chloroplast ribosomes are involved in making a regulatory protein involved in repression of the synthesis of membrane polypeptides, cells treated with chloramphenicol should resume synthesis of TMF*c* without exposure to light. As presented in figure 11, a polypeptide, which after electrophoresis was in the position expected for TMF*c*, was synthesized to a much greater extent in cells treated in the dark with chloramphenicol than in control cells. In these experiments, in which cells were labeled with [^{3}H]arginine, the extent of synthesis of TMF*b* could not be clearly determined after electrophoresis of total cellular protein because of the extensive labeling of other polypeptides that migrated through the gels at approximately the same rate.

Chlorophyll itself does not seem to be required for synthesis of the membrane polypeptides. Several agents that inhibited, in cells exposed to light, the production of chlorophyll at a stage earlier than protochlorophyllide did not interfere with synthesis of the membrane polypeptides (Hoober and Stegeman, 1973). These results suggest that chlorophyll is not an inducer of the synthesis of these polypeptides. However, since photoreduction of protochlorophyll was necessary for the synthesis of TMF*b* and TMF*c*, the evidence indicates that protochlorophyll is an inhibitor of their synthesis. And from the affects of chloramphenicol it was concluded that a protein is required for the inhibitory action of protochlorophyll to be expressed. In this broad outline form the regulatory mechanism resembles the corepressor-repressor mechanisms described for bacteria (Epstein and Beckwith, 1968; Martin, 1969; Chadwick et al., 1970).

Although the above results suggest that synthesis of TMF*b* and TMF*c* is controlled by the level of precursors of chlorophyll, the accumulation of chlorophyll, in turn, is dependent upon the production of these polypeptides. After etiolated cells are exposed to light, a lag is observed in both the accumulation of chlorophyll and in the synthesis of TMF*b* and TMF*c*, with the maximal rate of chlorophyll

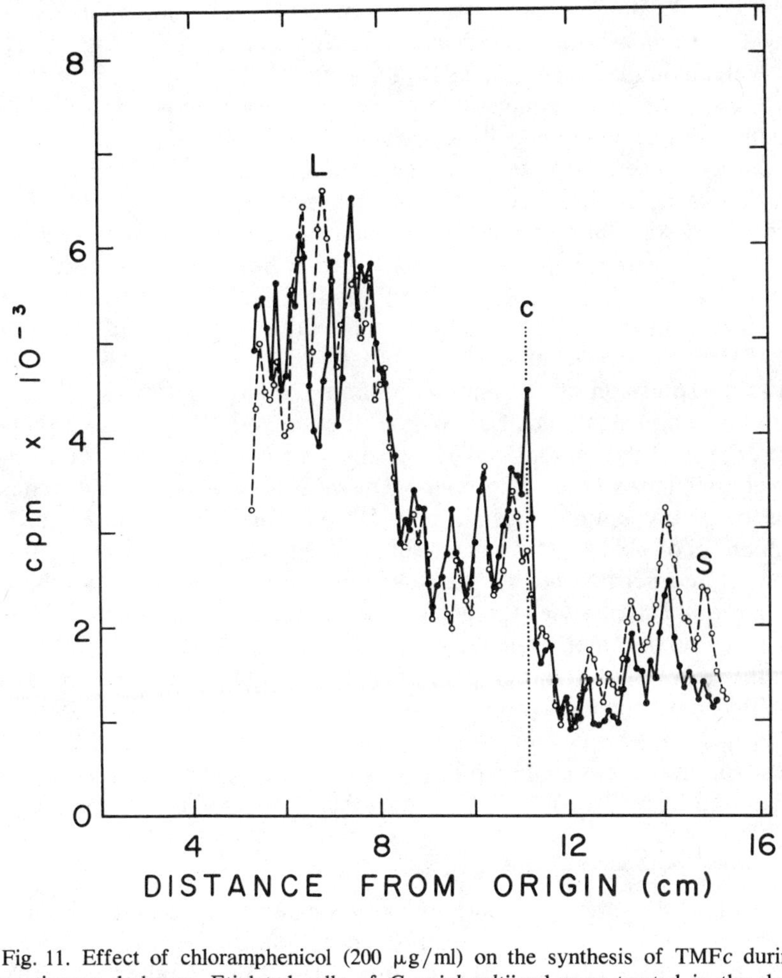

Fig. 11. Effect of chloramphenicol (200 μg/ml) on the synthesis of TMF*c* during continuous darkness. Etiolated cells of *C. reinhardtii* y-1 were treated in the dark with or without chloramphenicol for 6 hours. [^{3}H]Arginine then was added, and the cells were labeled for an additional hour. Total cellular protein was subjected to electrophoresis. O, control cells; ●, chloramphenicol-treated cells. (From Hoober and Stegeman, 1973).

synthesis achieved concomitantly with the maximal rate of synthesis of these polypeptides (Hoober, unpublished data). As greening proceeds, the amounts of chlorophyll and of these polypeptides in thylakoid membranes increase in parallel (Eytan and Ohad, 1970, 1972). If cycloheximide is added to cell cultures to inhibit cytoplasmic protein synthesis during greening, chlorophyll synthesis stops abruptly but resumes immediately when the drug is removed (Hoober et al., 1969). Although these results suggest a specific association between chlorophyll and the polypeptides, the existence and type of this association are not known. But both types of substances are required for membrane assembly, and the synthesis of each appears to be governed by synthesis of the other. Thus, an explanation for the lag when etiolated cells are placed in the light may be that this is the time required for cells to achieve a significant rate of synthesis of membrane polypeptides after derepression by the photoreduction of protochlorophyll.

As mentioned above (fig. 6), the site suggested by these results for control of the synthesis of TMF*b* and TMF*c* is at the transcriptional level, presumably on chloroplast DNA. Localization of the genes for these polypeptides to chloroplast DNA has not been accomplished. Inhibitors of prokaryotic RNA synthesis, such as rifampicin, streptolydigin and acriflavine, do not selectively inhibit synthesis of these membrane polypeptides (Stegeman and Hoober, 1974). Whether RNA synthesis in the chloroplast is refractory to the action of these inhibitors or the genes for these polypeptides are not contained in chloroplast DNA is not known. A genetic analysis would be the most satisfactory approach to this question, but mutant strains, in which the structural gene for one of these polypeptides is affected, have not been identified. The control mechanism proposed must therefore remain circumstantial.

ACKNOWLEDGEMENTS

Work in the authors' laboratory was supported by grants GB-8031 and GB-32411 from the National Science Foundation. Dr. Stegeman was the recipient of National Institutes of Health Postdoctoral Fellowship 5-F02-GM52669, 1972–74.

LITERATURE CITED

Anderson, J. M., and R. P. Levine. 1974. Membrane polypeptides of some higher plant chloroplasts. Biochim. Biophys. Acta 333:378–87.

Bar-Nun, S., D. Wallach, and I. Ohad. 1972. Biogenesis of chloroplast membranes.

X. Changes in the photosynthetic specific activity and the relationship between the light harvesting system and photosynthetic electron transfer chain during greening of *Chlamydomonas reinhardi* y-1 cells. Biochim. Biophys. Acta 267:138–48.

Bishop, N. I. 1971. Photosynthesis: The electron transport system of green plants. Ann. Rev. Biochem. 40:197–226.

Boardman, N. K. 1968. The photochemical systems of photosynthesis. Adv. Enzymol. Relat. Areas Mol. Biol. 30:1–79.

Boardman, N. K. 1970. Physical separation of the photosynthetic photochemical systems. Ann. Rev. Plant Physiol. 21:115–40.

Bogorad, L. 1966. The biosynthesis of chlorophyls, pp. 481–510. *In* L. P. Vernon and G. R. Seely (eds.), The chlorophylls. Academic Press, New York.

Borisov, A. Yu., and V. I. Godik. 1973. Excitation energy transfer in photosynthesis. Biochim. Biophys. Acta 301:227–48.

Boynton, J. E., N. W. Gillham, and J. F. Chabot. 1972. Chloroplast ribosome deficient mutants in the green alga *Chlamydomonas reinhardi* and the question of chloroplast ribosome function. J. Cell Sci. 10:267–305.

Branton, D., and R. B. Park. 1967. Subunits in chloroplast lamellae. J. Ultrastruct. Res. 19:283–303.

Chadwick, P., V. Pirrotta, R. Steinberg, N. Hopkins, and M. Ptashne, 1970. The λ and 434 phage repressors. Cold Spring Harbor Symp. Quant. Biol. 35:283–94.

Cheniae, G. M. 1970. Photosystem II and O_2 evolution. Ann. Rev. Plant Physiol. 21:467–98.

Egan, J. M., Jr., D. Dorsky, and J. A. Schiff. 1974. Action spectra for various processes associated with light induced chloroplast development in *Euglena gracilis* var. bacillaris. Plant Physiol. 53 (Supplement): 4.

Epstein, W., and J. R. Beckwith. 1968. Regulation of gene expression. Ann. Rev. Biochem. 37:411–36.

Eytan, G., and I. Ohad. 1970. Biogenesis of chloroplast membranes. VI. Cooperation between cytoplasmic and chloroplast ribosomes in the synthesis of photosynthetic lamellar proteins during the greening process in a mutant of *Chlamydomonas reinhardi* y-1. J. Biol. Chem. 245:4297–4307.

Eytan, G., and I. Ohad. 1972. Biogenesis of chloroplast membranes. VIII. Modulation of chloroplast lamellae composition and function induced by discontinuous illumination and inhibition of ribonucleic acid and protein synthesis during greening of *Chlamydomonas reinhardi* y-1 mutant cells. J. Biol. Chem. 247:122–29.

Fairbanks, G., T. L. Steck, and D. F. H. Wallach. 1971. Electrophoretic analysis of the major polypeptides of the human erythrocyte membrane. Biochemistry 10:2606–17.

Goldberg, I., and I. Ohad. 1970. Biogenesis of chloroplast membranes. IV. Lipid and pigment changes during synthesis of chloroplast membranes in a mutant of *Chlamydomonas reinhardi* y-1. J. Cell Biol. 44:563–71.

Goodenough, U. W., and L. A. Staehelin. 1971. Structural differentiation of stacked and unstacked chloroplast membranes. J. Cell Biol. 48:594–619.

Gornall, A. G., C. J. Bardawill, and M. M. David. 1949. Determination of serum proteins by means of the biuret reaction. J. Biol. Chem. 177:751–66.

Hoober, J. K. 1970. Sites of synthesis of chloroplast membrane polypeptides in *Chlamydomonas reinhardi* y-1. J. Biol. Chem. 245:4327–34.

Hoober, J. K. 1972. A major polypeptide of chloroplast membranes of *Chlamydomonas reinhardi.* Evidence for synthesis in the cytoplasm as a soluble component. J. Cell Biol. 52:84–96.

Hoober, J. K., and G. Blobel. 1969. Characterization of the chloroplastic and cytoplasmic ribosomes of *Chlamydomonas reinhardi.* J. Mol. Biol. 41:121–38.

Hoober, J. K., and W. J. Stegeman. 1973. Control of the synthesis of a major polypeptide of chloroplast membranes in *Chlamydomonas reinhardi.* J. Cell Biol. 56:1–12.

Hoober, J. K., P. Siekevitz, and G. E. Palade. 1969. Formation of chloroplast membranes in *Chlamydomonas reinhardi* y-1: Effects of inhibitors of protein synthesis. J. Biol. Chem. 244:2621–31.

Jennings, R. C., and G. Eytan. 1973. Biogenesis of chloroplast membranes. XIV. Inhomogeneity of membrane protein distribution in photosystem particles obtained from *Chlamydomonas reinhardi* y-1. Arch. Biochem. Biophys. 159:813–20.

Kahn, A., and O. F. Nielsen. 1974. Photoconvertible protochlorophyll(ide)$_{635/650}$ *in vivo:* A single species or two species in dynamic equilibrium? Biochim. Biophys. Acta 333:409–14.

Kahn, A., N. K. Boardman, and S. W. Thorne. 1970. Energy transfer between photochlorophyllide molecules: Evidence for multiple chromophores in the photoactive protochlorophyllide-protein complex *in vivo* and *in vitro.* J. Mol. Biol. 48:85–101.

Ke, B. 1973. The primary electron acceptor of photosystem I. Biochim. Biophys. Acta 301:1–33.

Kellems, R. E., and R. A. Butow. 1974. Cytoplasmic type 80 S ribosomes associated with yeast mitochondria. III. Changes in the amount of bound ribosomes in response to changes in metabolic state. J. Biol. Chem. 249:3304–10.

Kellems, R. E., V. F. Allison, and R. A. Butow. 1974. Cytoplasmic type 80 S ribosomes associated with yeast mitochondria. II. Evidence for the association of cytoplasmic ribosomes with the outer mitochondrial membrane *in situ.* J. Biol. Chem. 249: 3297–3303.

Klein, S. M., and L. P. Vernon. 1974. Protein composition of spinach chloroplasts and their photosystem I and photosystem II subfragments. Photochem. Photobiol. 19:43–49.

Koski, V. M., C. S. French, and J. H. C. Smith. 1951. The action spectrum for the transformation of protochlorophyll to chlorophyll *a* in normal and albino corn seedlings. Arch. Biochem. Biophys. 31:1–17.

Levine, R. P., W. G. Burton, and H. A. Duram. 1972. Membrane polypeptides associated with photochemical systems. Nature New Biol. 237:176–77.

Martin, R. G. 1969. Control of gene expression. Ann. Rev. Genetics 3:181–216.

Matsuda, Y., T. Kikuchi, and M. R. Ishida. 1971. Studies on chloroplast development in *Chlamydomonas reinhardtii.* Effect of brief illumination on chlorophyll synthesis. Plant Cell Physiol. 12:127–35.

McLeod, G. C., G. A. Hudock, and R. P. Levine. 1963. The relation between pigment concentration and photosynthetic capacity in a mutant of *Chlamydomonas reinhardi,* pp. 400–408. *In* Photosynthetic mechanisms of green plants. Publication 1145. National Acad. Sci.-National Res. Council, Washington, D.C.

Nielsen, O. F., and A. Kahn. 1973. Kinetics and quantum yield of photoconversion of protochlorophyll(ide) to chlorophyll(ide) *a*. Biochim. Biophys. Acta 292:117–29.

Ohad, I. 1975. Cytoplasm-chloroplast interrelationship during biogenesis of chloroplast membranes in *Chlamydomonas reinhardi*. *In* S. Puiseux-Dao (ed.), Nucleocytoplasmic relationships during cell morphogenesis in some unicellular organisms. Elsevier Publishing Co., Amsterdam (in press).

Ohad, I., P. Siekevitz, and G. E. Palade. 1967a. Biogenesis of chloroplast membranes. I. Plastid dedifferentiation in a dark-grown algal mutant (*Chlamydomonas reinhardi*). J. Cell Biol. 35:521–52.

Ohad, I., P. Siekevitz, and G. E. Palade. 1976b. Biogenesis of chloroplast membranes. II. Plastid differentiation during greening of a dark-grown algal mutant (*Chlamydomonas reinhardi*). J. Cell Biol. 35:553–84.

Rebeiz, C. A., and P. A. Castelfranco. 1973. Protochlorophyll and chlorophyll biosynthesis in cell-free systems from higher plants. Ann. Rev. Plant Physiol. 24:129–72.

Sadler, D. M., M. Lefort-Tran, and M. Pouphile. 1973. Structure of photosynthetic membranes of *Euglena* using x-ray diffraction. Biochim. Biophys. Acta 298:620–29.

Singer, S. J. 1974. The molecular organization of membranes. Ann. Rev. Biochem. 43:805–33.

Stegeman, W. J., and J. K. Hoober. 1974. Mitochondrial protein synthesis in *Chlamydomonas reinhardtii* y-1. Polypeptide products of mitochondrial transcription and translation *in vivo* as revealed by selective labeling with [^{3}H]leucine. J. Biol. Chem. 249:6866–73.

Torres-Pereira, J., R. Mehlhorn, A. D. Keith, and L. Packer. 1974. Changes in membrane lipid structure of illuminated chloroplasts—studies with spin-labeled and freeze-fractured membranes. Arch. Biochem. Biophys. 160:90–99.

Vernon, L. P. 1960. Spectrophotometric determination of chlorophylls and pheophytins in plant extracts. Anal. Chem. 32:1144–50.

Wallach, D., S. Bar-Nun, and I. Ohad. 1972. Biogenesis of chloroplast membranes. IX. Development of photophosphorylation and proton pump activities in greening *Chlamydomonas reinhardi* y-1 as measured with an open-cell preparation. Biochim. Biophys. Acta 267:125–37.

Wang, A. Y.-I., and L. Packer. 1973. Mobility of membrane particles in chloroplasts. Biochim. Biophys. Acta 305:488–92.

Wolken, J. J., and A. D. Mellon. 1956. The relationship between chlorophyll and the carotenoids in the algal flagellate, *Euglena*. J. Gen. Physiol. 39:675–85.

Zacharius, R. M., T. E. Zell, J. H. Morrison, and J. J. Woodlock. 1969. Glycoprotein staining following electrophoresis on acrylamide gels. Anal. Biochem. 30:148–52.

RUTH SAGER

Patterns of Inheritance of Organelle Genomes: Molecular Basis and Evolutionary Significance

7

Chloroplasts and mitochondria contain unique genomes located in the unique DNAs of these organelles (review in Sager, 1972a). Chloroplast genes were discovered in the decade following the rediscovery of Mendel's laws. Correns (1909) described the maternal inheritance of a chloroplast mutation in the four o'clock flower, *Mirabilis*, and Baur (1909) reported the biparental but non-Mendelian inheritance of a chloroplast mutation in *Pelargonium*. Since that time, organelle genes have been identified primarily by their non-Mendelian behavior in crosses, and also by the high frequency of segregation they show in clonal growth of hybrids.

In separate but related studies, high molecular weight double-stranded DNAs of unique base composition were identified in chloroplasts (Sager and Ishida, 1963; Chun et al., 1963) and in mitochondria (Luck and Reich, 1964). Subsequently, the mitochondrial DNA was shown to be the carrier of the mitochondrial genome (Mounolou et al., 1966; Michaelis et al., 1973), and the chloroplast genome of *Chlamydomonas* was associated with chloroplast DNA (Sager, 1972a; Schlanger and Sager, 1974a,b). These findings have been generalized to other organisms, in part on the basis of experiments and in part by inference. Thus, at present, newly discovered non-Mendelian genes are being attributed to particular organelles largely on the basis of phenotype, except in yeast and in *Chlamydomonas*, where linkage can be tested directly.

Department of Biological Sciences, Hunter College of the City University of New York.

The possibility remains open that cells may contain additional classes of cytoplasmic DNAs containing unique coding sequences. Cytoplasmic genes that appear unlinked to mitochondrial genes have been described in yeast (e.g., Lacroute, 1971), and cytoplasmic DNAs of unknown function have been reported (Clark-Walker, 1973; Meinke et al., 1973).

In the organisms so far studied, the pattern of non-Mendelian inheritance has been largely uniparental. The occurrence of biparental inheritance, foreshadowed by the findings of Baur (1909), has been documented in a few higher plants, of which the most thoroughly investigated are *Oenothera* (Renner, 1936; Schotz, 1954) and *Pelargonium* (Tilney-Bassett, 1970a; 1970b; 1973). Biparental inheritance, when it occurs, seems to be influenced by, or under the control of, specific nuclear genes (Tilney-Bassett, this volume).

This article will review the present knowledge about the molecular basis of inheritance of the chloroplast genome in *Chlamydomonas*, and then discuss the recent findings in yeast and in *Pelargonium* (presented elsewhere in this volume) in the light of our results. We have obtained strong evidence that the inheritance of the chloroplast genome is controlled by a DNA modification-restriction system analogous to those described in bacteria (Boyer, 1974). We will propose that similar modification-restriction systems regulate the transmission of organelle DNAs in other organisms.

THE MOLECULAR BASIS OF MATERNAL INHERITANCE IN *CHLAMYDOMONAS*

Chlamydomonas reinhardi is heterothallic with two mating types, determined by the nuclear alleles mt^+ and mt^-, and isogamous, the gametes of both mating types being of equal size. In higher plants simple physical exclusion of male cytoplasm from the fertilized egg has long been postulated as the effective basis of maternal inheritance. Clearly, in plants showing biparental inheritance, simple exclusion is not a sufficient explanation; and as discussed below, it may not be sufficient in any system. In *Chlamydomonas* the gametes fuse completely in zygote formation, so that the two parents contribute equal amounts of cytoplasm. Thus, events occurring after zygote formation must determine the pattern of maternal inheritance.

Three lines of experimentation carried out in our laboratory have provided evidence on the molecular basis of uniparental inheritance of chloroplast DNA and the associated chloroplast genome in *Chlamydomonas:*

1. Action of UV irradiation and of inhibitors of protein synthesis upon the pattern of chloroplast gene transmission (Sager and Ramanis, 1967, 1973).

2. Effects of mutations of two nuclear genes, *mat−1* and *mat−2*, upon the pattern of chloroplast gene transmission (Sager and Ramanis, 1974).

3. Behavior in zygotes of chloroplast DNAs, differentially labeled, from the two parental strains (Sager and Lane, 1972; Sager, 1972a); and effects of UV irradiation and of the *mat−1* gene upon this behavior (Schlanger and Sager, 1974b).

These lines of experimentation will now be briefly reviewed.

Action of UV Irradiation and of Inhibitors of Protein Synthesis

UV irradiation of the mt^+ (female) parent just before mating increases the frequency of exceptional zygotes from the spontaneous level of about 0.1% in our strains up to 40–100%, depending upon the UV dose (Sager and Ramanis, 1967). Exceptional zygotes are those that transmit chloroplast genes from the mt^- (male) parent to progeny in crosses; the term includes biparental zygotes, in which chloroplast genes from both parents are transmitted, and paternal zygotes, in which only those from the mt^- parent appear in the progeny. In some strains of this species spontaneous frequencies of exceptional zygotes as high as 20% have been reported (Gillham, 1969).

The ratio of maternal:biparental:paternal zygotes is influenced by the UV dose, and furthermore the UV effect is partially photoreactivable (Sager and Ramanis, 1967). UV irradiation of the mt^- gametes has very little if any effect upon the pattern of transmission, and irradiation of the zygotes after fusion is highly lethal. The UV dose that gives about 50% biparental zygotes (with photoreactivation) and little if any lethality, is a dose that gives about 10% survival of unmated irradiated (not photoreactivated) gametic controls. Thus, viability is restored when irradiated gametes are mated with unirradiated cells of the opposite mating type.

The UV experiments demonstrate that a factor produced by the mt^+ (female) parent is required for maternal inheritance, i.e., for loss of the chloroplast genome coming from the mt^- (male).

Analogous experiments were carried out in which gametes, either mt^+ or mt^-, were pretreated with drugs that block chloroplast DNA replication, transcription, or protein synthesis (Sager and Ramanis, 1973). The pretreatments were done either for the full time of gametogenesis (24 hours) or for a six-hour period at the end of gametogenesis, during which mature gametes were treated. In general, treatments during gametogenesis were difficult to interpret, since numerous processes were simultaneously affected. However, treatments of mature gametes were quite successful, influencing the pattern of maternal inheritance but not survival, indicating that few essential processes were operative at that time. Only the results with mature gametes will be described here. Ethidium bromide (10 $\mu g/ml$) was found to produce an effect similar to that of UV: when female gametes were treated for six hours just prior to zygote formation, a great increase (about 70-fold) in exceptional zygotes resulted, whereas a similar treatment of male gametes had little effect. Rifamycin SV, erythromycin, and spiramycin were each more effective (about 30-fold increase in exceptional zygotes over controls) when the male gametes were treated than when the female gametes were treated (ca. 5-fold increase over controls). Effects of cycloheximide were inconclusive, since extensive lethality occurred at all effective concentrations.

These studies provide evidence that both parental strains contribute different but essential components to the process of maternal inheritance. The UV and ethidium bromide effects suggest that for maternal inheritance to result, transcription of DNA must occur in the female gametes; and the inhibitor studies indicate that protein synthesis must occur in the male gametes, as well as in the females.

Effects of the mat−1 and mat−2 Mutations

Two nuclear mutations were recently discovered that dramatically alter the pattern of transmission of chloroplast genes. The *mat−1* mutation was identified by the very *high* frequency of exceptional zygotes occurring in crosses that were *not* UV-irradiated. The *mat−2* mutation was identified by the very *low* frequency of exceptional zygotes occurring in crosses that *were* UV-irradiated. The principal properties conferred by these mutations are listed in table 1.

The *mat* genes are each closely linked to mating type: *mat−1* is linked to mt^- and *mat−2* is linked to mt^+. This linkage is not surprising, since the mating type locus governs the differential behavior of gametes in sexual fusion. As first demonstrated 20 years ago (Sager, 1954) and documented in several subsequent studies (Gillham, 1969; Sager, 1972), maternal inheritance of the chloroplast genome is regulated

TABLE 1

PROPERTIES OF *MAT-1* AND *MAT-2* MUTANT GENES

	Wild Type	*mat-1*	*mat-2*
1. Phenotype			
a.	Frequency of spontaneous exceptional zygotes 0.1–1.0%	High frequency of spontaneous exceptional zygotes (20–100%)	Lower than wild-type frequency of spontaneous exceptional zygotes (0.02–0.09%)
b.	UV increases frequency of exceptional zygotes, dose dependent	UV low dose further increases frequency of exceptional zygotes to 100%	UV high dose slightly increases frequency of exceptional zygotes to about 10%
c.	Both biparental and paternal among exceptional zygotes (paternal rare)	Both biparental and paternal among exceptional zygotes (paternal frequent)	Only biparental; no paternal zygotes
d.		Expressed only in zygotes with *mat-1* mt^- parents	Expressed only in zygotes with *mat-2* mt^+ parents
2. Inheritance		Linked to mt^-	Linked to mt^+
3. Zygote viability	Close to 100%	Variable	Close to 100%

Sager and Ramanis, 1974

by the mating type locus. As proposed by Gillham (1969), it seems likely that a cluster of genes is located in the mating type region, governing different steps in the complex set of events that characterize the sexual cycle.

The *mat–1* gene in the mt^- parent increases the frequency of exceptional zygotes, both biparental and paternal, up to 100%. Thus, it is at least as effective as UV irradiation of the mt^+ parent. In crosses in which the effect of *mat–1* is about 50–60%, the yield of exceptional zygotes can be further increased by UV irradiation of the mt^+ gametes, indicating that the effects can be additive and therefore that *mat–1* and UV act at different steps in the over-all process.

The *mat–2* mutation in the mt^+ parent effectively lowers the frequency of exceptional zygotes. The spontaneous frequency is already quite low, i.e., about 0.1% in the wild-type, and the frequencies seen in the presence of *mat–2* are about 20-fold lower. The effect of the *mat–2* mutation is seen more strikingly after UV irradiation.

Instead of frequencies in the range of 50-60% exceptionals, as in the wild-type controls, the highest frequencies after UV irradiation of the *mat−2* parent were only about 10%, and higher doses of UV did not increase the yield. Also, paternal zygotes have been found only rarely and at the highest UV doses, among the exceptionals recovered in the presence of *mat−2*. Survival curves examining the UV sensitivity of vegetative cultures carrying *mat−2* show that the mutation does not confer increased UV resistance in terms of viability.

In crosses between strains carrying the *mat−1* and *mat−2* mutations, *mat−1* is expressed in the zygote regardless of the presence of *mat−2*. Also paternal zygotes are found in crosses of *mat−2* × *mat−1*, but none in crosses of *mat−2* × *mat−2*$^{+}$. Thus, the principal effects of the *mat−2* mutation are not seen when both mutant genes are acting in the same zygote.

We have previously proposed a model (Sager and Ramanis, 1973) to explain the mechanism of maternal inheritance in *Chlamydomonas*. The model is based on our hypothesis that maternal inheritance of cytogenes is regulated by modification and restriction of chloroplast DNA occurring in zygotes. In this model an inactive modification enzyme (M) is postulated to be present in the chloroplast of *mt*$^{+}$ gametes, and an inactive restrictive enzyme (R) in the chloroplast of *mt*$^{-}$ gametes. Both enzymes are postulated to be activated at the time of mating; their effects are seen only in zygotes. The fact, noted in light microscope (Sager, unpublished) and electron microscope (Friedmann et al., 1968; Cavalier-Smith, 1970) studies, that chloroplast fusion occurs six hours after zygote formation provides time for modification to occur within the chloroplast from the female parent, and degradation of chloroplast DNA to occur within the plastid from the male parent, while both coexist within the zygote.

Before discussing the proposed functions of the *mat* genes in terms of this model, we will consider the evidence from physical studies of chloroplast DNA that bears on the subject.

Differential Fates of Chloroplast DNAs in Zygotes

The different fates of chloroplast DNAs coming from the two parents were followed by prelabeling the parental DNAs differentially with $^{15}N/^{14}N$ in some experiments (Sager and Lane, 1969; 1972) and with $^{3}H/^{14}C$ adenine in others (Schlanger and Sager, 1974a,b). Two remarkable events were discovered to occur soon after zygote formation. (1) Chloroplast DNA from the female parent was found to undergo

TABLE 2

BUOYANT DENSITIES OF CHLOROPLAST DNAS FROM ZYGOTES

Cross[a]	Observed density[b]	Computed[c]	Discrepancy
$^{14}N \times {}^{14}N$	1.690 ± 0.0009	1.690	0
$^{14}N \times {}^{15}N$	1.692 ± 0.001	1.6935	0.0015
$^{15}N \times {}^{14}N$	1.698 ± 0.0009	1.7005	0.0025
$^{15}N \times {}^{15}N$	1.7005	1.704	0.0035
^{14}N vegetative cells	1.695 ± 0.0005		
^{14}N gametes	1.695 ± 0.0005		
^{15}N vegetative cells	1.709 ± 0.0005		
^{15}N gametes	1.709 ± 0.0005		

Source: Sager and Lane, 1972, and Sager, unpublished data.

[a] Female (mt^+) parent given first

[b] Data from five or more preparations except $^{15}N \times {}^{15}N$ from two preparations

[c] Computation assumes one round of semi-conservative replication: one strand conserved and one strand replicated from equal pools of ^{14}N and ^{15}N precursors in reciprocal crosses; both strands modified (0.005 g/cm^3 lighter).

a density shift, within six hours after zygote formation, from the buoyant density of 1.695 g/cm^3 in CsCl, characteristic of vegetative cells and of gametes, to the lighter density of 1.690 g/cm^3 seen only in zygotes. (2) Chloroplast DNA from the male parent was found to disappear from CsCl gradients within the first six hours after zygote formation. After replication, the chloroplast DNA returned to its usual buoyant density.

What causes the density shift? We infer that it results from covalent addition of some component with a lighter buoyant density in CsCl than chloroplast DNA, covalent because the density shift withstands detergent extraction with sodium lauryl sulfate, sarkosyl, Triton X-100, and deoxycholate, as well as enzymatic digestion with pronase, ribonucleases, and amylase. If the density shift results from methylation, as in bacterial modification, about 5% methylation would be required (Szybalski and Szybalski, 1971) and should be detectable; experiments are in progress to look for it.

In the reciprocal crosses between parents grown in ^{15}N-medium and in ^{14}N-medium, a single chloroplast peak was found in DNA from zygotes sampled at 6 hours and at 24 hours after mating: at 1.692 g/cm^3 in the $^{14}N(mt^+) \times {}^{15}N(mt^-)$ cross, and 1.698 g/cm^3 in the reciprocal cross. As listed in table 2, the densities were similar to, but not identical with, those from $^{15}N \times {}^{15}N$ and from $^{14}N \times {}^{14}N$ crosses. The $^{14}N \times {}^{15}N$ average value was about 0.002 g/cm heavier than the $^{14}N \times {}^{14}N$ average, and similarly, the $^{15}N \times {}^{14}N$ value was about 0.002 g/cm^3 lighter than the $^{15}N \times {}^{15}N$ value. These small but significant differences could be the result of limited recombination, except that we have genetic evidence indicating that recombination of our markers, situated all around the map, does not occur in unirradiated crosses at a frequency that would be detectable in the DNA. Another possibility, which we favor, is that the 0.002 g/cm^3 density difference reflects either repair or partial replication in which the nucleotides come from a common $^{15}N - {}^{14}N$ pool.

Another point to be noted in table 2 is that the density shift in the $^{15}N \times {}^{15}N$ chloroplast DNA is greater than in the $^{14}N \times {}^{14}N$ DNA and similarly in the pairs of reciprocal crosses. This difference could be the result of a heavy isotope effect, in which the presence of ^{15}N influences the extent of covalent addition of whatever component is responsible for the density shift of maternal chloroplast DNA in zygotes.

Methods have been developed recently in our laboratory for examining the differential fates of parental chloroplast DNAs in zygotes by growing parental cultures with either ^{3}H-adenine or ^{14}C-adenine before gametogenesis. After zygote formation, lysis and separation of nuclear and chloroplast DNAs in preparative CsCl density gradients has given high yields of both DNA fractions. Mitochondrial DNA, banding at a buoyant density of 1.706 (Ryan et al., 1973) was rarely seen in these preparations.

The question posed was whether the transmission of chloroplast genes from the mt^- parent following UV irradiation of mt^+ gametes, or in the presence of the *mat−1* mutant allele, would be paralleled by transmission of chloroplast DNA from the mt^- parent. The results were unambiguous. Chloroplast DNAs of the mt^- (male) origin were recovered from 24-hour zygotes in both classes of experiments: those in which the mt^+ parent had been UV-irradiated before mating, and those involving crosses of $mt^+ \times mat{-}1^- \; mt^-$ strains. The genetic behavior was examined in the same experiments: transmission of male chloroplast DNA paralleled transmission of chloroplast genes from the male parent (Schlanger and Sager, 1974b).

The three lines of evidence summarized above have provided the experimental basis for our proposal that the modification and restriction of chloroplast DNA is the molecular mechanism that regulates the pattern of inheritance of the chloroplast genome. The detailed mechanism by which the modification (M) and restriction (R) enzymes are regulated is not yet well understood. However, the effects of the *mat* genes, of UV irradiation, and of the inhibitors of macromolecular synthesis have led us to postulate two factors that regulate M and R enzymes: G_1 and G_2. G_1 is produced in the mt^+ gametes at the time of mating, and activates both enzymes, M located in the chloroplast from the mt^+, and R in the chloroplast from the mt^-. G_2 is produced in mt^- gametes at the time of mating, and acts to inhibit G_1. Thus, the activation of M and R is regulated by the balance between G_1 and G_2. In our wild-type strains, G_1 is much more effective than G_2, and most zygotes show maternal inheritance. UV irradiation blocks G_1 activity, leading to the absence of both M and R. We postulate that the *mat−1* mutation increases the effectiveness of G_2 vis-à-vis G_1, whereas the *mat−2* mutation makes G_1 even more effective than it is in the wild-type.

The identities of G_1 and G_2 are not defined in this hypothesis. Since they are tightly linked to mating type, it seems likely that they are part of the modification-restriction system, and may represent controlling elements associated with the genes coding for these enzymes.

PLASTID INHERITANCE IN PELARGONIUM

Studies of plastid inheritance in higher plants have shown that most plants exhibit strict uniparental transmission, following the pattern first described by Correns (1909), whereas certain genera, such as *Oenothera* and *Pelargonium*, follow the pattern of biparental but non-Mendelian inheritance first described by Baur (1909) (see Kirk and Tilney-Bassett, 1967, for a thorough review). Intensive studies of plastid inheritance in *Oenothera* (Renner, 1936; Schotz, 1954) and in *Pelargonium* (Tilney-Bassett, 1963; 1965; 1970a; 1970b; 1973) have documented the occurrence of extensive male transmission of plastid phenotypes (i.e., plastid DNA) in these genera. The present status of the *Pelargonium* work is reviewed by Tilney-Bassett in this volume. Similarities between the *Pelargonium* and *Chlamydomonas* systems will be considered here.

In *Pelargonium*, crosses can be made between isogenic plants that differ only in the presence or absence of periclinal chimeras. The chimeric plants produce some white flowers as a result of a non-Mendelian plastid mutation present in white tissues of the germ line. In reciprocal crosses (G × W and W × G) three classes of progeny are recovered: green, variegated, and white. The frequencies of these classes, and of the green and white sectors in variegated progeny plants, provide the data for evaluating the ratios of parental inputs of normal and mutant plastids (presumably, normal and mutant plastid DNA). In studies previously reviewed (Sager, 1972b), Tilney-Bassett found that the pattern of transmission was always biparental, but that the genotype of the female parent was the principal determinant of the precise G:W ratio in reciprocal crosses. More recently, Tilney-Basset (1973) has been able to demonstrate that the major effect upon the G:W ratio is controlled by a single Mendelian gene with two alleles that can be distinguished in the cultivars (varieties) used: Dolly Varden (DV) and Flower of Spring (FS). The allele Pr_1 is proposed to control one typical pattern called Type I: most offspring are green, fewer are variegated, and fewer yet are white. Cultivar DV is considered to be homozygous $Pr_1\ Pr_1$. The cultivar FS shows a Type II pattern in which green and white seedlings are approximately equally frequent, and variegated seedlings are few. FS is considered to be a $Pr_1\ Pr_2$ heterozygote. To account for the data, the genotype $Pr_2\ Pr_2$ is considered to be lethal. The allele Pr_2 is proposed to favor the replication of the white plastids (i.e., mutant plastid DNA), and Pr_1 to favor the green plastids (i.e., normal plastid DNA). The Type II pattern is the result of half the eggs carrying the Pr_1 allele and giving rise mainly to green seedlings, and half carrying the Pr_2 allele and giving rise to white seedlings. The nuclear genotype of the male parent affects the outcome only to a minor extent.

Although Tilney-Bassett has not suggested any particular molecular model to explain his data, the formal resemblance to the *Chlamydomonas* data merits some consideration. The G:V:W plastid ratios in *Pelargonium* are formally analogous to the maternal:biparental:paternal ratios of chloroplast genes in *Chlamydomonas*. The excess of green plastids, like the excess of maternal zygotes, reflects maternal transmission, and the appearance of white plastids, like the appearance of chloroplast genes from the *mt*- parent, reflects the extent of male transmission. The *Pr* gene in *Pelargonium* seems to be the counterpart of the mat^{-1} gene in *Chlamydomonas*. Thus, Pr_1 would be equivalent to the wild-type $mat{-}1^{+}$, and Pr_2 would be the analog of the $mat{-}1^{-}$ mutant allele.

In both systems it was necessary to choose the most favorable material in which the noise level from secondary effects of other genes was minimized. In *Chlamydomonas*, for example, strains have been examined in which the spontaneous frequency of exceptional zygotes is as high as 30% (Sager, unpublished data). Similarly, in other cultivars of *Pelargonium*, the difference between Type I and

II patterns is somewhat blurred. Thus, in neither system have all the genetic factors influencing the inheritance of chloroplast DNA been identified. What is striking, however, is the similarity in the over-all patterns determined by the major genetic influences.

On this basis I consider it warranted to postulate that there is a common molecular basis in the control of plastid DNA inheritance in these two organisms, and by inference in all eukaryotic organisms with chloroplasts. The molecular basis that I propose is the modification and restriction of chloroplast DNA by site-specific enzymes similar to the methylases and endonucleases that have been described in bacteria (Smith et al., 1972; Meselson et al., 1972; Boyer, 1974). The application of this model to the *Chlamydomonas* data has been discussed above. Here I would simply point out that the secondary genetic influences upon the systems may be organism-specific, depending upon the mode of sexual differentiation (e.g., mating type, flower development, and so on) as well as the rate of attack by less specific nucleases upon DNAs following the initial cut by the restriction endonucleases.

With respect to plants showing strict maternal inheritance, Tilney-Bassett points out that "examples of purely maternal inheritance may be due to the failure of male plastids to replicate rather than their failure to enter the egg as is usually assumed" (Tilney-Bassett, 1973). If by plastid replication is meant replication of plastid DNA, then modification-restriction provides the best known mechanism to achieve *selective* replication of one DNA in the presence of its homologue, which is discarded (Sager and Kitchin, 1975).

TRANSMISSION OF MITOCHONDRIAL DNAS IN YEAST CROSSES

In recent studies of the mitochondrial genetics using mutations to drug-resistance as markers, two phenomena have been described that complicate genetic analysis: *bias* and *polarity*. According to Dujon et al. (1974), *bias* is determined by two separate and distinguishable phenomena: *asymmetry* and *polarity*. *Asymmetry* refers to the difference in frequencies of parental alleles and of recombinant classes recovered as a result of the action of nuclear genes that influence mitochondrial DNA transmission in crosses. Asymmetry affects all mitochondrial markers equally. *Polarity*, on the other hand, has only been seen in three of the six loci so far identified genetically in yeast mitochondrial DNA. *Polarity* is the difference in recovery frequencies of these alleles and of recombinants that results from action at a specific site in mitochondrial DNA called *omega.* Yeast strains fall into two classes, ω^+ and ω^-, defined by their behavior in crosses. In general, mitochondrial alleles from the ω^+ parent are found in excess in recombinant progeny and those from the ω^- parent are deficient in $\omega^+ \times \omega^-$ crosses. The polarity is not seen in $\omega^+ \times \omega^+$ or in $\omega^- \times \omega^-$ crosses. The extent of polarity has provided the basis for mapping these loci in linear order, starting at omega (ω). The closer a locus is to ω, the greater is its polarity (see Birky, this volume).

In a recent review Dujon et al. (1974) have proposed that polarity is the result of a gene conversion process, in which the mitochondrial DNA is preferentially degraded starting at ω on the ω^- molecule, and the strand is restored by copying from the intact ω^+ DNA. Differences in gene frequencies would depend upon the length of the DNA segment destroyed before repair processes took over. They do not propose a molecular basis for the enzymatic attack on ω^- DNA.

In our view polarity can be explained by the action of modification-restriction enzymes. Thus, the ω region in mitochondrial DNA would represent a recognition site for a pair of modification-restriction enzymes coded by nuclear genes. Formally, the difference between ω^+ and ω^- strains could result either from differences in nuclear gene control of modification-restriction enzyme expression, or from differences in the ω site in mitochondrial DNA. The latter possibility is supported by the finding that the ω^- property is lost in crosses: the diploid clones coming from vegetative multiplication of zygotes are ω^+ as are the four products of meiosis recovered after germination of diploid spores. This result suggests that the ω^- property is a special feature of the mitochondrial DNA of these strains that renders it vulnerable to the modification-restriction enzymes.

Perlman and Birky (1974) have suggested that differences in length of the ω^+ and ω^- molecules, as a result of a deletion in ω^+ and a consequent looping-out of a segment in the ω region, might initiate endonuclease activity, followed by exonuclease or endonuclease digestion until stopped by repair processes.

We favor the modification-restriction hypothesis primarily because the uniparental inheritance of organelle DNAs seems to be a fundamental feature of organelle genetics, of some evolutionary importance, and would therefore be likely to be regulated by similar molecular mechanisms in all organisms. Also, Bolotin et al. (1971) have reported that the polarity of transmission of mitochondrial markers in yeast crosses can be partly reversed by a low dose of UV irradiation, a finding that parallels our results with UV irradiation in *Chlamydomonas*, and suggests a common mechanism.

CONCLUDING REMARKS

The pattern of transmission of nuclear genes—equal contribution from the two parents and extensive recombination in meiosis—represents the genotypic basis of evolutionary genetics. Whereas mutation is the ultimate source of genetic variation, recombination provides the proximate genetic basis of the variability upon which natural selection operates. The science of genetics, both molecular and evolutionary, has been built upon the study of variation; and genetic stability, though of obvious consequence, has been taken for granted. Recently, Lewontin has shown that in polymorphic populations highly restricted recombination maximizes fitness at the population level (Lewontin, 1971).

In cytoplasmic genetics, however, stability plays a far more eye-catching role than in nuclear genetics. The studies of cytoplasmic gene transmission, discussed here and elsewhere (Sager, 1972) reveal uniparental transmission as the predominant mode. When biparental inheritance occurs, its frequency is found to be under nuclear gene control. In crosses between species of *Neurospora* (Reich and Luck, 1966) and of *Xenopus* (Dawid and Blackler, 1972) the inheritance of mitochondrial DNA is maternal. The mitochondrial DNA of cell hybrids, produced by interspecies somatic cell fusion, parallels that of the surviving nuclear DNA (Coon et al., 1973).

Thus, uniparental inheritance is the prevailing pattern of organelle DNA transmission in crosses. Special enzymatic mechanisms are required to maintain this pattern, since the organelle DNA of one parent must be maintained and replicated while its homolog is destroyed. The modification and restriction of DNA by site-specific enzymes provides the best-known molecular model to meet these biological requirements. Both strict uniparental and nuclear gene regulated biparental patterns can be readily accommodated within this model. The application of the modification-restriction model to other eukaryotic systems including heterochromatization and elimination of chromosomes in insects, higher plants, and mammals has recently been discussed (Sager and Kitchin, 1975).

The evolutionary significance of uniparental inheritance would seem to be primarily in the restriction of recombination. Recent evidence has shown that some subunits of key organelle enzymes including the RUdP carboxylase of chloroplasts (Wildman, this volume), cytochromes *a* and *b*, and the F_1-ATPase of mitochondria (Mahler et al., this volume) are coded by organelle DNAs and other subunits of the same enzyme are coded by nuclear DNAs. This division of labor would seem to require genetic stability for its maintenance, and the inhibition of genetic variation inherent in uniparental transmission may provide the requisite stability. The hydrophobic, membrane-bound condition of organelle-coded proteins may also require mutational stability, and this too may somehow be abetted by the uniparental transmission pattern.

In summary, the study of patterns of transmission of cytoplasmic genes has brought us to propose two generalizations:

1. that the evolutionary significance of uniparental inheritance of organelle DNAs lies in the maintenance of genetic stability by blocking recombination and perhaps by limiting allelic polymorphism as well; and
2. that the molecular basis of uniparental inheritance is similar throughout the eukaryotes, namely, the modification and restriction of organelle DNAs.

ACKNOWLEDGMENTS

This research was supported by grants from the National Institutes of Health and the American Cancer Society.

LITERATURE CITED

Baur, E. 1909. Das Wesen und die Erblichkeitsverhaltnisse der "Varietates albomarginatae hort" von *Pelargonium zonale*. Z. Verebungsl. 1:330–51.

Bolotin, M., D. Coen, J. Deutsch, B. Dujon, P. Netter, E. Petrochilo, and P. P. Slonimski. 1971. La recombinaison des mitochondries chez *Saccharomyces cerevisiae*. Bull. Inst. Pasteur. 69:215–39.

Boyer, H. W. 1974. Restriction and modification of DNA: Enzymes and substrates. Fed. Proc. 33:1125–27.

Cavalier-Smith, T. 1970. Electron microscope evidence for chloroplast fusion in zygotes of *Chlamydomonas reinhardii*. Nature (London) 228:333–35.

Chun, E. H. L., M. H. Vaughan, and A. Rich. 1963. The isolation and characterization of DNA associated with chloroplast preparations. J. Mol. Biol. 7:130–41.

Clark-Walker, G. D. 1973. Size distribution of circular DNA from petite-mutant yeast lacking ρDNA. Eur. J. Biochem. 32:263–67.

Coon, H. G., I. Horak, and I. B. Dawid. 1973. Propagation of both parental mitochondrial DNAs in rat-hamster and mouse-human hybrid cells. J. Mol. Biol. 81:285–98.

Correns, C. 1909. Vererbungsversuche mit blass (gelb) grunen und bluntblattrigen Sippen bei *Mirabilis jalapa, Urtica pilulifera* and *Lunaria annua.* Z. Vererbungsl. 1:291–329.

Dawid, I. B., and A. W. Blackler. 1972. Maternal and cytoplasmic inheritance of mitochondrial DNA in *Xenopus.* Dev. Biol. 29:152–61.

Dujon, B., P. P. Slonimski, and L. Weill. 1974. Mitochondrial genetics. IX. A model for recombination and segregation of mitochondrial genomes in *Saccharomyces cerevisiae.* Genetics 78:415–37.

Friedmann, I., A. L. Colwin, and L. H. Colwin. 1968. Fine-structural aspects of fertilization in *Chlamydomonas reinhardi.* J. Cell Sci. 3:115–28.

Gillham, N. W. 1969. Uniparental inheritance in *Chlamydomonas reinhardi.* Amer. Nat. 103:355–88.

Haberman, A., J. Heywood, and M. Meselson. 1972. DNA modification methylase activity of *Escherichia coli* restriction endonucleases K and P. Proc. Nat. Acad. Sci. USA. 69:3138–41.

Kirk, J. T. O., and R. A. E. Tilney-Bassett. 1967. The plastids: Their chemistry, structure, growth, and inheritance. W. H. Freeman, London. 608 pp.

Lacroute, F. 1971 Non-mendelian mutation allowing ureidosuccinec acid uptake in yeast. J. Bact. 106:519–22.

Lewontin, R. 1971. The effect of genetic linkage on the mean fitness of a population. Proc. Nat. Acad. Sci. USA. 68:984–86.

Luck, D. J. L., and E. Reich. 1964. DNA in mitochondria of *Neurospora crassa.* Proc. Nat. Acad. Sci. USA. 52:931–38.

Meinke, W., M. R. Hall, and D. A. Goldstein. 1973. Physical properties of cytoplasmic membrane-associated DNA. J. Mol. Biol. 78:45–56.

Meselsen, M., R. Yuan, and J. Heywood. 1972. Restriction and modification of DNA. Ann. Rev. Biochem. 41:447–66.

Michaelis, G., E. Petrochilo, and P. P. Slonimski. 1973. Mitochondrial genetics. III. Recombined molecules of mitochondrial DNA obtained from crosses between cytoplasmic petite mutants of *Saccharomyces cerevisiae:* Physical and genetic characterization. Molec. Gen. Genet. 123:51–65.

Mounolou, J. C., H. Jakob, and P. P. Slonimski. 1966. Mitochondrial DNA from yeast "petite" mutants: Specific changes of buoyant density corresponding to different cytoplasmic mutations. Biochem. Biophys. Res. Commun. 24:218–44.

Perlman, P. S., and C. W. Birky, Jr. 1974. Mitochondrial genetics in baker's yeast: A molecular mechanism for recombinational polarity and suppressiveness. Proc. Nat. Acad. Sci. USA. 71:4612–16.

Renner, O. 1936. Zur Kenntnis der nichtmendelnden Buntheit der Laubblatter. Flora (Jena) N.F. 30:218–90.

Reich, E., and D. J. L. Luck. 1966. Replication and inheritance of mitochondrial DNA. Proc. Nat. Acad. Sci. USA. 55:1600–1608.

Ryan, R. S., D. Grant, K.-S. Chiang, and H. Swift. 1973. Isolation of mitochondria and characterization of the mitochondrial DNA of *Chlamydomonas reinhardii*. J. Cell Biol. 59:297a. (Abstr.)

Sager, R. 1954. Mendelian and non-Mendelian inheritance of streptomycin resistance in *Chlamydomonas reinhardi*. Proc. Nat. Acad. Sci. USA. 40:356–63.

Sager, R. 1972a. Cytoplasmic genes and organelles. Academic Press, New York. 405 pp.

Sager, R. 1972b. Evolution of preferential transmission mechanisms in cytoplasmic genetic systems. Brookhaven Symp. 23:495–505.

Sager, R., and M. R. Ishida. 1963. Chloroplast DNA in *Chlamydomonas*. Proc. Nat. Acad. Sci. USA. 50:725–30.

Sager, R., and D. Lane. 1969. Replication of chloroplast DNA in zygotes of *Chlamydomonas*. Fed. Proc. 38:347. (Abstr.)

Sager, R., and D. Lane. 1972. Molecular basis of maternal inheritance. Proc. Nat. Acad. Sci. USA. 69:2410–13.

Sager, R., and R. Kitchin. 1975. Selective silencing of eukaryotic DNA. Science (in press).

Sager, R., and Z. Ramanis. 1967. Biparental inheritance of non-chromosomal genes induced by ultraviolet irradiation. Proc. Nat. Acad. Sci. USA. 58:931–37.

Sager, R., and Z. Ramanis. 1973. The mechanism of maternal inheritance in *Chlamydomonas:* Biochemical and genetic studies. Theor. Appl. Genet. 43:101–8.

Sager, R., and Z. Ramanis. 1974. Mutations that alter the transmission of chloroplast genes in *Chlamydomonas*. Proc. Nat. Acad. Sci. USA. 71:4698–4702.

Schlanger, G., and R. Sager. 1974a. Localization of five antibiotic resistances at the subunit level in chloroplast ribosomes of *Chlamydomonas*. Proc. Nat. Acad. Sci. USA. 71:1715–19.

Schlanger, G., and R. Sager. 1974b. Correlation of chloroplast DNA and cytoplasmic inheritance in *Chlamydomonas* zygotes. J. Cell. Biol. 63:301a (abstr.).

Schotz, F. 1954. Uber Plastidenkonkurrenz bei *Oenothera*. Planta (Berlin) 43:182–240.

Smith, J. D., W. Arber, and V. Kuhnlein. 1972. Host specificity of DNA produced by *Escherichia coli*. XIV. J. Mol. Biol. 63:1–8.

Szybalski, W., and E. H. Szybalski. 1971. Equilibrium density gradient centrifugation. Prog. Nucleic Acid Res. 2:311–54.

Tilney-Bassett, R. A. E. 1963. Genetics and plastid physiology in *Pelargonium*. Heredity 18:485–504.

Tilney-Bassett, R. A. E. 1965. Genetics and plastid physiology in *Pelargonium* II. Heredity 20:451–66.

Tilney-Bassett, R. A. E. 1970a. Genetics and plastid physiology in *Pelargonium*. III. Effect of cultivar and plastids on fertilization and embryo survival. Heredity 25:89–103.

Tilney-Bassett, R. A. E. 1970b. The control of plastid inheritance in *Pelargonium*. Genet. Res. Camb. 16:49–61.

Tilney-Bassett, R. A. E. 1973. Control of plastid inheritance in *Pelargonium* II. Heredity 30:1–13.

R. A. E. TILNEY-BASSETT

Genetics of Variegated Plants

8

Variegated plants are of so many different kinds, and are based on such a wide variety of genetic mechanisms, that I do not wish to review the whole group. Instead, I should like to limit my discussion to developments in our understanding of variegation involving normal and mutant plastids; that is to say, plants in which the green and white variegated tissues are caused by plastids of different genotype, not just different phenotype. In particular, I should like to consider variegated plants of two different origins: first, cases of nuclear gene-induced plastid mutations, and, second, studies of plastids that have originated by spontaneous mutations.

GENE-INDUCED PLASTID MUTATIONS

The Basic Behavioral Pattern

Gene-induced plastid mutations fall into a characteristic pattern of behavior (Kirk and Tilney-Bassett, 1967; Sager, 1972). A recessive mutant allele, when homozygous, induces sporadic plastid mutation that is visually marked by a change in plastid phenotype; most conspicuously the plastid changes from a structurally and physiologically normal green chloroplast to an underdeveloped and physiologically inactive white plastid. During the ensuing cell divisions each mutated plastid sorts out from the wild-type plastids in the mixed cells to give pure mutant cells, and eventually early mutant lineages may even give rise to pure mutant sectors of leaves and vegetative and flowering shoots. The variegated shoot is a mixture of many sorting-out lineages with a new lineage beginning in a dividing cell

Department of Genetics, University College of Swansea, Wales whenever and wherever a new plastid mutation occurs. Light and electron microscopical investigations during the early stages of sorting-out have proved the existence of these mixed cells in which the markedly distinct normal and mutant plastids have been found together in the same cell (table 1).

On selfing the variegated plants there is usually a mixture of green, variegated, and white offspring in which their highly variable, non-Mendelian frequencies bear a close correlation with the degree of variegation of the flowering shoot from whence they arose, and individually the seedling phenotype bears an even closer correlation with the degree of variegation of the individual flower or fruit parts from where the seed was harvested.

On crossing the homozygous-recessive variegated plant back to the wild type, the F_1 hybrid progeny are usually all green, if the variegated plant is used as the male parent, or a mixture of green, variegated, and white, when the variegated plant is the female parent. The difference between the reciprocal crosses is caused by the maternal inheritance of the mutated plastids (table 1); occasionally the mutant plastids are transmitted by both parents as found in *Nepeta cataria* and *Oenothera hookeri*, but even in these instances the female influence predominates. The behavior of *Oenothera hookeri* crosses (Epp, 1973) depends on the nature of the plastids in the two parents. When the slower-multiplying type (plastome) IV wild-type plastids (table 4) are present in the egg cells, the same type IV mutant plastids succeed in being transmitted by the male parent; but when the plastids in the egg are the faster-multiplying type I wild-type plastids, the mutant type IV plastids fail to get through to the next generation. Within the variegated F_1 plants the normal and mutant plastids continue to sort-out during development. Nevertheless, a few mixed cells may still exist in the flowering shoot and so give rise to eggs with both kinds of plastids again. In this way variegated plants may be obtained in which the mutation-inducing allele is no longer present even in the heterozygous state. Such a plant is really no different from a spontaneously occurring variegation with a mixture of normal and mutant plastids as, in the absence of the mutant allele, there are no new plastid mutations.

These observations illustrate three fundamental principles of extranuclear inheritance:

1. Apart from the original induction, the subsequent inheritance of the variegation is non-Mendelian.

TABLE 1

EXAMPLES OF VARIEGATED PLANTS IN WHICH REPEATED PLASTID MUTATIONS (G → W) ARE INDUCED BY A NUCLEAR GENE WHEN HOMOZYGOUS-RECESSIVE

Plant and Form	Gene Symbol	Mixed Cells	Plastid Mutations G → W		Plastid Restitution W → G	Other Effects	Plastid Inheritance	Authors	Dates
			One Kind	Many Kinds					
Arabidopsis thaliana albomaculans	*am*	E	+	–	+	–	Maternal	Röbbelen	1966
Arabidopsis thaliana	*chm*[1] *chm*[2]	L	–	+	+	–	Maternal	Redei	1973
Capsicum annuum	–	–	+	–	–	–	Maternal	Hagiward & Oomura	1939
Epilobium hirsutum	mp_1	L	–	+	+	+	Maternal	Michaelis	1965
									1968a,b
									1969
								Michaelis & Fritz	1966
Epilobium hirsutum	mp_2	L	–	+	–	–	Not maternal	Michaelis	1968c
Hordeum vulgare Okina-mugi and Okina-mugi tricolor	–	–	+	–	–	–	Maternal	So	1921
								Imai	1928
									1935
									1936a,b
Hordeum vulgare	*w*	L	+	–	–	–	Maternal	Arnason et al.	1946
								Arnason & Walker	1949
Hordeum vulgare albostrians	*as*	–	+	–	–	–	Maternal	Hagemann and Scholz	1962
Hordeum vulgare striata-4	–	–	+	–	–	–	Maternal	Wettstein and Eriksson	1965
Nepeta Cataria	*m*	L	–	+	+	+	Biparental	Woods and DuBuy	1951
Oenothera hookeri plastome mutator	*pm*	–	–	+	–	–	Biparental	Epp	1973
Oryza sativa	–	–	+	–	–	–	Maternal	Kondo etal.	1927
								Morinaga	1932
Oryza sativa	–	–	+	–	–	–	–	Pal and Ramanujam	1941
Petunia hybrida	a^-	L	–	+	+	–	Maternal	Potrykus	1970
Zea mays iojap	*ij*	E	+	–	–	–	Maternal	Jenkins	1924
								Rhoades	1943
								Rhoades	1946
								Rhoades	1950
								Shumway and Weier	1967
Zea mays chloroplast mutator	*cm*	–	+	–	–	–	Maternal	Stroup	1970

Mixed cells were viewed by light (L) or electron microscopy (E). Other effects include sterility and developmental abnormalities not attributed to plastid mutation. A – is a negative observation. It is frequently impossible to know whether the appropriate behaviour was not observed, or whether it was overlooked, or whether it was simply not recorded.

2. Reciprocal crosses between green and variegated plants do not give identical results but are governed by the direction of the cross and the degree of variegation of the female parent.
3. The alternative plastid phenotypes, green and white, sort-out from one another during development.

The characteristic pattern of sorting-out and the obvious effect on plastid phenotype strongly suggest that the mutation has occurred in the plastid itself, which is confirmed by the finding of mixed cells. Finally, the permanency of the altered plastid phenotype in the absence of the nuclear mutation-inducing gene confirms the permanent change in plastid genotype.

Of course, there is a Mendelian component to the inheritance of the variegation. The green F_1 heterozygote on selfing again segregates in a typical 3:1 Mendelian ratio, but strictly speaking this is Mendelian inheritance of the plastid mutation-inducing gene, not of the variegation.

Up to now attention has been focused on accumulating the evidence and establishing the concept that a nuclear gene induces plastids to mutate, and that thereafter the plastids behave autonomously just like many examples of spontaneously occurring mutant plastids. In my opinion we should now become more attentive to the problem of the mechanisms by which the nuclear gene induces the plastids to mutate. At the moment we certainly do not know the answer, but we can hope to make some progress by assembling all the known observations, many of which have not seemed important for establishing the basic behavioral pattern.

Variation in Mutator Genes

The plastid mutation-inducing gene is derived by mutation, either spontaneously or following mutagen treatment, from a wild-type gene that does not induce plastid mutation. In every case the mutant allele is recessive so that there is no, or very little, inducing ability in heterozygotes. It is therefore the complete absence of the wild-type allele that produces the loss of normal function that results in the widespread expression of plastid mutation. As similar mutants have arisen in a variety of flowering plants, including monocotyledons and dicotyledons, it seems probable that the wild-type gene or genes is common and serves an important function in the plant. At present, however, we can say little about its structure.

In five species (table 1) there have been two or more separate occurrences of nuclear gene mutants with the power to induce plastid mutations, but for the most part it is not known whether these represent mutations of different genes or of the same gene. Only in *Zea mays* has Stroup (1970) been able to say that the gene *chloroplast mutator* (*cm*) is not allelic to *iojap* (*ij*) because F_1 hybrids between the two variegated stocks were green. The location of the two genes in *Arabidopsis thaliana* (*am* and *chm*) is not stated. One of these two genes exists as two alternative mutant alleles (chm^1 and chm^2). The two alleles induce more or less similar phenotypic alterations, though it appears that chm^1 often causes a diffuse type of sectoring in advanced stages of the ontogenetic development of the plants, while chm^2 is characterized generally by a sharp demarcation of sectors. It is not known if the two *Epilobium* genes (mp_1 and mp_2) are independent or allelic. The *plastome mutator* gene (*pm*) in *Oenothera hookeri* that induces mutation of the type I plastids can also be transferred by crossing to *O. parviflora,* where it induces mutation of the type IV plastids. The penetrance of the plastid mutator genes varies in different individuals so that variegation may not develop in all homozygotes; but where it does develop, mutation frequency per plant varies from as low as once to much higher values that are difficult to estimate.

Multiple Mutation Sites

It is apparent from the literature that more than one change can be induced. Whether this is not always so, or whether in some instances the authors simply failed to report other changes, is an open question. Woods and DuBuy (1951) reported that in *Nepeta cataria* the gene induces many different kinds of plastid mutation, either to a new color—pale green, yellow, or white—or to a structural alteration—macrograna or vacuolate. Sometimes these changes even proceed in steps—green, yellow-green, white, or normal, macrograna, vacuolate—as if a first mutation gives one change and a second mutation the next. Similar observations were made in *Epilobium hirsutum* (Michaelis, 1965). Redei (1973) reported many kinds of plastid mutant in *Arabidopsis* (*chm*), and Potrykus (1970) even found three types of plastid (green, yellow-green, and white) all within the same living mixed cell of *Petunia* (table 1). At least six different plastid mutants were induced by the plastome mutator gene of *Oenothera hookeri* (Epp, 1973). These differed in their color during the cotyledon and mature leaf stages, in sensitivity to light intensity, and in their effect on the expansion of the leaf lamina; some of them also differ in their fine structural phenotype.

Unfortunately there have been only a few statements on plastid ultrastructure. Röbbelen (1966) is fairly definite that the *am* gene in *Arabidopsis* induces only one specific phenotype in which the plastid mutant is blocked in thylakoid differentiation at an early phase. The plastids contained only some vesicles and osmiophilic globules in their matrix; there is no mention of the presence or absence of ribosomes. Their development is somewhat improved in mixed cells. Shumway and Weier (1967) also describe only one type of mutant plastid induced by the *iojap* gene in *Zea mays*. The mutant plastids are aberrant at all developmental stages, they lack the normal grana system as well as a normal prolamellar body; in this plant, plastid ribosomes are absent.

Another kind of alteration appears to occur outside of the plastids. In *Arabidopsis thaliana* besides variegation the *chm* mutant also induces changes to asymmetric growth and a rough, uneven leaf surface as well as sterility in either sex of some plants. Moreover, Redei (1973) was able to show that the rough leaf shape character is based on a cytoplasmic alteration, independent of the color variegation, which can be isolated and maintained separately. Electron microscope investigations reveal that the rough leaves contain morphologically different, though photosynthetically active, green chloroplasts in these cells. The cells often deviate significantly in size and or shape from normal. The fact that the various expressions of the cytoplasmically transmitted traits are all associated with detectable changes in the plastids, and no alterations are observed in other cell organelles, suggests that the plastids harbor the genetic determinants. Naturally, because the two types of genetic alteration are separable, their determinants must be located in different plastids. Michaelis (1965) also found other kinds of alterations to the growth of *Epilobium hirsutum* plants producing stunting and narrowing of the leaves and both male and female sterility. Again these abnormalities occur either in combination with variegation or independently; Michaelis describes many examples in great detail (1968a,b). Michaelis did not consider that these too were due to plastid mutation but rather to mutation of some other non-specified cytoplasmic component.

During an electron microscope examination of a cytoplasmic alteration, produced after hybridization between *Epilobium hirsutum*-Essen ♀ × *Epilobium parviflorum*-Tübingen ♂, Anton-Lamprecht (1967) found in the shoot apex mixed cells for mitochondria. Besides the normal mitochondria, the mixed cells contained mitochondria with an atypical development of their tubular inner membranes lying in parallel rows. This observation rather suggests that there was a connection between the cytoplasmic alteration and the existence of mutant mitochondria, so that the possibility of gene-induced mitochondrial mutation leading to developmental disturbances is by no means unthinkable. The mitochondrial abnormalities observed by Wettstein and Eriksson (1965) appear more in the nature of pleiotropic effects as they were only found in cells with mutant plastids. On this hypothesis both plastid and mitochondrial changes are caused by a primary alteration in some other cytoplasmic component. If they represented distinct mutational events, then we should expect to find mitochondrial abnormalities in cells either in combination with, or independently of, mutant plastids. Alternatively the primary mutation could be in one organelle, and the other main organelle secondarily changed following a metabolic interaction between the two.

Carry-over of Mutation-Inducing Principle

Michaelis (1968a), whose studies with *Epilobium hirsutum* are far more extensive than those of any other worker, considers that the nuclear gene might alter the plastid through an intermediate variegation-inducing principle. When he crossed a variegated female (mp_1 mp_1) with the wild-type *E. hirsutum*-Essen male, besides some maternally inherited variegated progeny, there were also 1-3% newly induced variegations even though these hybrid plants were now heterozygous (Mp_1 mp_1). No such variegated offspring are obtained in the reciprocal cross so we cannot argue that the mp_1 gene is still working in the heterozygote. He then took wholly green sisters of the variegated F_1 plants and once again crossed these with *E. hirsutum*-Essen males, and again obtained between 2-3% variegated progeny. Finally he took the variegated heterozygous plants, which had sorted-out to pure green, crossed these with the wild-type males, and once more obtained 2-3% variegated offspring. Hence the carrying-over of the mp_1 variegation-inducing ability is always observed in crosses in which the female cytoplasm had contained the mp_1 gene for one or two generations, suggesting that the cytoplasm itself is sufficiently changed to be capable of inducing plastid mutations. The objection that this might really reflect still incomplete sorting-out is not valid owing to examples in which the original sorting-out variegation is yellow-variegated and the newly induced white-variegated; prolonged sorting-out would not produce a color change. Arnason and Walker (1949) are of the opinion that the *w* mutant allele in *Hordeum vulgare* is still active in the heterozygote. Other workers have recorded no carry-over effect or gene activity in the heterozygotes of their plants.

Plastid Restitution

The gene-induced mutant plastids, like some spontaneously induced mutant plastids (Kirk and Tilney-Bassett, 1967), are not always permanently altered but are sometimes capable of recovery. Both Röbbelen (1966) and Redei (1973) noticed green islands in white sectors as though there had been a back-mutation from white to green in *Arabidopsis thaliana,* and Potrykus (1970) observed similar islands in *Petunia hybrida.* Most workers do not record these green islands (table 1); either they do not occur, or the islands may be so small as to be only visible through the microscope and therefore not apparent to the eye. Michaelis (1969) found many islands in *E. hirsutum* to be no larger than one or two cells. Another possibility is that the islands were present but were mistaken for the background sorting-out. Michaelis (1969) prefers to call the mutations restitutions in recognition of the fact that the newly originated green plastids are frequently different in structure and behavior from the original green plastids and are therefore not the result of true back-mutations. In my opinion the findings of Michaelis in regard to these restituted plastids are so interesting that I should like to summarize his observation in some depth, although I can not hope to do full justice to his very detailed paper.

Proof of plastid restitution. The first problem in studying restituted plastids is to be certain that the occurrence of a green island in white tissue really is caused by a plastid mutation and is not the result of another phenomenon. Michaelis (1969) recognized that plastid restitution might be mistaken for:

1. Sorting-out between green and white plastids
2. Modifications of the plastids
3. Mutation of another cell component
4. Duplication of a genetically green epidermis

He was able to overcome these sources of error in *Epilobium* by a detailed analysis that I shall briefly outline. Plastid restitution is easily mistaken for, or confused with, sorting-out, especially toward the end of sorting-out when green flecks may suddenly appear. It is therefore important to be certain that sorting-out is over. This depends on the number of plastids per cell, the kind of plastid division and multiplication, and the cell-to-cell plastid distribution. There are at least twenty successive cell divisions required to produce an *Epilobium* leaf with up to one million cells. If at the start there is a mixed cell with only a single green plastid, then Michaelis calculates, with random sorting-out, that after twenty successive divisions there should be 2-3% pure green cells in the leaf; similarly from two successive leaf whorls requiring about 40–55 cell divisions he calculates that there should be 6-9% pure green cells. By a one-sided sorting-out or an uneven plastid mixing the appearance of pure green cells would be speeded up. Hence if no visible green flecks appear through three leaf whorls of a white shoot, one can safely conclude that sorting-out is complete and that green plastids from sorting-out no longer exist. Michaelis kept a white shoot alive for 21 months while it passed through at least 150 successive cell divisions. During this period he never found any cluster larger than a group of four cells containing up to 16 green plastids between them, too small to be visible to the eye. Most frequently he found only single cells containing one green restituted plastid. If these green plastids had existed earlier in development, there would have been a prompt increase in green cells to produce an easily visible green fleck. Their presence is best explained by restitution in relatively late leaf development that left little time for sorting-out since few or no cell divisions remained.

The pattern of normal sorting-out is also rather different from the restitution pattern. In mixed cell descendents of sorting-out plants where there are pure green and pure white cells, it is exceptionally rare to actually find a mixed cell with only one green plastid; but among the small group of cells with restituted plastids, mixed cells with one or two green plastids are particularly frequent. Moreover, the ratios in which restituted plastids appear in mixed cells of restitution groups correspond closely to expectation on the basis that the green plastids have arisen by a chance mutation of a single plastid. For example, Michaelis and Fritz (1966) in a sample of 93 isolated single cells observed 70.3% of cells with 1-2 green plastids, 26.5% with 3-8 green plastids and 3.2% with 9-16 green plastids, which compares well with the theoretically expected values of 78.6%, 21.4% and 0.001% respectively on the basis of a random distribution after the original plastid mutation. By contrast the majority of mixed cells following the original green-to-white plastid mutation are expected to have predominantly green plastids with few white plastids. Redei (1973) also observed single cells containing from one to a majority, or all, green plastids surrounded by thousands of defective cells, but he has no data on the frequency of mixed cells with different plastid ratios. As additional proof against sorting-out Michaelis was often able to distinguish features of the restituted green plastids that led him to conclude that they are not identical to the original green wild-type plastids.

The objection that restituted plastids could be modifications of mutant plastids is invalidated by the very clear distinction between the two, and by their occurrence in sister cells in similar ratios that agree with the expectations of true mixed cells. Plastid restitution also occurs in some plastid mutants that originate spontaneously, and in such a variegated plant Anton-Lamprecht (1966) observed mixed cells containing one or two green restituted plastids along with many mutant ones. The differences in plastid structure were very clear, and there were no transitional forms. In the mature leaf the plastid had lost all grana structure and had become highly vacuolated whereas the restituted plastid showed well-developed grana formation.

The presence of mixed cells with plastids in expected ratios also invalidates the objection that plastid restitution might be due to a mutation of another cell component. Finally, although it is true that the duplication of a genetically green epidermis covering underlying white tissues could produce a green fleck, Michaelis was able to rule this out on several counts, especially as restitution cells were found in all cell layers and frequently had no contact with the epidermis, which in any case was pure white without the mixed cells needed to give rise to the mixed cells in the restitution group.

Restitution frequency. The frequency of restitution varies from not detectable to too numerous to be possible to estimate accurately. The highest countable frequency by microscopical observation in variegated *Epilobium* plants was about 1 cell with a restituted plastid among 60 wholly colorless cells. Many factors appear to influence the restitution frequency, and Michaelis (1969) obtained highly significant differences in frequency as between seasonal effects, day length, developmental stage, differences between rosette and stem leaves and between the tips and bases of the leaves. For example, in one variegated plant, the restitution frequency found in leaves developed in July was as high as 1:57, but by the following January this had fallen to 1:1,537; and by February not one restitution cell could be found in a sample of 56,680 colorless cells. In another variegated plant, however, the difference between summer and winter leaves was almost negligible. It seems probable that only by a thorough comparison of the many variables by the analysis of variance technique, with the plants grown under carefully controlled environmental conditions, could one hope to recognize the important causes of variation in restitution frequency. Redei (1973) recorded a restitution frequency of the order of 10^{-5} per cell in the *chm* mutant of *Arabidopsis*, or of the order of 10^{-6} per plastid, but he did not experiment with varying environmental conditions.

Inheritance of restitution frequency. Among variegated plants derived from spontaneous plastid mutants some show restitution cells and some do not. Moreover in the *Epilobium* studies, whereas the mp_1 plants produce some macroscopically visible restitution flecks, restitution groups from spontaneous mutants are only visible microscopically. Michaelis was able to use these differences to determine the inheritance of the restitution ability. Using plants in which sorting-out had produced pure mutant shoots, he made reciprocal crosses between an ivory-white plastid mutant incapable of plastid restitution and a pale golden plastid mutant that gave macroscopically visible restitution flecks. The ivory ♀ × pale golden ♂ cross gave 66 ivory seedlings with no restitution flecks, whereas the reciprocal pale golden ♀ × ivory ♂ cross gave 91 pale golden seedlings of which 45 possessed relatively large restitution flecks. The result clearly shows that restitution ability is tied to the female plastids and is maternally inherited.

Isolation of pure restituted plants. All seedlings from white shoots in which restitutions occur very late in development die, as do many seedlings from white shoots with macroscopically visible restitution groups; but in some, early restitution can produce a sufficiently large green restitution zone in the cotyledons that they survive. From these seedlings pure green shoots may arise from which eventually one can isolate clones pure for the new green restitution plastids.

In some instances restitution appears to occur actually under the influence of the mp_1 gene. Michaelis made a backcross between a pale yellow ♀ (Mp_1 mp_1) × a white ♂ (mp_1 mp_1). Both parent branches were completely mutant, and initially the seedlings were all colorless. The heterozygous seedlings died, but, as the cotyledons expanded, the homozygous seedlings developed restitutions producing a marbled variegation that kept them alive and enabled them to be grown to maturity. In turn, crossing these plants, in which sorting-out to pure green cells occurs, produces a mixture of white seedlings, white seedlings with green restitution flecks, and also pure green restitution seedlings developed from eggs containing only green restituted plastids. New green seedlings without the mp_1 gene remain stable, and such plants can be used to compare the behavior of the new green plastids with that of the original wild-type plastids.

Comparison between restituted and original wild-type plastids. The pigment content of the new clones is not qualitatively different from the wild type (Michaelis and Fritz, 1966), but other features of their growth, especially under different temperature regimes, suggest real differences. Michaelis (1969) observed four distinct clones. Clone I had dark green, thick leaves that contained plastids bigger than normal, but the plants died early in development. Clones II, III, and IV all survived and showed morphological and physiological differences that could be criticized as being not very conclusive, so Michaelis compared their behavior with the wild type *Epilobium hirsutum*-Essen by combining them in crosses with six different test genomes: *Epilobium hirsutum*-Attika, -Kanada, -München, -Kew *albiflorum*, -Coimbra, and *Epilobium parviflorum* resistens, a derivative of *E. parviflorum*-Tübingen. A number of statistically significant growth differences as measured by the height of the plants were observed with several of the test genomes of *E. hirsutum*, suggesting that the clones were not the same as wild type nor identical to each other. Similar conclusions could be drawn from measurement of leaf lengths. By far the most impressive differences, however, were with the species *E. parviflorum*. Hybrids between *E. hirsutum*-Essen ♀ × *E. parviflorum* ♂ are characterized by many developmental disturbances such as stunting, leaf deformities, deficiency in flower formation, and sterility. Yet when the *E. hirsutum*-Essen clones II, III, and IV were tested with *E. parviflorum*, Michaelis found the restitution clones II and IV produced normal fertile hybrids, and a part of the disturbances were removed with restitution clone III. This complete deviation from the normal reaction of the hybrid proved conclusively that restitution really had produced a new type of plastid with features quite distinct from the wild type. The new type of plastids showed the usual maternal inheritance and remained constant for the two generations tested.

Link with Pattern Genes

A second nuclear gene capable of inducing plastid mutation in *E. hirsutum* was called mp_2 by Michaelis (1968c). Selfing homozygous

$mp_2\,mp_2$ plants produced 81.7% clearly variegated offspring; in some plants variegation is limited to a few flecks and it seems probable that variegation is not always manifested. Selfing green or variegated branches of variegated plants made no difference to the offspring, and reciprocal crosses between $mp_2\,mp_2$ variegated plants and wild-type $Mp_2\,Mp_2$ green plants produced only green F_1 progeny whichever way the cross was made. Hence the maternal effect of the variegated phenotype upon the offspring, as found in $mp_1\,mp_1$ plants, was completely lacking in $mp_2\,mp_2$ plants. In other words the variegation behaves as a mosaic pattern gene plant (Kirk and Tilney-Bassett, 1967) with purely Mendelian inheritance. Detailed analysis, however, convinced Michaelis that the nuclear gene really was inducing plastid mutation, but maternal inheritance of the mutated plastids could not be demonstrated because the mutations were occurring too late in leaf development to exert an effect on the shoots and cell lines giving rise to the egg cells. Moreover, the mutation frequency appeared to be at a maximum in the cotyledon and early rosette leaves, gradually decreasing to a minimum in the flowering shoot.

His evidence for plastid mutation is based on the characteristic multiple cell lineage sorting-out pattern in the leaves, and the finding of mixed cells. In addition he found several different kinds of plastid alteration, which could be expected following plastid mutation rather than any physiological change in the leaf. The existence of mixed cells containing both kinds of plastid and with similar mixture proportions in sister cells excluded pseudo-mixed cells and firmly supported plastid mutation. The evidence is therefore quite convincing that this is an example of a mosaic pattern gene that works by causing plastid mutation. It seems probable that many other mosaic or striped pattern genes might behave in a similar fashion.

Thus a nuclear gene that induces plastid mutation early in development can lead to sorting-out of mutant plastids, their maternal or biparental inheritance, and, following separation of the mutant plastids from the mutator gene, may even lead to a stable chimera. Whereas, if the plastid mutations only occur late in development the end product is a simple mosaic or striped pattern. Viewed in this manner the link between the two types of variegation is indeed very close. At the molecular level, however, it seems unlikely that quite the same mechanism could be the cause of the unrestricted timing of plastid mutation, on the one hand, and mutation restricted to rather late in plastid development, on the other. But, as I shall argue shortly, the difference might be only a matter of the frequency of plastid mutation.

Mechanisms

In the past the similarity of the basic behavioral pattern of gene-induced plastid mutation has led to the belief that this is a homogeneous group of plants presumably sharing a common induction mechanism (Kirk and Tilney-Bassett, 1967). An appreciation of the finer observations of the many examples now studied forces one to the inescapable conclusion that this is an oversimplification. Instead of one, there may be as many as three distinct types of behavior:

1. A nuclear gene induces plastid mutation to a single, specific mutant phenotype.
2. A nuclear gene induces plastid mutation to a wide variety of mutant phenotypes.
3. A nuclear gene induces plastid mutation to a wide variety of mutant phenotypes and also induces independent cytoplasmic mutations.

Examples of all three of these types are present in table 1, but, as there are reservations about some of the original observations, this is no guarantee that all three types really exist.

In the first two types the nuclear gene has a specific effect on the plastids alone, whereas in the third type mutation must have occurred in a nuclear gene controlling processes in the cytoplasm as well as the plastids. Hence the difference between the two groups suggests that one type of nuclear gene controls a function specifically limited to the plastid, and that the other type of nuclear gene controls a more general function. In the examples where only one kind of plastid mutation occurs, as in the *am* mutant of *Arabidopsis*, the barley mutants, and the *ij* and *cm* mutants of maize, it seems probable that plastid mutation is limited to one specific site in the plastid DNA, although not necessarily an equivalent site in each species. A strictly localized, repetitive mutation of this kind might indeed be gene-induced and is reminiscent of the behavior of controlling elements (Whitehouse, 1973).

When a wide variety of plastid mutations occur, as in the *chm* mutant of *Arabidopsis*, *Oenothera*, and *Petunia*, lesions must be occurring in many different regions of the plastid DNA, possibly quite randomly. Controlling elements are not expected to behave in this way; instead we must look for an indiscriminate kind of error. A likely candidate would be a mutation in a nuclear-transcribed DNA polymerase enzyme so that plastid mutations arise by faulty replication of plastid DNA (Epp, 1973). As mitochondria do not necessarily replicate at the same time as the plastids, we may expect differences in their polymerases thereby allowing errors of plastid DNA replication without affecting mitochondrial DNA replication. There are far fewer significantly placed cells in the apical meristem than in the leaves, and moreover these apical cells have only about 10 plastids per cell compared with 30–50 plastids in leaf mesophyll tissue (Michaelis, 1969). Now, assuming that the probability of a mutation is the same every time the plastid DNA replicates, it follows that there will be many more mutations in the leaves than in the apical cells. A very inefficient enzyme allowing a high frequency of mistakes could provide a significant probability of mutations occurring in the apical cells thereby allowing the opportunity for an eventual sorting-out in the vegetative and flowering shoot. On the other hand, a rather less inefficient, although still mutant, enzyme allowing a lower frequency of mistakes might reduce the probability of a mutation occurring in the apical cells to an insignificant level, yet still permit mutations to appear in the leaves; the mp_2 mutant in *Epilobium* may be just such a case.

In the case of plastid and cytoplasmic mutations, as in the mp_1 mutant of *Epilobium* and in *Nepeta,* in which mutation has occurred in a nuclear gene of more general function, one possibility would be mutation of a repair enzyme. One may expect natural errors to occur during replication of both plastid and mitochondrial DNAs; these might normally be repaired by a common repair enzyme coded by nuclear genes. A slight change in the code could produce an enzyme that was rather poor at repair, hence allowing plastid and mitochondrial mutations to be expressed. Mutations would be expected at many different sites in both organelle DNAs, thus accounting for the wide variety of phenotypes expressed.

Potrykus (1970) interprets the variegation in *Petunia* in terms of two models. Model I is the classical one of nuclear gene-induced plastid mutations. In model II the characteristics of nuclear gene-induced plastid mutations are not caused by independent, self-replicating factors but by feed-back through extrachromosomal products of chromosomal gene action. As with all the models discussed, we are particularly in need of further analyses especially at the ultrastructural and biochemical levels. Redei (1973) stresses the need for further knowledge of the degree of polyteny or polyploidy of individual chloroplasts. At the moment we do not know whether a plastid mutation is expressed immediately, or whether there are many copies of the plastid DNA per organelle so that there could be a lag period between mutation and expression owing to the need for an internal segregation of mutant and wild-type plastid DNA strands.

Restitution changes may be in the nature of suppressor mutations. This is certainly a possible explanation for the restitutions that occur under the influence of the mp_1 gene in *Epilobium;* here it looks as if a second mutation induced in the mutant plastid DNA strand may have corrected the first mutation, thus restoring the plastid to normal chloropast development. Whether suppressor mutations can also account for the very high frequency of restitutions in some leaves of the order of 1 plastid per 60 cells, that is about 1 plastid per 2,000–3,000 plastids, is another matter. Whatever the mechanism, we have to marvel at the effect; for according to Michaelis, some newly restituted plastids in *Epilobium hirsutum*-Essen have so changed from the wild type that they have completely overcome the incompatibility that previously existed in hybrids between *E. hirsutum*-Essen ♀ × *E. parviflorum* ♂. This implies that the former incompatibility between these species was actually brought about by the unfavorable reaction between the *E. hirsutum* plastids and the hybrid nucleus so that speciation itself must have been initiated either by a change in a plastid to make it incompatible with the nucleus or by a change in the nucleus to make it incompatible with the plastid. Whichever way round evolution worked, it is nonetheless remarkable to observe the restoration of harmony following the two successive plastid mutations. The very fact of restitution ability makes it seem unlikely that the primary mutation could have altered the plastid DNA very much, and to be fully functional it seems unlikely that the restituted plastid could be very different from wild type, yet these small changes are enough to restore harmonious relations between the plastids and the hybrid nucleus. This demonstration of plastid evolution in experimental material is also highly significant for our appreciation of the possible role of plastid evolution in nature.

SPONTANEOUS PLASTID MUTATIONS

The Distinction between Maternal and Biparental Plastid Inheritance

Following spontaneous plastid mutation, cells pass through the characteristic stages of sorting-out of plastids from mixed cells, followed by the development of pure mutant and pure green cells and tissues and finally pure or chimerical shoots, which can then be used for breeding experiments. Unfortunately sorting-out is frequently too slow to give pure shoots, so that many observers have been obliged to use variegated shoots in their experiments. Reciprocal crosses between green and white shoots or, less favorably, between green and variegated shoots have been used to determine the inheritance pattern of the plastids. When all the offspring have the same phenotype as the mother, irrespective of the direction of the cross, the inheritance is maternal. When the offspring have a mixed phenotype, partly following the maternal parent and partly the paternal, the inheritance is biparental. In a comprehensive and critical review of such studies (Kirk and Tilney-Bassett, 1967) I listed 29 genera for which there was some evidence of maternal inheritance; this list has been reprinted by Sager (1972). I should like now to take the opportunity of updating these lists by adding some more recent examples for which there is sufficient published data to justify a claim to maternal or biparental inheritance (table 2). This list includes several genera previously known; new genera are *Allium* and *Gossypium* with maternal inheritance, *Fagopyrum* and *Solanum*, which may have a trace of paternal inheritance, and *Secale*, the first monocotyledon genus to show biparental inheritance. In addition, *Triticum aestivum*, with maternal inheritance, follows the behavior of *Triticum vulgare*. The discovery of a very strong paternal bias to the biparental transmission in *Cryptomeria japonica*, the first gymnosperm to be tested genetically, is particularly interesting, especially as the use of pure crosses, uncomplicated by sorting-out, and the large number of progeny scored, make the data very reliable.

The apparent trace of paternal inheritance in *Solanum* and *Fagopyrum* after green ♀ × variegated ♂ crosses illustrates that the division between biparental and maternal inheritance should be thought of as a probability rather than categorically one way or the other. This has been particularly clearly shown in new experiments with *Antirrhinum*, formerly one of the classic examples of maternal inheritance. By using a rare and very characteristic plastid mutant called *prasinizans*, which could be grown as a free-living plant under reduced daylight, Diers (1967, 1971) was able to make crosses between pure plants free from the problems of sorting-out and in which the mutant plastids could be clearly recognized in the progeny with no confusion with newly originating spontaneous plastid mutants. In this way he could be certain that the low level of male transmission was not

TABLE 2

RECENT EXAMPLES OF HIGHER PLANTS WITH MATERNAL OR BIPARENTAL TRANSMISSION OF THEIR PLASTIDS

Plant	Demonstration of mixed cells		Results of Reciprocal Crosses							Authors	Dates
	By Light or Electron Microscopy	By Segregation of Seedlings	Type of Cross[a]	Ia G × W IIa G × V			Ib W × G IIb V × G				
				G	V	W	G	V	W		
Gymnosperms											
Cryptomeria japonica	—	G V W	Ia,b	111	284	3386	8618	49	22	Ohba et al.	1971
Dicotyledons											
Antirrhinum majus	L	— — —	Ia,b	9481	3	—	—	1	6531	Diers	1967
Antirrhinum majus	—	— — —	Ia,b	41121	14	—	—	6	40456	Diers	1971
Fagopyrum esculentum	—	G V W	IIa,b	161	16	—	37	3	1	Tatebe	1972
Gossypium hirsutum	—	G V W	IIa,b	202	—	—	52	24	72	Kohel	1967
Gossypium hirsutum	—	G V W	IIa,b	G	—	—	G	V	W	Krishnaswami	1968
Nicotiana tabacum	E	— — —	—	—	—	—	—	—	—	Shumway and Kleinhofs	1973
Nicotiana tabacum	L	G V W	Ia,b	G	—	—	—	—	W	Dulieu	1967
Nicotiana tabacum	L	— — —	Ia,b	500	—	—	—	—	307	Wildman et al.	1973
Pelargonium zonale	L	G V W	Ia,b	29	1	—	277	356	11	Herrmann, & Hagemann	1971
Pelargonium zonale	L	G V W	Ia,b	106	31	3	319	205	5	Börner et al.	1972
Pelargonium zonale	—	G V W	Ia,b	G	V	W	G	V	W	Tilney-Bassett	1963–74
Solanum tuberosum	—	G V W	IIa,b	1868	1	4	79	56	576	Simmonds	1969
Monocotyledons											
Allium cepa	—	G V W	IIa,b	226	—	—	401	18	44	Tatebe	1968
Allium fistulosum	—	G V W	IIa,b	898	—	—	160	19	137	Tatebe	1961
Secale cereale	—	G V W	IIa,b	124	50	—	3	5	—	Fröst et al.	1970
Triticum aestivum	—	G V W	IIa,b	49	—	—	25	66	11	Briggle	1966
Zea mays	—	G V —	IIa,b	1219	—	—	2865	3318	—	Shumway and Bauman	1967

[a] Ia: G × W; Ib: W × G; IIA: G × V; IIB: V × G

TABLE 3

LIST OF SPECIES SHOWING THEIR MODE OF PLASTID INHERITANCE
BASED ON ANALYSIS OF VARIOUS KINDS OF PLASTID VARIEGATION

Species	Method of Analysis[a]				Mode of Inheritance[b]
Ferns					
Scolopendrium vulgare		2			Bip
Gymnosperms					
Cryptomeria japonica	1				Bip
Dicotyledons					
Acacia decurrens				4	Bip
Acacia mearnsii				4	Bip
Antirrhinum majus	1				Bip (trace)
Arabidopsis thaliana	1		3		Mat
Arabis albida	1				Mat
Aubrietia graeca	1				Mat
Aubrietia purpurea	1				Mat
Beta vulgaris	1				Mat
Borrago officinalis		2			Bip
Capsicum annuum	1		3		Mat
Cucurbita maxima	1				Mat
Epilobium hirsutum	1		3		Mat
Fagopyrum esculentum	1				Bip (trace)
Geranium bohemicum				4	Bip
Gossypium hirsutum	1				Mat
Humulus japonica		2			Mat
Hydrangea hortensis	1				Mat
Hypericum acutum	1			4	Bip
Hypericum montanum				4	Bip
Hypericum perforatum	1				Bip
Hypericum pulchrum				4	Bip
Hypericum quadrangulum				4	Bip
Lactuca sativa	1				Mat
Lycopersicum esculentum	1	2			Mat
Medicago truncatula				4	Bip
Mesembryanthemum cordifolium	1				Mat
Mimulus quinquevulnerus	1				Mat
Mirabilis jalapa	1				Mat
Nepeta cataria			3		Bip
Nicotiana colossea	1				Mat
Nicotiana tabacum	1	2			Mat
Oenothera (*Euoenothera*):					
28 species listed table 4	1		3	4	Bip
Oenothera (*Raimannia*):					
Oenothera berteriana				4	Bip
Oenothera odorata				4	Bip
Pelargonium denticulatum				4	Bip
Pelargonium filicifolium				4	Bip
Pelargonium × *Hortorum* (*zonale*)	1				Bip
Petunia hybrida			3		Mat
Petunia violacea	1				Mat
Pharbitis nil	1				Mat
Phaseolus vulgaris		2			Bip
Pisum sativum	1				Mat
Plantago major		2			Mat
Primula sinensis	1				Mat
Primula vulgaris	1				Mat

TABLE 3 (*continued*)

Species	Method of Analysis[a]	Mode of Inheritance[b]
Rhododendron hortense	4	Bip
Rhododendron japonicum	4	Bip
Rhododendron kaempferi	4	Bip
Rhododendron mucronatum	4	Bip
Rhododendron obtusum	4	Bip
Rhododendron pulchrum	4	Bip
Rhododendron ripense	4	Bip
Rhododendron serpyllifolium	4	Bip
Rhododendron sublanceolatum	4	Bip
Rhododendron transiens	4	Bip
Rhododendron yedoense	4	Bip
Silene otites	4	Bip
Silene pseudotites	4	Bip
Solanum tuberosum	1	Bip (trace)
Stellaria media	1	Mat
Trifolium pratense	1	Mat
Viola tricolor	1	Mat
Monocotyledons		
Allium cepa	1	Mat
Allium fistulosum	1	Mat
Avena sativa	1	Mat
Chlorophytum comosum	1	Bip (trace)
Chlorophytum elatum	1	Bip (trace)
Hordeum vulgare	1 3	Mat
Hosta japonica	1	Mat
Oryza sativa	3	Mat
Secale cereale	1	Bip
Sorghum vulgare	1	Mat
Triticum aestivum	1	Mat
Triticum vulgare	1 2	Mat
Zea mays	1 3	Mat

Source: Kirk and Tilney-Bassett (1967) and present text, tables 1, 2, and 4.

[a]The kinds of genetic analysis used are the results of crosses between normal plastids and their spontaneous mutants, which may be stable (1) or unstable (2); the results of crosses between normal plastids and their gene-induced plastid mutants (3); and the results of crosses between plants containing normal plastids in both parents and producing hybrid variegation among the offspring (4).

[b]The plastid inheritance is classified as maternal (Mat) or biparental (Bip). The examples should not be taken as rigorously proved in all instances.

due to any artefact, a possibility that cannot be ruled out for other examples showing a trace of male transmission. The maximum observed frequency of about 1 variegated:3,000 non-variegated seedlings is so low that nearly all other studies in over thirty genera would not have been expected to show any paternal transmission in the small numbers of progeny scored; only in maize and *Epilobium*, and perhaps tobacco, have there been sufficiently large populations scored to be able to eliminate the possibility of male transmission at a frequency as low as 0.03%. Since this frequency is of the same order of magnitude as the occurrence of new variegation by spontaneous plastid mutation it will usually be difficult to recognise in G × W crosses, although possible in W × G crosses so long as the white plastids are not liable to restitution changes. Hence for the majority of species at present classified as having a maternal inheritance of plastids, we have not yet eliminated the possibility of a low level of paternal transmission, a reservation that should be kept in mind in assessing the summary list of species presented in table 3. Included in this list are examples of biparental transmission classified by hybrid variegation. This is a phenomenon in which, following crosses between two normal green plants of different races or different species, the plastids of one species are compatible with the hybrid nucleus but not the plastids of the other species. Hence when the plastids are transmitted by both parents, one develops into a normal green chloroplast and the other remains white. The two types of plastid then sort-out to produce a variegated seedling, a witness to the biparental transmission of their plastids. There appear to be no new examples of hybrid variegation.

Gradually, as our list of case histories expands, we can hope to gain some insight as to when the distinction between maternal and biparental plastid inheritance developed. At present maternal inheritance appears to be more common than biparental, but both types are found in the two major groups of angiosperms, the dicotyledons with 24 genera maternal and 14 genera with at least a trace of biparental inheritance, and the monocotyledons with 8 genera maternal and 2 genera with at least a trace or suspicion of biparental inheritance (table 3). At the family level, examples of both biparental and maternal are found in the Leguminosae and Gramineae, but so far only maternal in the Cruciferae. At the level of the genus, species are either all maternal or all biparental. The one example from ferns shows biparental inheritance. In the case of gymnosperms, the breeding evidence in *Cryptomeria japonica,* which shows biparental with a strong paternal bias (table 2), is supported by detailed electron microscopical observations of other gymnosperms—*Larix decidua, Biota orientalis* (Chesnoy and Thomas, 1971), and *Chamaecyparis lawsoniana* (Chesnoy, 1973), which show that the plastids of the proembryo are mostly, if not all, derived from the male parent. Similarly the mitochondria are mostly derived from the male parent in *Biota* and *Chamaecyparis,* but mostly from the female in *Larix.* These gymnosperms have a very large egg, which only goes to show how trivial can be the common argument that the female plastid contribution is so much more important than the male, because of the large size difference between their gametes. We also find a lack of correlation between egg size and the number of plastids in *Quercus gambelii,* because Morgensen (1972) states that he found plastids to be very scarce in the egg and zygote of this species. Although the information is still minimal, these examples are a warning that we should be extremely wary of making any assumptions about the male and female plastid contributions to the zygote.

The importance of knowing more about events at fertilization has been emphasized by the demonstration that viable plastids are still present in the pollen of tobacco, which follows a strictly maternal inheritance. This demonstration was achieved by the successful growth of green, variegated, and white haploid plantlets from the anthers of corresponding phenotype on a variegated plant of *Nicotiana tabacum* cv. Samsun (Nilsson-Tillgren and von Wettstein-Knowles, 1970). Hence in tobacco the male plastids must be eliminated after the first mitotic division of the pollen grain and before the young embryo, that is during pollen transmission, pollen tube growth, fertilization, or the first zygotic division. Lombardo and Gerola (1968) examined the ultrastructure of the generative cell of the pollen grain in *Mirabilis jalapa,* with maternal plastid inheritance, and could find only a very small number of membrane-bound vesicles that they tentatively interpreted as either very young mitochondria or proplastids. In *Pelargonium,* with biparental plastid inheritance, the generative cells contained a great number of proplastids packed close together and filling the whole cytoplasm. Diers (1963) similarly found proplastids in the generative cells of *Oenothera,* which also has a biparental plastid inheritance. The scantiness, or total absence, of proplastids in the male generative cell in *Mirabilis* may indicate that the main block to male transmission of plastids is at this stage, which it clearly is not in tobacco, nor is it likely to be the main block in *Antirrhinum* since some male plastids succeed in overcoming all the hurdles. The problem of the control of maternal inheritance of plastids in higher plants is still a long way from being solved. Fortunately, we understand a little more about the control of biparental plastid inheritance based on extensive studies in two genera, *Oenothera* and *Pelargonium.*

The ever increasing number of plastome mutants now described has led Hagemann (1971) to propose a system of nomenclature whereby one might avoid confusion with nuclear mutants. In his scheme all extranuclear mutants are prefixed with the symbol *en* followed by the symbol representing the diagnostic feature of the phenotype, for example, gilva (*gil*) for yellow leaves. The whole symbol *en:gil* would thus characterize the particular genetical factor with regard to its extranuclear location as well as its influence on the phenotype. As yet, however, most higher plant mutants have not been given any symbol. This is not entirely a question of carelessness. The normal breeding tests for allelism of phenotypically similar nuclear mutants are not available for distinguishing the many extremely similar plastid mutants of higher plants, and only by detailed structural and biochemical analysis may we hope to distinguish them. Although considerable progress has been made on the structure and biochemistry of plastome mutants as, for example, the detection of a change in the base composition of chloroplast DNA in a mutant of *Nicotiana tabacum* (Wong-Staal and Wildman, 1973), this seems to have little direct bearing on the problem of their genetical behavior, which enables me to excuse myself from further consideration of these very interesting aspects in the present review.

Genetic Control of Biparental Plastid Inheritance

The Oenothera system: hybrid variegation. One aspect of research into the genus *Oenothera* has been based on extensive studies of hybrid variegation. I do not wish to enter into the details here as it has been extensively reviewed by Kirk and Tilney-Bassett (1967) and by Sager (1972), and there have been no more recent developments. The outcome has been to show the great importance of the plastids in the evolution of the genus. By making a large number of crosses between fourteen species, Stubbe (1959a,b, 1960, 1963, 1964) was able to combine as far as possible each plastome with all available nuclear genomes, and each nuclear genome with all available plastomes. He found that in about one-third of the combinations the plastids were able to develop into normal green chloroplasts, but in the remainder a wide range of chlorophyll deficiences were produced. A comparison of the behavior of the plastids from the fourteen species with different test genomes showed that some plastids were identical, but not others, so that they fell into five groups. These five different wild-type plastomes Stubbe designated as *hookeri* plastome (I), *suaveolens* plastome (II), *lamarckiana* plastome (III), *parviflora* plastome (IV), and *argillicola* plastome (V). Similarly a comparison of varying nuclear genomes with different test plastomes indicated that the nuclear genomes had evolved three main haploid types, A, B, and C, which combine in pairs to give six different diploid genomes. Now the compatibility relations between the different genomes and plastomes shows that, of the five plastome types, only plastome IV is compatible with both the B and C genome and part of the A genome. It therefore seemed plausible that the evolution of the three main genomes could have arisen from a common ancestor containing plastome IV. It so happens that this is the plastome with the slowest rate of multiplication (see next section). Hence one is led to the conclusion that plastome IV is the most ancient plastome from which the other types evolved by increases in their multiplication rates and by changes leading to incompatibility between them. This vivid demonstration of plastid evolution, in parallel with the evolution of the nuclear genome, is particularly useful in that it shows that wild-type plastids can alter, and that it is therefore sensible to look for differences in the plastids themselves as the basis for any significant variation in the pattern of biparental plastid inheritance.

The Oenothera system: plastid competition. The probelm of the control of biparental plastid inheritance in *Oenothera* has been studied by Schötz (1954) and summarized by Kirk and Tilney-Bassett (1967) and by Sager (1972). His technique was to cross a dozen species of *Oenothera,* containing green plastids, by two chimeras containing mutant white plastids in their germ cells; the mutant plastids were derived by spontaneous mutation from green plastids of *O. biennis* and *O. lamarckiana.* The essential features of his results were as follows:

1. The G × W crosses produced green and variegated progeny, but no white seedlings, and the W × G crosses produced white and variegated progeny, but no green seedlings. This result shows the predominant effect of the maternal parent on the phenotype and segretation of the offspring.
2. The percentage of variegated seedlings differed in each cross, and species could be arranged according to an increasing percentage. This order could be defined more precisely by taking into account the proportion of green and white tissue in each variegated seedling and so calculating a variegation rating.
3. The ordering of the green species was essentially the same in G × W crosses whether the mutant plastids originated from *O. biennis* or *O. lamarckiana,* but the percentage of variegated seedlings and the variegation rating were somewhat higher with the *lamarckiana* plastids. The same order in both sets demonstrates the predominant role of the female parent, and the difference between the two sets indicates the effect of the male.

4. The ordering of the green species in W × G crosses roughly corresponded to that in G × W crosses, but in decreasing order. That is to say, species giving the lowest frequency of variegation in G × W crosses tended to give the highest frequency in W × G crosses. The relationship between reciprocal crosses shows that the species with the most successful green plastids in G × W crosses is also the species with the most successful green plastids in W × G crosses, and similarly, species whose green plastids succeed less well in G × W crosses also succeed less well in W × G crosses.

As the performance of the green plastids of one species relative to that of another is unaltered by the direction of the cross, it followed that the relative performance of green plastids of all species could be satisfactorily assessed without making reciprocal crosses each time. Hence Schötz was able to concentrate his attention on G × W crosses, which had the great advantage of no seedling lethality compared with the high death rate of white seedlings from W × G crosses.

Differences between the success or competitiveness of green plastids from varying species in conjunction with the same white plastids might be due either to an effect of the different hybrid nucleus in each cross or to a difference between the green plastids. Schötz was able to distinguish between these alternatives of nuclear or plastome control by comparing the effects of identical plastid pairs with different nuclear genomes, and conversely, by comparing the effects of identical nuclear genomes with varying plastid pairs. Although he could not rule out that the nucleus had a minor modifying effect, there was no doubt that the major changes in segregation frequencies were caused by changes in the plastid pairs. He was therefore able to interpret his observations of the different success of green plastid transmission as evidence of the differential multiplication rates of the plastids from varying species.

More recently Schötz (1968) has extended his analysis of G × W crosses to cover as many as 28 green species of *Oenothera* (subgenus Euoenothera), as females, with the same two different sources of mutant white plastids, as males. He distinguishes three categories of green plastid, with a strong ability, with an average ability, and with a weak ability to compete with the mutant plastids. Moreover, these three groups correspond closely with the five plastome types recognized by Stubbe. The group 1, with the fastest-multiplying plastids, corresponds to plastomes I and III; group 2, with an average multiplication rate, corresponds to plastome II; and group 3, with the slowest multiplication rate, corresponds to plastomes IV and V (table 4). An analysis of the ratios of green:white plastids in mixed cells, following many plastid crosses, generally upheld this grouping (Schötz and Heiser, 1969), although it did raise problems as to the causes of some disagreement between observed and expected curves of mixed cell frequencies.

The classification of the species into three groups also reveals why there are always more variegated seedlings and a higher variegation rating when the mutant plastids in G × W crosses are derived from *O. lamarckiana* (group 1) than when they are derived from *O. biennis* (group 2). Evidently this is because the *lamarckiana* plastids multiply faster than the *biennis,* and this property is not altered by the green to white plastid mutation.

When Schötz made crosses between female plants with the fastest-multiplying plastids and male plants with the slowest-multiplying plastids, he found a purely maternal inheritance. Evidently the male plastids could not overcome the double disadvantage of entering from the male and having the slower multiplication. When, on the other hand, he crossed females with the slowest-multiplying plastids by males with the fastest-multiplying plastids, the male plastid transmission was at its best; even so, the males could not transmit their plastids beyond a certain level owing to the great initial advantage enjoyed by plastids from the female parent.

One may therefore recognize two main factors in the control of biparental inheritance in *Oenothera* species through plastid competition:

1. Sexual differentiation. The major sex control is by the female parent, ensuring that all offspring obtain at least some, and sometimes all, plastids from the mother. It seems probable that this is a passive control in which the female advantage is gained through contributing many more plastids to the zygotes than the male. Conversely, the minor sex control is by the male parent, ensuring that male plastids enter at least some zygotes. Again, this is probably a passive control in which the disadvantage of the male is through contributing far fewer plastids to the zygote than the female.

2. Plastid multiplication. Given that the average zygote has, immediately after fertilization, a ratio of male to female plastids greatly

TABLE 4

SUMMARY OF OENOTHERA SPECIES CLASSIFIED ACCORDING TO THEIR RATES OF PLASTID MULTIPLICATION (GROUPS 1–3), THEIR PLASTID COMPATIBILITY RELATIONSHIPS (PLASTOMES I–V), AND THEIR BASIC NUCLEAR GENOMES A (A_1, A_2), B, AND C.[a]

Group 1 : Fast		Group 2 : Medium		Group 3 : Slow	
Plastome I		Plastome II		Plastome IV	
O. bauri	$A_1 A_1$	*O. purpurata*	$A_? A_?$	*O. ammophila*	A_1 C
O. cockerelli	$A_1 A_1$	*O. biennis*	A_1 B	*O. germanica*	A_1 C
O. mollis	$A_1 A_1$	*O. hölscheri*	A_1 B	*O. syrticola*	A_1 C
O. franciscana	$A_2 A_2$	*O. nuda*	A_1 B	*O. pictirubata* (hybrid)	B B
O. hookeri	$A_2 A_2$	*O. rubricaulis*	A_1 B	*O. atrovirens*	B C
Plastome III		*O. suaveolens*	A_1 B	*O. flexirubata* (hybrid)	B C
O. blandina	$A_2 A_2$	*O. conferta*	A_2 B	*O. parviflora*	B C
O. deserens	$A_2 A_2$	*O. coronifera*	A_2 B	*O. rubricuspis*	B C
O. lamarckiana	A_2 B	*O. grandiflora*	B B	*O. silesiaca*	B C
O. chicaginensis	B A_1			Plastome V	
				O. argillicola	C C

[a]Data taken from the text of Schötz (1968).

> favoring the female, an active control now comes into operation in which the success of the male and female plastids depends on their competitive ability as determined by their relative rates of plastid multiplication.

These two factors bring into play a balance between the physical advantage to the female of a higher initial proportion of plastids in the zygote and the effect of any differences in multiplication rates. Although there is no evidence of any difference between the species in regard to the physical advantage of the female, and therefore of the starting ratio in the zygote, the nine permutations of plastids with three multiplication rates and their association with male or female parents produce a variety of shifts in the plastid ratios by the time of seed germination, and in consequence a change in the progeny segregation frequencies and the variegation rating.

Schötz (1974) has now shown in a very detailed study that these variegation ratings can indeed be modified by the nuclear background. In general, tests of plastomes I to IV showed an increasing competitive ability when combined with the diploid genomes of the type $BA_2 < A_1A_2 < A_2A_2$ in this succession. Hence the combination of genome A_2A_2 with the fastest plastome I, as it exists in *O. hookeri* or *O. fransciscana,* is the most fit of all for plastid competition. Similarly, in competition between *biennis* (II) and *atrovirens* (IV) plastids, the strength of the *atrovirens* plastids increases with the genome succession $BA_2 < BB < BC$, showing that the *atrovirens* plastids are most able to compete with *biennis* plastids in their natural genome environment, BC. We may therefore add "Hybrid Genome" as a third—but compared with the first two, minor—factor in the control of biparental plastid inheritance in *Oenothera* species.

Although these explanations seem fairly conclusive, it still is too early to close the story. In particular, we do not know why the female contributes more plastids to the zygote than does the male. Does this bear a direct relationship to the numbers of plastids contributed by the two sexes at fertilization? Or could there be competition at the level of plastid DNA replication that always favors the female? Or could there even be a recombination process as Sager (1972) has suggested? Clearly at the ultrastructural and molecular levels, there are still many important questions awaiting an answer.

The Pelargonium system. One might expect that the control of plastid inheritance in *Pelargonium zonale* would be similar to that in *Oenothera* as both share the property of biparental plastid transmission. My studies over the last decade clearly show that the position is not as simple as this. Moreover, the relationships of the plants studied are rather different. The *Oenothera* experiments examined the nature of variation produced by crosses between many species; Schötz (1954, 1968, 1969) did not comment on any major variation in behavior within a single species. By contrast, the *Pelargonium* experiments analyze the extensive variations in behavior exhibited by different cultivars within a species.

The *Pelargonium* results are based on two kinds of crosses. In one kind of cross cultivars of unknown genotypic relationship are used. One parent contains mutant, white plastids in the germ layers, and the other parent, the wild type, contains normal, green plastids (fig. 1). The source of mutant plastids is always a stable mesochimera, GWG, in which the mutant germ layer is sandwiched between two green layers. Extensive observation shows that the variegated chimera breeds very true to the character of its sub-epidermal layer, L II, which besides producing the germ cells is also responsible for producing the white marginal tissue so characteristic of the white-over-green, variegated leaves (Tilney-Bassett, 1963a; Kirk and Tilney-Bassett, 1967). The second kind of cross is basically the same, but the parents are isogenic. This is achieved by obtaining a pure green clone from a green bud variant of a GWG mesochimera. Hence when a green clone of cv. (cultivar) Flower of Spring is crossed with the original chimera cv. Flower of Spring, the nucleus is effectively selfed, and this is equally true with W × G as with G × W crosses, or with a G × G or W × W self. These isogenic crosses are particularly useful, as it means that any differences in the segregation ratios after reciprocal crosses cannot be attributed to possible differences in the nucleus, as the nucleus is unchanged. The essential features of the *Pelargonium* results are as follows:

1. The early experimentalists (see summaries by Tilney-Bassett, 1963b; Hagemann, 1964) showed that the G × W plastid crosses produced varying proportions of green, variegated, and white progeny, and the W × G crosses behaved similarly. This result shows that all the plastids of some of the progeny come from the female, or from the male, or, in the case of variegated seedlings, from both parents, Strictly speaking, therefore, the inheritance is a mixture of maternal, biparental, and parental, whereas in *Oenothera* it is a mixture of maternal and biparental plastid inheritance. In some crosses the *Pelargonium* plastid

inheritance is biparental plus maternal, or biparental plus paternal (Tilney-Bassett, 1963b), or in extreme cases only maternal (Tilney-Bassett, 1973) or only paternal (Tilney-Bassett, 1964).

2. Comparison between pairs of reciprocal crosses showed that green or white plastids are more successfully transmitted by the female parent in some crosses, or by the male parent in others (Tilney-Bassett, 1965; Kirk and Tilney-Bassett, 1967). Evidently, the universal maternal bias found in *Oenothera* is absent from *Pelargonium*, which has no constant sex effect.
3. Reciprocal crosses frequently produced neither the same nor reciprocal proportions of green, variegated, or white seedlings (Tilney-Bassett, 1963b; 1965). Apparently, the plastid contribution from each parent is not simply dependent on the sex difference, but also on whether the plastids are normal or mutant.
4. It used to seem as if green plastids were always more successfully transmitted than white (Tilney-Bassett, 1965), but a wider range of cultivars showed that this need not be so (Tilney-Bassett, 1970c). Hence there is no universal plastid effect separating normal from mutant.
5. The segregation ratios were not significantly altered by fairly extreme changes in the environment (Tilney-Bassett, 1970b). Plainly, significant differences cannot be explained away by environmental effects.
6. The combination, in all G × W crosses, within and between two isogenic cultivars from Flower of Spring and cv. Dolly Varden, of no differences in mean fertilization, no difference in embryo survival between fertilization and three weeks after, and no change in the proportion of green, variegated, and white embryos, demonstrated that there is no preferential elimination of white embryos (Tilney-Basset, 1970a). Plainly, significantly different segregation ratios cannot be explained away by selective effects.

The proof that differing segregation ratios could not be explained by environmental or selective effects and must therefore have a genetic basis was a major step in the understanding of the control of plastid inheritance in *Pelargonium*. Yet it had not been possible to make any permanent generalization in regard to either a sex effect or a

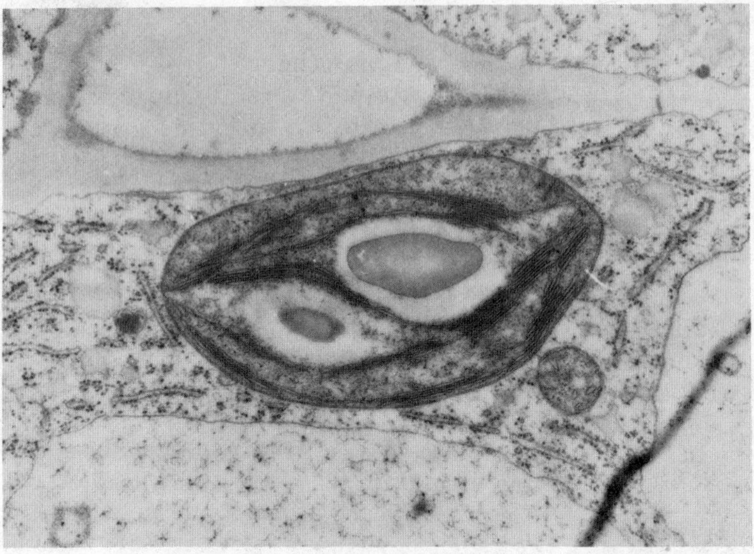

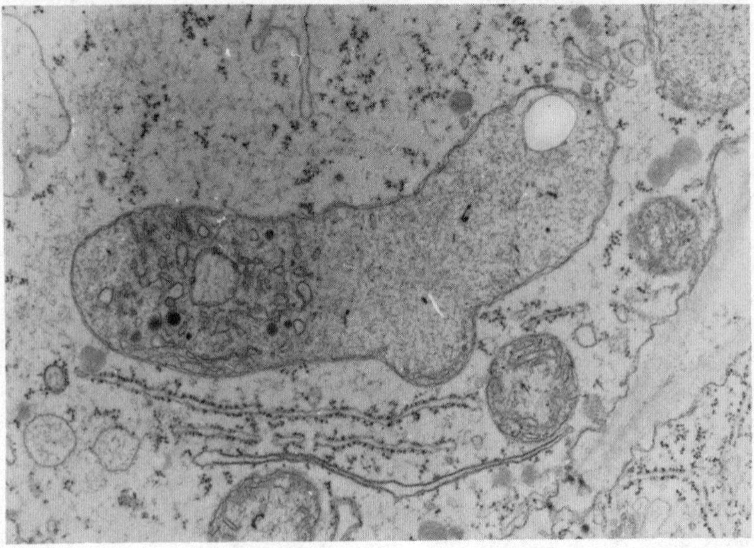

plastid effect, so how could this genetic control be operating? One answer came with the realization that the major control could still be by the female parent, even when most plastids came from the male. To examine this possibility six variegated cultivars containing mutant white plastids were each crossed reciprocally with the two green cultivars F. of Spring and D. Varden, which were known to differ significantly in their behavior (Tilney-Bassett, 1970c). The results of these twenty-four crosses showed that in the six G × W crosses with F. of Spring as the female parent the segregation ratios were very similar with a consistent 50–56% green plastid transmission, whereas with D. Varden as female parent, the segregation ratios altered markedly, for all six crosses together, changing to a new level of 77–93% green plastid transmission, and with a different segregation pattern. In the reciprocal crosses the six different white females could be arranged in an identical sequence, with each green male, in which white plastids were increasingly successful, and white and variegated embryos increasingly frequent. Hence in both G × W and W × G sets of crosses, the influence of the female parent was decisive, irrespective of whether the male or female contributed the most plastids, and the male had only a minor, modifying effect.

The important role of the female parent in the control of plastid inheritance still left undecided the question of whether the nucleus or the plastid was the more important factor. This was resolved by inbreeding the two cultivars F. of Spring and D. Varden, and by determining the behavior of hybrids between them (Tilney-Bassett, 1973). The results showed that when the cultivars were used as female parents in G × W crosses there were two distinct segregation patterns: these are called type I (G > V > W), in which most offspring are green, rather less variegated, and a few are white, or occasionally the variegated are more frequent than green; and type II (G > V

Fig. 1. *Above:* Section through a cotyledon cell from a 2-week-old green embryo of *Pelargonium zonale* cv. Flower of Spring. The section shows a fully developed chloroplast containing starch grains, a thylakoid-grana membrane system, small osmiophilic globules, and plastid ribosomes. (Magnification × 29,400.) *Below:* Section through a cotyledon cell from a 2-week-old variegated embryo of *Pelargonium zonale* cv. Flower of Spring. The section shows a mutant plastid containing a starch grain and small osmiophilic globules, but there are only vesicles with no development of a thylakoid-grana membrane system, and plastid ribosomes are probably absent. (Magnification × 29,400.)

The cytoplasm is normal in both sections and shows endoplasmic reticulum, cytoplasmic ribosomes, mitochondria, and the early development of storage lipids.

Photographs kindly supplied by Mrs. Paramjit Khera.

< W), in which green and white offspring are approximately equally frequent and variegated few. The cultivar D. Varden was true breeding for the type I pattern, while F. of Spring segregated for the two patterns (table 5). Hybrids with green plastids derived from either F. of Spring or D. Varden behaved either as type I or type II. These important observations showed, for the first time, that although the two cultivars differed in their segregation pattern, this is not due to differences in their plastids, and must therefore be due to differences in their nuclear genotypes.

It was suggested that the two patterns are controlled by alternative alleles of a nuclear gene, named Pr_1 and Pr_2, with an effect on plastid replication at the time of fertilization. Females whose offspring segregate with a type I pattern are homozygous $Pr_1 Pr_1$, and females whose offspring segregate with a type II pattern are heterozygous $Pr_1 Pr_2$. We therefore have in *Pelargonium* a nuclear control of the segregation pattern of the extranuclear plastid inheritance. The assumption that the mutant plastids arose through spontaneous mutation in their own plastid DNA, and not in that of any other cytoplasmic organelle, has been greatly strengthened by observations of mixed cells (Herrmann and Hagemann, 1971; Börner, Knoth, Herrmann, and Hagemann, 1972) and by the finding that at least some mutations have a block in plastid rRNA formation (Börner, Knoth, Herrmann, and Hagemann, 1972; Börner, Herrmann, and Hagemann, 1973), which is known to be coded for by plastid DNA (Börner, 1973).

Selfing the heterozygote, and backcrosses between the heterozygote and homozygote, produced deviations from expected monohybrid and

TABLE 5

SUMMARY OF THE SEGREGATIONAL PATTERNS OF SELECTED PROGENY PRODUCED BY INBREEDING F. OF SPRING AND D. VARDEN, AND BY HYBRIDS BETWEEN THEM, AND OF THE UNSELECTED GREEN PROGENY OF SELFED F_1 HYBRIDS[a]

Parents	Presumed Genotypes of Parents	Segregations Patterns		
		Type I	Type II	Total
D. Varden selfed	$Pr_1 Pr_1 \times Pr_1 Pr_1$	17		17
F. of Spring selfed	$Pr_1 Pr_2 \times Pr_1 Pr_2$	14	11	25
D. Varden ♀ × F. of Spring ♂	$Pr_1 Pr_1 \times Pr_1 Pr_2$	29	18	37
F. of Spring ♀ × D. Varden ♂	$Pr_1 Pr_2 \times Pr_1 Pr_1$	24	1	25
Green F_1 selfed	$Pr_1 Pr_2 \times Pr_1 Pr_2$	20	18	38

[a]In this analysis one has to determine the segregational phenotype of the green progeny, by crossing them as females to a mutant plastid male, in order to deduce the genotypes of their parents. An independent test had shown that the green F_1 was a heterozygous, type II plant before it was selfed.

backcross ratios (table 5) that, if the ratios are unbiased, can be explained on the assumption that the rare Pr_2 allele (Tilney-Bassett, 1974a) is almost completely lethal on the female side, and slightly lethal on the male side. These initial observations were based on the analysis of G × W and W × G plastid crosses and their variegated hybrids. This meant that the Mendelian ratios might be biased by selecting for analysis only the variegated hybrids, while ignoring the green and inviable white progeny. These objections have recently been overcome by selfing heterozygous green plants and testing their solely green F_2 progeny (Tilney-Bassett, 1974b). The results again show an F_2 ratio of 1 homozygote $Pr_1 Pr_1$:1 heterozygote $Pr_1 Pr_2$ instead of the expected Mendelian 3:1 or 1:2:1 ratio. The fact that these F_2 progeny were all green rules out the possibility that the ratio could be biased through loss of white or variegated embryos, the fact that the mean fertilization and embryo survival is the same for families of homozygotes and heterozygotes contradicts zygotic selection, and the fact that embryo death is insufficient to swing the ratio from 3:1 to 1:1 all point to gametophytic selection against the Pr_2 allele, particularly on the female side. These conclusions imply that the conspicuous differences between the segregation patterns from heterozygous and homozygous families after G × W plastid crosses occur despite the fact that the vast majority of the female eggs that are fertilized must be of uniform genotype Pr_1. Hence the differences between the segregation patterns must be determined by the female parent before meiosis when still heterozygous and diploid.

The major, maternal control of plastid inheritance is now being verified with six different variegated cultivars reciprocally crossed by the six isogenic green clones derived from them, to make a total of 36 W × G crosses and 36 G × W crosses (Tilney-Bassett, unpublished data). The results are now sufficiently complete to be able to state that the rule of maternal control holds equally well if the female carries the mutant plastids, W × G crosses, or if it carries the green plastids, G × W crosses; and it holds equally well if the female is homozygous, $Pr_1 Pr_1$, or heterozygous, $Pr_1 Pr_2$. To this extent the behavior is similar to *Oenothera;* the difference lies in the fact that in *Oenothera* maternal control is synonymous with maternal predominance, whereas in *Pelargonium* maternal control permits either male or female predominance.

The *Pelargonium* behavior is evidently more complicated than that of *Oenothera*, but is made even worse by very significant differences between normal and mutant plastids. For example, in the isogenic

F. of Spring reciprocal crosses the nucleus remains constant, yet in G × W crosses the female transmission is about 55% (G plastids), whereas in W × G crosses the female transmission is about 15% (W plastids). The inferior performance of white plastids causes a switch from female predominance in the former cross to male predominance in the latter. By contrast, the isogenic Miss Burdette-Coutts favours the female parent in both directions to give an approximately 95% green plastid transmission in G × W crosses with a 65% white plastid transmission after W × G crosses. Hence it appears that the white plastids of F. of Spring do better when they enter through the male, and the white plastids of Miss Burdette-Coutts do better when they enter through the female. In table 6 I have compared two whole sets of reciprocal crosses using first F. of Spring and then D. Varden as the constant source of green plastids in crosses with six different variegated cultivars as the source of white plastids. The results show an interesting change in polarity from strongly paternal to strongly maternal, depending on the source of the white plastids. Apart from slight alterations, which are probably not significant, both sets follow the same sequence. The difference between the two green cultivars is vividly expressed by the shift, through all six crosses, of 15–25% toward a greater maternal polarity with D. Varden compared with F. of Spring.

The transition from a strongly paternal to a strongly maternal polarity vividly expresses the effect of the different sources of white plastids upon the results of W × G crosses. The question that must now be answered is whether these differences are caused by varying nuclear genotypes associated with the white female parents, or whether their varying behavior reflects real differences between the mutant plastid genotypes. Our electron microscope studies of the ultrastructural development of the mutant plastids of these cultivars do not suggest any major differences between their structural capabilities (Khera and Tilney-Bassett, unpublished data), but this does not prove that they are identical. A more appropiate analysis would be to test the behavior of the plastid mutants, after crossing them into a similar nuclear background, in the same manner as was used to confirm the similarity of the green plastids of D. Varden and F. of Spring. An associated experiment is to compare in more detail the behavior of white plastids from a homozygous female parent with that from a heterozygous female parent after W × G crosses. Experiments now under way should soon make it possible to determine whether or not there is a distinction between type I and type II segregation

TABLE 6

COMPARISON BETWEEN TWO SETS OF SIX PAIRS OF RECIPROCAL CROSSES TO SHOW THE GRADATION IN THE POLARITY OF PLASTID TRANSMISSION FROM STRONGLY PATERNAL TO STRONGLY MATERNAL

Cross[a]		Percentage of Plastid Distribution		Polarity (Green Female minus Green Male)
		Female	Male	
FS (Green Parent)[b]				
FS × FoS	G × W	56.3	43.7	−32.8 (paternal)
	W × G	10.9	89.1	
FS × FS	G × W	54.6	45.4	−30.6 (paternal)
	W × G	14.8	85.2	
FS × JCM	G × W	55.0	45.0	−22.3 (paternal)
	W × G	22.7	77.3	
FS × DV	G × W	54.5	45.5	−17.1 (paternal)
	W × G	28.4	71.6	
FS × LG	G × W	50.1	49.9	−1.2 (paternal)
	W × G	48.7	51.3	
FS × MBC	G × W	55.0	45.0	20.4 (maternal)
	W × G	65.4	34.6	
DV (Green Parent)[b]				
DV × FoS	G × W	89.1	10.9	−9.2 (paternal)
	W × G	1.7	98.3	
DV × FS	G × W	83.8	16.2	−11.7 (paternal)
	W × G	4.5	95.5	
DV × JCM	G × W	93.6	6.4	5.5 (maternal)
	W × G	11.9	88.1	
DV × DV	G × W	77.2	22.8	−1.1 (paternal)
	W × G	21.7	78.3	
DV × LG	G × W	80.3	19.7	17.7 (maternal)
	W × G	37.4	62.6	
DV × MBC	G × W	87.6	12.4	45.6 (maternal)
	W × G	58.0	42.0	

[a]FoS = cv. Foster's Seedling; FS = cv. Flower of Spring; JCM = cv. J. C. Mappin; DV = cv. Dolly Varden; LG = cv. Lass O' Gowrie; MBC = cv. Miss Burdette-Coutts

[b]Green Dolly Varden is a type I, and Green Flower of Spring a type II, cultivar.

patterns after W × G crosses, corresponding to, although not identical with, the two patterns after G × W crosses.

Until these questions are answered, it is premature to speculate on any hypotheses that assume genetic differences between the plastids; nevertheless it is tempting to wonder whether there might be a recombination of organelle DNAs in higher plants comparable to that found in the chloroplasts of *Chlamydomonas* (Sager, 1972), or the mitochondria of yeast (Coen et al., 1970) or paramecia (Adoutte, 1974). In these organelles it has been possible to demonstrate both a recombination of organelle genes and a polarity of their transmission, dependent on the varying distances of the genes from a "centromere-like" origin or attachment point. Since the *Pelargonium* mutants are phenotypically alike, there is little immediate prospect of establishing the necessary conditions for demonstrating recombination should it exist. On the other hand, if it should turn out that the white mutants are genetically different, and have therefore probably arisen by mutation at different sites along the plastid genome, then their varying behaviour in W × G crosses might be interpreted as demonstrating a real polarity of transmission, in which the probability of inheritance for a particular mutant is, at least in part, related to its distance along the plastid genome. *Pelargonium* has already proved to be an important organism for studying the genetic controls of plastid inheritance; it may also be the most suitable higher plant for studying the possibility of recombination in a cytoplasmic organelle, and certainly, for the moment, it appears to be the only plant in the running.

LITERATURE CITED

Adoutte, A. 1974. Mitochondrial mutations in *Paramecium:* Phenotypical characterization and recombination, pp. 263–71. *In* A. M. Kroon and C. Saccone (eds.), The biogenesis of mitochondria. Academic Press, London.

Anton-Lamprecht, I. 1966. Beitrage zum Problem der Plastidenäbanderung. III. Uber das Vorkommen von "Ruckmutationen" in einer spontan enstandenen Plastidenscheсke von *Epilobium hirsutum.* Z. Pflanzenphysiol. 54:417–45.

Anton-Lamprecht, I. 1967. Elektronenmikroskopische Untersuchungen an Plasmonabänderungen von *Epilobium*-Bastarden. IV. Mischzellen mit normalen und atypischen Chondriosomen. Z. Naturf. 22b:1340–42.

Arnason, T. J., J. B. Harrington, and H. A. Friesen. 1946. Inheritance of variegation in barley. Canad. J. Res. 24:145–57.

Arnason, T. J., and G. W. R. Walker. 1949. An irreversible gene-induced plastid mutation. Canad. J. Res. 27c:172–78.

Börner, Th. 1973. Struktur und Funktion der genetischen Information in den Plastiden. Biol. Zentralbl. 92:545–61.

Börner, Th., R. Knoth, F. Herrmann, and R. Hagemann. 1972. Struktur und Funktion der genetischen Information in den Plastiden. V. Das Fehlen von ribosomaler RNS in den Plastiden der Plastommutante "Mrs. Parker" von *Pelargonium zonale* Ait. Theoret. Appl. Genet. 42:3–11.

Börner, Th., F. Herrmann, and R. Hagemann. 1973. Plastid ribosome deficient mutants of *Pelargonium zonale.* FEBS Letters 37:117–19.

Briggle, L. W. 1966. Inheritance of a variegated leaf pattern in hexaploid wheat. Crop Sci. 6:43–45.

Chesnoy, L. 1973. Sur l'origine paternelle des organites du proembryon du *Chamaecyparis lawsoniana* A. Murr. (Cupressacées). Carylogia 25:223–32.

Chesnoy, L., and M. J. Thomas. 1971. Electron microscopy studies on gametogenesis and fertilization in gymnosperms. Phytomorphology 21:50–63.

Coen, R., J. Deutsch, P. Netter, E. Petrochilo, and P. Slonimski. 1970. Mitochondrial genetics. I. Methodology and phenomenology. Symp. Soc. Exp. Biol. 24:449–96.

Diers, L. 1963. Elektronenmikroskopische Beobachtugen an der generativen Zelle von *Oenothera hookeri* Torr et Gray. Z. Naturf. 18b:562–66.

Diers, L. 1967. Übertragung von Plastiden durch den Pollen bei *Antirrhinum majus.* Molec. Gen. Genet. 100:56–62.

Diers, L. 1971. Übertragung von Plastiden durch den Pollen bei *Antirrhinum majus.* II. Der Einfluss verschiedener Temperaturen auf die Zahl der Schecken. Molec. Gen. Genet. 113:150–53.

Duhlieu, H. 1967. Sur les différents types de mutations extranucléaires induites par le méthane sulfonate d'éthyle chez *Nicotiana tabacum* L. Mut. Res. 4:177–89.

Epp, M. D. 1973. Nuclear gene-induced plastome mutations in *Oenothera hookeri.* I. Genetic analysis. Genetics 75:465–83.

Fröst, S., L. Vaivars, and C. Carlbom. 1970. Reciprocal extrachromosomal inheritance in rye (*Secale cereale* L.). Hereditas 65:251–60.

Hagemann, R. 1964. Plasmatische Vererbung. Veb Gustav Fischer Verlag, Jena.

Hagemann, R. 1971. Struktur und Funktion der genetischen Information in den Plastiden. I. Die Bedeutung von Plastommutanten und die genetische Nomenklatur extranukleärer Mutationen. Biol. Zentralbl. 90:409–18.

Hagemann, R., and F. Scholz. 1962. Ein Fall geninduzierter Mutationen des Plasmotypus bei Gerste. Züchter 32:50–59.

Hagiwara, T., and Y. Oomura. 1939. Plastid inheritance of variegation in *Capsicum annuum.* Jap. J. Genet. 15:328–30.

Herrmann, F., and R. Hagemann. 1971. Struktur und Funktion der genetischen Information in den Plastiden. III. Genetik, Chlorophylle und Photosyntheseverhalten der Plastommutante "Mrs. Pollock" und der Genmutante "Cloth of Gold" von *Pelargonium zonale.* Biochem. Physiol. Pfl. 162:390–409.

Imai, Y. 1928. A consideration of variegation. Genetics 13:544–62.

Imai, Y. 1935. Variation in the rate of recurring plastid mutations in *Hordeum vulgare* caused by differences in the sowing times. Genetics 20:36–41.

Imai, Y. 1936a. Recurrent auto- and exomutation of plastids resulting in tricolored variegations of *Hordeum vulgare.* Genetics 21:752–57.

Imai, Y. 1936b. Chlorophyll variegations due to mutable genes and plastids. Z. VererbungsLehre 71:61–83.

Jenkins, M. T. 1924. Heritable characters of maize (Iojap striping, a chlorophyll defect). J. Hered. 15:467–72.

Kirk, J. T. O., and R. A. E. Tilney-Bassett. 1967. The plastids: Their chemistry, structure, growth, and inheritance. W. H. Freeman and Co., London and San Francisco.

Kohel, R. J. 1967. Variegated mutants in cotton, *Gossypium hirsutum* L. Crop Sci. 7:490–92.

Kondo, M., M. Takeda, and S. Fujimoto. 1927. Untersuchungen über die weissgestreifte Resipflanze (Shimaine). Ber. Ohara Inst. 3:291–317.

Krishnaswami, R. 1968. Occurrence of plastid mutation and its inheritance in *Gossypium*. Curr. Sci. 37:294–95.

Lombardo, G., and F. M. Gerola. 1968. Cytoplasmic inheritance and ultrastructure of the male generative cell of higher plants. Planta 82:105–10.

Michaelis, P. 1965. Uber kerninduzierte Plasma- und PlastidenAbänderungen. Z. Naturforsch. 20b:264–67.

Michaelis, P. 1968a. Beiträge zum Problem der Plastidenabänderung. IV. Uber das Plasma- und Plastidenabänderungen auslösende, isotopen-(^{32}p)-induzierte Kerngen mp_1 von *Epilobium*. Molec. Gen. Genet. 101:257–306.

Michaelis, P. 1968b. Uber eine vermutliche Kombination von Plasma- und Plastidenabänderungen mit abnormen Scheckungsmustern und abnormen Erbgang. Flora 159:307–38.

Michaelis, P. 1968c. Beiträge zum Problem der Plastidenabänderungen. V. Uber eine weitere isotopen-(^{35}S)-induzierte Kernmutante, die Plastidenabänderungen hervorruft. Theoret. Appl. Genet. 38:314–20.

Michaelis, P. 1969. Über Plastiden-Restitutionen (Rückmutationen). Cytologia 34 Suppl.:1–115.

Michaelis, P. and H. G. Fritz. 1966. Beiträge zum Problem der Plastidenabänderungen. II. Chlorophyllbestimmungen an Pflanzen mit Plastiden- "Rückmutationen". Z. Naturforsch. 21b:66–71.

Morgensen, H. L. 1972. Fine structure and composition of the egg apparatus before and after fertilization in *Querqus gambelii:* The functional ovule. Amer. J. Bot. 59:931–41.

Morinaga, T. 1932. The chlorophyll deficiencies in rice. Bot. Mag. Tokyo 46:202–7.

Nilsson-Tillgren, T., and P. von Wettstein-Knowles. 1970. When is the male plastome eliminated? Nature 227:1265–66.

Ohba, K., M. Iwakawa, Y. Ohada, and M. Murai. 1971. Paternal transmission of a plastid anomaly in some reciprocal crosses of Suzi, *Cryptomeria japonica* D. Don. Silvae Genet. 20:101–7.

Pal, B. P., and S. Ramanujam. 1941. A new type of variegation in rice. Indian J. Agric. Sci. 11:170–76.

Potrykus, I. 1970. Mutation und Rückmutation extrachromosomal verbter Plastidenmerkmale von Petunia. Z. Pfl. Zücht. 63:24–40.

Redei, G. P. 1973. Extra-chromosomal mutability determined by a nuclear gene locus in *Arabidopsis*. Mut. Res. 18:149–62.

Rhoades, M. M. 1943. Genic induction of an inherited cytoplasmic difference. Proc. Nat. Acad. Sci. USA. 29:327–29.

Rhoades, M. M. 1946. Plastid mutations. Cold Spring Harbor Symp. Quant. Biol. 11:202–7.

Rhoades, M. M. 1950. Gene induced mutation of a heritable cytoplasmic factor producing male sterility in maize. Proc. Nat. Acad. Sci. USA. 36:634–35.

Röbbelen, G. 1966. Chloroplastendiffenenzierung nach geninduzierter Plastommutation bei *Arabidopsis thaliana* (L.) Heynh. Z. Pflphysiol. 55:387–403.

Sager, R. 1972. Cytoplasmic genes and organelles. Academic Press, New York.

Schötz, F. 1954. Über Plastidenkonkurrenz bei *Oenothera*. Planta 43:182–240.

Schötz, F. 1968. Über Plastinenkonkurrenz bei *Oenothera*. II. Biol. Zentralbl. 87:33–61.

Schötz, F. 1974. Untersuchungen über die Plastidenkonkurrenz bei *Oenothera*. IV. Der Einfluss des Genoms auf die Durchsetzungsfähigkeit der Plastiden. Biol. Zentralbl. 93:41–64.

Schötz, F., and F. Heiser. 1969. Über Plastidenkonkurrenz bei *Oenothera*. III. Zahlenverhältnisse in Mischzellen. Wiss. Z. Pädagogische Hochschule Potsdam 13:65–89.

Shumway, L. K., and L. F. Bauman. 1967. Nonchromosomal stripe of maize. Genetics 55:33–38.

Shumway, L. K., and A. Kleinhofs. 1973. Aspects of the biochemistry and ultrastructure of a cytoplasmically inherited plastid defect (D1) of tobacco. Biochem. Genet. 8:271–80.

Shumway, L. K., and T. E. Weier. 1967. The chloroplast structure of iojap maize. Amer. J. Bot. 54:773–80.

Simmonds, N. W. 1969. Variegated mutant plastid chimeras of potatoes. Heredity 24:303–6.

So, M. 1921. On the inheritance of variegation of barley. Jap. J. Genet. 1:21–36.

Stroup, D. 1970. Genic induction and maternal transmission of variegation in *Zea mays*. J. Hered. 61:139–41.

Stubbe, W. 1959a. Genetische Analyse des Zusammenwirkens von Genom und Plastom bei *Oenothera*. Z. VererbungsLehre 90:288–98.

Stubbe, W. 1959b. Genetic analysis of the correlation between different genomes and plastomes of *Oenothera*. Recent Adv. Bot. 2:1439–42.

Stubbe, W. 1960. Untersuchungen zur genetischen Analyse der Plastoms von *Oenothera*. Zeit. Bot. 48:191–218.

Stubbe, W. 1963. Die Rolle des Plastoms in der Evolution der Oenotheren. Ber. deut. bot. Ges. 76:154–67.

Stubbe, W. 1964. The role of the plastome in evolution of the genus *Oenothera*. Genetica 35:28–33.

Tatebe, T. 1961. Genetic studies on the leaf variegation of *Allium fistulosum* L. Jap. J. Genet. 36:151–56.

Tatebe, T. 1968. Genetic studies on the leaf variegation of *Allium cepa* L. J. Jap. Soc. Hort. Sci. 37:345–48.

Tatebe, T. 1972. Inheritance of leaf variegation in buckwheat (with English summary, p. 45) Jap. J. Breed. 22:40–45.

Tilney-Bassett, R. A. E. 1963a. The structure of periclinal chimeras. Heredity 18:265–85.

Tilney-Bassett, R. A. E. 1963b. Genetics and plastid physiology in *Pelargonium*. Heredity 18:485–504.

Tilney-Bassett, R. A. E. 1964. Failure to transmit mutant plastids in a *Pelargonium* cross. Heredity 19:516–18.

Tilney-Bassett, R. A. E. 1965. Genetics and plastid physiology in *Pelargonium*. II. Heredity 20:451–66.

Tilney-Bassett, R. A. E. 1970a. Genetics and plastid physiology in *Pelargonium*. III. Effect of cultivar and plastids on fertilisation and embryo survival. Heredity 25:89–103.

Tilney-Bassett, R. A. E. 1970b. Effect of environment on plastid segregation in young embryos of *Pelargonium* × *Hortorum* Bailey. Ann. Bot. 34:811–16.

Tilney-Bassett, R. A. E. 1970c. The control of plastid inheritance in *Pelargonium.* Genet. Res. Camb. 16:49–61.

Tilney-Bassett, R. A. E. 1973. The control of plastid inheritance in *Pelargonium.* II. Heredity 30:1–13.

Tilney-Bassett, R. A. E. 1974a. A search for the rare type II (G > V < W) plastid segregation pattern among cultivars of *Pelargonium* × *Hortorum* Bailey. Ann. Bot. 38:333–35.

Tilney-Bassett, R. A. E. 1974b. The control of plastid inheritance in *Pelargonium.* III. Heredity (in press).

Wettstein, D. von, and G. Ericksson. 1965. The genetics of chloroplasts. Proc. 11 Int. Cong. Genet. 3:591–612.

Whitehouse, H. L. K. 1973. Towards an understanding of the mechanism of heredity. 3d ed. Edward Arnold (Publishers) Ltd., London.

Wildman, S. G., C. Lu-Liao, and F. Wong-Staal. 1973. Maternal inheritance, cytology, and macromolecular composition of defective chloroplasts in a variegated mutant of *Nicotiana tabacum.* Planta 113:293–312.

Wong-Staal, F., and S. G. Wildman. 1973. Identification of a mutation in chloroplast DNA correlated with formation of defective chloroplasts in a variegated mutant of *Nicotiana tabacum.* Planta 113:313–26.

Woods, M. W., and H. G. DuBuy. 1951. Hereditary and pathogenic nature of mutant mitochondria in *Nepeta.* J. Nat. Cancer Inst. 11:1105–51.

S. G. WILDMAN, K. CHEN, J. C. GRAY, S. D. KUNG, P. KWANYUEN, AND K. SAKANO

Evolution of Ferredoxin and Fraction I Protein in the Genus *Nicotiana*

9

Fraction I protein is the most abundant protein in the world. It is found in all organisms containing chlorophyll *a*, including the procaryotic blue-green algae. Fraction I and protein is identical to ribulose diphosphate carboxylase (RuDPCase), the enzyme that catalyzes the crucial step of carbon dioxide fixation during photosynthesis. Since RuDPCase is also found in photosynthetic bacteria, it is reasoned that this protein must be of very ancient origin and had already evolved into a vital catalyst in those primeval organisms that had only just developed a mechanism for transducing light into chemical energy. Evidently RuDPCase appeared on the biological scene one or two billion years before the oxygen-producing function of photosynthesis had evolved and created an atmosphere suitable for evolution of animals. If, as seems probable, chemoautotrophic bacteria were the progenitors of the photosynthetic organisms, we can also surmise that the ancestor of Fraction I proteins evolved along with the proteins capable of electron transport or shortly thereafter. In this sense we can order the appearance of primordial Fraction I protein as not much later than cytochrome *c* and ferredoxin, and we can imagine, therefore, that it has been buffeted by the forces of evolution for more than 3 billion years.

In the case of cytochrome *c* and ferredoxin, there is a great reservoir of knowledge about the sequences of amino acids constituting the primary polypeptide chains of these proteins. Indeed, the information

Department of Biology, Molecular Biology Institute, University of California, Los Angeles, California 90024 is already sufficient to be the foundation for the new discipline of chemical paleogenetics, whereby phylogenetic trees have been constructed solely by comparing amino acid sequences from proteins isolated from diverse species of biological organisms. Thus, the probable number of mutational steps, for example, separating the cytochrome *c* of man from apes, or even wheat, can be estimated, as well as the probable time of evolution that separated one mutation in the sequence from another. The phylogenetic trees constructed from amino acid sequences are remarkably faithful to their predecessors built by comparing morphological properties of living organisms with those of fossils. In the case of Fraction I proteins, not enough is known about amino acid sequences to permit construction of such phylogenetic trees. However, Fraction I proteins isolated from different species of the genus *Nicotiana* have been shown to differ in physicochemical properties, which has made it possible to describe the evolution of this protein during the evolution of a new species of *Nicotiana*. Moreover, there is also the knowledge of the geological time when the ancestors of some *Nicotiana* species became geographically separated from other species. This information makes possible a meaningful comparison of the degree of mutation of Fraction I protein with that of ferredoxin during the same span of evolutionary time. Both proteins are uniquely necessary to the photosynthetic function of chloroplasts. The genes coding for ferredoxin are contained in nuclear DNA (Kwanyuen and Wildman, 1975), whereas the genes coding for Fraction I protein are contained in both nuclear and chloroplast DNAs (Chan and Wildman, 1972; Kawashima and Wildman, 1972). Therefore, it will be possible to compare nuclear *vs.* chloroplast DNA genes in regard to mutability as it pertains to two proteins of vital concern to the plant's existence as a photoautotrophic organism.

THE GENUS *NICOTIANA*

The pioneering cytogenetic and taxonomic research of Thomas Harper Goodspeed and his students is the main source of our extensive knowledge of the interrelationships and evolution of the more than 60 well-defined species that constitute the genus *Nicotiana* (Goodspeed, 1954). Apparently, the stimulus for working on *Nicotiana* in the first place was the ability to produce viable F_1 interspecific hybrids between many of the species. *Nicotiana* is one of the few genera in the entire biological world affording the opportunity to produce such a wide range of interspecific hybrids. According to Goodspeed, whose ideas have been supported and refined by the more recent investigations of Dr. N. Burbidge in Australia (Burbidge, 1960), the large number of species of *Nicotiana* in nature arose primarily by hybridization encounters between different species. Most such encounters were non-productive in regard to speciation because of the failure of the two complements of parental chromosomes to pair properly during meiosis. The F_1 hybrids were self-infertile, could not produce seeds, and, therefore, could not survive alternation of generations. However, on extremely rare occasions, spontaneous doubling of the somatic F_1 hybrid chromosomes occurred and permitted satisfactory pairing in meiosis and consequent self-fertility. Self-fertility was the evolutionary necessity for a new species of *Nicotiana* to be capable of self-perpetuation by seeds and thereby complete alternation of generations.

It is Goodspeed's view that the ancestors of the *Nicotiana* species arose before the present day continents of the earth began to drift apart. He places the locale of these *Nicotiana* ancestors near to what is now the northern half of South America. As evolution ensued, the ancestors that were to evolve into present-day *Nicotiana* species radiated into what has now become North and South America. The ancestors also radiated by the Antarctica land bridge to what later became the island continent of Australia. However, no ancestors of *Nicotiana* had spread into those regions of the earth that were to later separate into Europe, Asia, and Africa when continental drift began during the Cretaceous period. The separation of continents was, therefore, the starting point of the entirely separate pathways of evolution of the Australian *Nicotiana* species and those of the Western Hemisphere. Evolution during the ensuing period of 150–250 million years has resulted in the present day occurrence of about twenty different species of *Nicotiana* in Australia and nearby islands and about forty other species in North and South America.

Fraction I protein and ferredoxin are present in all species of *Nicotiana*. Consequently, the effect of 150–250 million years of evolution on the amino acid composition of these two different proteins can be compared. Of greater importance, however, is that this long period of separation has optimized the probability of finding differences in the composition of proteins contained in different species of *Nicotiana*. The ability to produce reciprocal, interspecific *Nicotiana* hybrids has provided a system with which to determine whether the genes coding for the amino acid sequences of chloroplast proteins are located in nuclear or extranuclear DNA.

THE *NICOTIANA* SYSTEM FOR DETERMINING THE LOCATION OF THE GENETIC INFORMATION CODING FOR CHLOROPLAST PROTEINS

The basis for using the *Nicotiana* system is that some heritable factors affecting the structure of chloroplasts are transmitted from generation to generation only via the maternal line. In fact, it was the genetic analysis of a variegated *N. tabacum* by Correns around 1908 that resulted in the first discovery of maternal, or what is often called cytoplasmic or extranuclear, inheritance. Defective chloroplasts lacking, for the most part, a thylakoid system of membranes containing chlorophyll are the fundamental cause of the variegation of *N. tabacum* leaves. Defective chloroplasts can exist in the same cell with perfectly normal chloroplasts (Wildman et al., 1973). Defective chloroplasts contain chloroplast DNA, which is slightly different in physicochemical properties from wild-type *N. tabacum* chloroplast DNA (Wong-Staal and Wildman, 1973). Thus, a positive correlation exists between a chemical mutation in chloroplast DNA and a profound defect in the structure of the chloroplast, the defect being inherited only by the maternal line (Wildman et al., 1973). There is thus some assurance that chloroplast DNA itself is transmitted from generation to generation exclusively by the maternal line in the genus *Nicotiana*.

To determine the location of the genetic information coding for a chloroplast protein, it is necessary to find two species of *Nicotiana* in which the protein under investigation can be demonstrated to differ in amino acid sequence, or some physicochemical property related to amino acid sequence. Reciprocal F_1 hybrids are then produced between these species, and the nature of the protein in the interspecific hybrids is examined. The outcome of experiments utilizing the *Nicotiana* system have been interpreted as follows:

1. If the genetic information for the protein in the F_1 interspecific hybrids is transmitted by the pollen of the male parent as well as by the egg cell of the female parent, then the protein is most probably coded by nuclear DNA.
2. If the genetic information for the protein in the F_1 interspecific hybrids is transmitted solely by the maternal parent, then the protein is coded in extranuclear DNA, most probably chloroplast DNA. In this case the maternal inheritance of the protein must be established by an analysis of the protein from both reciprocal interspecific F_1 hybrids.

A slight reservation is necessary, however, because the possibility that the genetic information is contained in mitochondrial DNA cannot be entirely ruled out. The mode of inheritance of mitochondrial DNA in higher plants is unknown. In animals mitochondrial DNA is inherited solely from the maternal parent (Dawid and Blackler, 1972), and this mode of inheritance probably also occurs in higher plants. However, mitochondria have been seen to migrate down pollen tubes in higher plants (Rosen et al., 1964), and, therefore, the possibility also exists that mitochondrial DNA is inherited from both parents in interspecific F_1 hybrids.

With this *Nicotiana* system, nuclear DNA has been found to contain the genetic code for two of the numerous ribosomal proteins composing the 50S subunit of the chloroplast ribosome (Bourque and Wildman, 1972), for the thylakoid membrane protein to which the chlorophyll pigments constituting Photosystem II are attached (Kung et al., 1972), for ferredoxin (Kwanyuen and Wildman, 1975), and for the small subunit of Fraction I protein (Kawashima and Wildman, 1972). Chloroplast DNA has been found to contain the genetic information for the large subunit of Fraction I protein (Chan and Wildman, 1972). Genetic analysis with the *Nicotiana* system has also shown that the information regulating the RuDPCase catalytic activity of Fraction I protein is inherited solely from the maternal parent (Singh and Wildman, 1973).

PHYSICOCHEMICAL NATURE OF FERREDOXIN AND FRACTION I PROTEIN

In *Nicotiana,* as in other plants, ferredoxins are proteins of about 12,000 daltons containing elemental iron and sulfur that are the functional moieties of their electron transport properties. The amino acid sequences of ferredoxins from five different kinds of plants, from the green alga *Scenedesmus* to spinach, show a high degree of similarity. About 60 percent of the residues are invariant in the five proteins (Hall et al., 1973). The amino acid sequences of ferredoxin from *Nicotiana* species are still unknown, but the proteins are all remarkably similar. Fingerprinting of tryptic peptides of ferredoxins from ten species of *Nicotiana* failed to reveal any differences (Kwanyuen and Wildman, 1975).

Fraction I protein of *Nicotiana* species has an $S^{\circ}_{20,w}$ of 18.3 equivalent to a molecular weight of about 550,000 daltons (Kawashima and Wildman, 1971a). Crystalline Fraction I protein, of constant specific RuDPCase activity and constant physicochemical composition, can be easily obtained from fifty species of *Nicotiana* and numerous interspecific F_1 hybrids (Chan et al., 1972). The pure protein is composed entirely of amino acids (Sakano et al., 1973).

By the use of hydrogen bond-disrupting agents, such as 8 M urea or dodecyl sulphate, Fraction I protein is dissociated into large and small subunits that can be physically separated from each other by column chromatography (Rutner and Lane, 1967; Kawashima, 1969). The large subunits make up 75 percent of the mass of Fraction I protein, with the small subunits making up the remaining 25 percent (Kawashima, 1969). In the monomeric form the large subunits have a molecular weight of about 56,000 daltons and the small subunits have a molecular weight of about 12,500 daltons (Rutner, 1970; Kawashima and Wildman, 1970). Hydrolysis of the isolated subunits of *N. tabacum* Fraction I protein with trypsin gives about 55 peptides for the large subunit and about 30 peptides for the small subunit (Kung et al., 1974). By fingerprinting the tryptic peptides of the large and small subunits of Fraction I proteins crystallized from two species of *Nicotiana* and their reciprocal F_1 hybrids, it was found that chloroplast DNA contains the genetic code for the large subunit, and nuclear DNA contains the code for the small subunit (Chan and Wildman, 1972; Kawashima and Wildman, 1972).

More recently, isoelectric focusing of S-carboxymethylated Fraction I protein in polyacrylamide gels containing 8 M urea has proved to be a remarkably simplified and reproducible method for resolving the large and small subunits (Kung et al., 1974) and for determining the location of the DNA coding for these subunits (Sakano et al., 1974). The method requires only microgram amounts of pure protein, compared with the milligram quantities of each subunit that were previously necessary to perform fingerprint analysis. Of more significance, electrofocusing uncovered differences in polypeptide composition that had gone undetected by fingerprinting. Not only did the method enlarge the number of reciprocal interspecific hybrids in which chloroplast DNA coding for the large subunit and nuclear DNA coding for the small subunit could be demonstrated, but the method revealed phenotypic markers for several different chloroplast DNA genes.

GENETIC ANALYSIS OF FERREDOXIN AND FRACTION I PROTEIN BY ISOELECTRIC FOCUSING

Genetic Analysis of Ferredoxin

Ferredoxin isolated from several species of *Nicotiana* gave a single polypeptide band on electrofocusing in polyacrylamide gel containing 8 M urea. The isoelectric point of ferredoxin from *N. glutinosa* was lower than the isoelectric point of all the other species examined, including *N. tabacum* and *N. glauca*. Analysis of the ferredoxin isolated from the hybrid *N. glauca* × *N. glutinosa* showed the presence of two polypeptide bands of equal intensity, corresponding to those of the parent species. This genetic analysis demonstrated that the genetic information for ferredoxin is located in nuclear DNA (Kwanyuen and Wildman, 1975).

Genetic Analysis of Fraction I Protein

The results of analysis of the polypeptide composition of the large and small subunits of a Fraction I protein from most species and interspecific hybrids of *Nicotiana* can be known within five days after harvest of leaves (Sakano et al., 1974). On the first day the protein crystallizes from extracts of leaves after the homogenates have been clarified by centrifugation and the remaining solution concentrated through molecular membranes and dialyzed to remove salt. During two more days the protein is brought to constant specific RuDPCase activity and physicochemical properties as well as amino acid composition by recrystallization on each day. On the fourth day the protein in 8 M urea is S-carboxymethylated and applied to prefocused polyacrylamide gels containing 8 M urea and ampholines to establish a pH gradient. The dissociated polypeptides resolve during overnight electrofocusing so that on the next morning staining the gel with bromophenol blue will cause the bands of polypeptides of different isoelectric points to become visible.

Figure 1 is a photograph of a result of simultaneously electrofocusing Fraction I proteins of *N. glauca, N. tabacum*, and the reciprocal hybrids, *N. glauca* × *N. tabacum*, in the same slab gel. The photograph shows that the large subunit of each species of protein consists of three polypeptides of different isolectric points. However, one of the three polypeptides of *N. tabacum* protein has a different isoelectric point from those of *N. glauca* protein. In the reciprocal hybrids the isoelectric points of the three large subunit polypeptides correspond to the isoelectric points of the polypeptides of the female parent. The three large subunit polypeptides are inherited and expressed together; there is no separation of the inheritance of the individual polypeptides. Thus the genetic analysis obtained by electrofocusing is entirely consistent with the genetic analysis derived from fingerprinting the tryptic peptides of the isolated large subunits (Chan and Wildman, 1972). Both of these methods demonstrate that the genes

Fig. 1. Genetic analysis of Fraction I protein by isoelectric focusing. Polypeptide composition of Fraction I proteins from reciprocal interspecific hybrids of *N. tabacum* and *N. glauca*. Left to right: (1) *N. tabacum;* (2) *N. glauca;* (3) *N. glauca* × *N. tabacum;* (4) *N. tabacum* × *N. glauca.*

that code for the primary structure of the large subunit are inherited only from the maternal line and are, therefore, located in chloroplast DNA.

Figure 1 also shows that the small subunit of *N. glauca* Fraction I protein is composed of a single polypeptide, whereas the small subunit of *N. tabacum* consists of two polypeptides with different isoelectric points. The small subunit polypeptides characteristic of both parents are found in each of the reciprocal, *N. glauca* × *N. tabacum* F_1 hybrids. Therefore, each hybrid contains genetic information for the small subunit that was brought to the female parent by pollen and was, therefore, contained in nuclear DNA.

Previous analyses by fingerprinting of tryptic peptides of the small subunits of Fraction I protein from the reciprocal F_1 hybrids of *N. glauca* × *N. tabacum* had failed to reveal the presence of the *N. glauca* small subunit, and it was suggested that this represented an example of dominance of the *N. tabacum* genome over that of *N. glauca* (Kawashima and Wildman, 1972). However, owing to the difficulties inherent in the fingerprinting technique used, it is believed that the isoelectric focusing method is more reliable. Examination of the inheritance of subunits by isoelectric focusing in all available interspecific hybrids has failed to reveal dominance (Sakano et al., 1974).

By the electrofocusing method, chloroplast DNA coding for the large subunit and nuclear DNA coding for the small subunit of Fraction I protein has been demonstrated by analysis of the following reciprocal hybrids: *N. tabacum* × *N. glutinosa, N. tabacum* × *N. sylvestris, N. tabacum* × *N. glauca, N. tabacum* × *N. gossei,* and *N. glauca* × *N. langsdorffii* (Sakano et al., 1974). The polypeptide patterns, typified by those shown in figure 1 are completely reproducible and have provided a remarkably simplified method for exploring the evolution and other expressions of the genetic behavior of Fraction I protein.

POLYPEPTIDE COMPOSITION OF FRACTION I PROTEINS FROM *NICOTIANA* SPECIES

Figure 2 is a diagram showing the polypeptide patterns produced by electrofocusing Fraction I proteins isolated from sixteen different species of *Nicotiana,* five being indigenous to Australia and the rest to the Western Hemisphere (unpublished data). The patterns are representative of many other species and provide an insight into the mode of evolution of Fraction I protein.

Composition of the Large Subunit Coded by Chloroplast DNA

The proteins from all species are composed of three polypeptides in about the same proportions and separated in isoelectric point to about the same extent. However, the large subunit polypeptides fall into four groups based on their isoelectric points. The five species

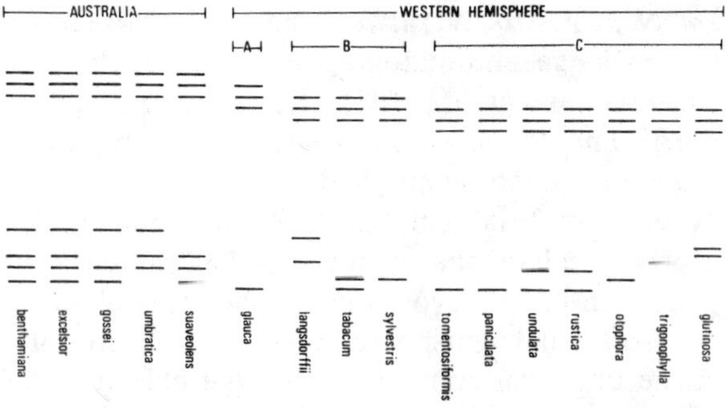

Fig. 2. Polypeptide composition of large and small subunits of Fraction I proteins from *Nicotiana* species.

of *Nicotiana* indigenous to Australia shown in figure 2 are in the same group, as are the following Australian species: *N. ingulba, N. amplexicaulis, N. hesperis, N. debneyi, N. goodspeedii, N. maritima, N. velutina, N. simulans, N. cavicola, N. occidentalis, N. megalosiphon, N. rotundifolia,* and *N. exigua* (unpublished data). The polypeptides of the large subunits of Fraction I proteins of the eleven species of *Nicotiana* indigenous to the Western Hemisphere shown in figure 2 also consist of three polypeptides, but their isoelectric points allow for classifying the species of protein under one of three groups: A, B, and C. *N. glauca* is the only species in Group A, which, on the basis of the isoelectric points of the large subunit polypeptides, has two polypeptides in common with the Australian species and also with the species in Group B. The Fraction I proteins from species in Group B have two large subunit polypeptides in common with species in Group A and with species in Group C. The Group C proteins do not have any polypeptides in common with the Australian species. The three polypeptides could be due to (1) three distinct polypeptides coded by three separate chloroplast DNA genes, or to (2) modifications, such as deamidation of glutaminyl- or asparaginyl-residues, of a single gene product producing three polypeptides of different charge. However, the differences between the three polypeptides were not reflected in an obvious change in the fingerprint pattern of the 55 tryptic peptides composing each of the three polypeptides of the large subunit of *N. tabacum* Fraction I protein. This problem is currently under investigation.

The nature of the differences between the large subunits of the proteins in Groups A, B, and C are also unknown. Analysis of the tryptic peptides of the large subunits of proteins from *N. glauca, N. tabacum, N. sylvestris, N. rustica,* and *N. glutinosa* by fingerprinting and by ion exchange chromatography has failed to reveal any differences (Kawashima et al., 1971; Kawashima et al., 1974). Only *N. langsdorffii* and the Australian species can be shown to differ from *N. tabacum* by these methods and then only by a very few peptides (Chan and Wildman, 1972; Kawashima et al., 1974). We can, therefore, conclude that in the 150–250 million years that have separated the continuing evolution of the Australian species from those in the Western Hemisphere, there have been only very small changes in the chemical composition of the chloroplast DNA genes that code for the large subunit.

The stability of the chloroplast DNA coding for the large subunit is also indicated by an investigation of the specific RuDPCase activity of the protein. The specific activity of Fraction I proteins is identical in all species examined, except for a single Australian species, *N. gossei,* indicating the stability of the structure of the protein during the 150–250 millions years of evolution of *Nicotiana* (Singh and Wildman, 1973). We can, therefore, imagine that virtually no mutation in chloroplast DNA can survive. Evidently, any mutation in chloroplast DNA that significantly alters the arrangement of amino acids of the large subunit of Fraction I protein responsible for RuDPCase activity is equivalent to a lethal mutation and is not sustained.

Composition of the Small Subunit Coded by Nuclear DNA

As also shown in figure 2, the number of peptides comprising the small subunit ranges from one to four depending on the species of Fraction I protein. However, the following features stand out: the Australian Fraction I proteins may contain up to four polypeptides, but no species has been found with less than two polypeptides; the Western Hemisphere species contain either one or two polypeptides. There are numerous differences in numbers of polypeptides as well as isoelectric points indicating that the nuclear DNA coding for the small subunit of Fraction I protein is much more susceptible to mutation than the chloroplast DNA coding for the large subunit or the nuclear DNA coding for ferredoxin. The function of the small subunit in Fraction I protein is unknown, but the large differences in the small subunits do not affect the enzymic properties of the protein (Singh and Wildman, 1973).

Polypeptide Composition of Fraction I Protein from N. glauca from Four Continents

Among the *Nicotiana* species, *N. glauca* is unusual for its widespread distribution in the world, and this provides an opportunity to analyze the Fraction I protein for possible mutations induced by the changes in environment. Seeds were collected from volunteer plants located in California, Mexico, Israel, Africa, and Australia, and the progeny were grown in Los Angeles. How *N. glauca* arrived at such diverse geographical locations, one being as remote as the shores of Lake Tanganyika, is something of a mystery. One guess is that they traveled as stowaway seeds in company with other agricultural products of value. In any event, there is no detectable difference in the isoelectric points of either the large or small subunit polypeptides of the Fraction I proteins regardless of the origin of *N. glauca* plants (Chen et al.,

1975). Since no evidence was obtained that the forces of natural selection had permitted a mutation to survive in either the chloroplast DNA or nuclear DNA genes coding for *N. glauca* Fraction I protein, an investigation was made to see whether mutations could accumulate as a result of the intensive breeding of *N. tabacum* by man.

Polypeptide Composition of Fraction I Protein from 13 Cultivars of N. tabacum

Because of the commercial importance of tobacco, which developed almost from the time of discovery of the New World by Columbus in 1492, the nuclear genes of *N. tabacum* have been under constant manipulation by plant breeders. The end result has been the development of numerous cultivars (cv) of distinctly different phenotypic appearance, for example, Turkish Samsun, Havana 38, Maryland Mammoth, Green Daruma, Burley, and many more. The electrofocusing patterns of Fraction I proteins obtained from ten different cultivars have shown no departure from that shown in figure 1 for the Turkish Samsun cultivar (Chen et al., 1975). The list of cultivars of identical polypeptide composition includes: a variegated strain of Turkish Samsun with the mutation in chloroplast DNA previously mentioned, the Fraction I protein having been isolated from the white tissue containing almost exclusively defective chloroplasts; an isogenic Turkish Samsun derived from doubling the chromosomes of a haploid plant produced by anther culture; and a White Daruma differing from a Green Daruma cultivar by containing only one-half the amount of chlorophyll and an accompanying reduction in size of the grana of the chloroplasts. The ten cultivars analyzed are representative products of the intensive efforts of man to deliberately change the anatomy, cytology, physiology, disease resistance, climate tolerance, biochemical quality of commercial product, and so on, of a plant that has been under continuous cultivation in most regions of the world for several hundred years. The breeding efforts have produced enough changes in the appearance of some cultivars to raise questions as to whether they should still be considered legitimate members of the *N. tabacum* species. Yet, in none of these ten cultivars could a change in the isoelectric points of the large or small subunits shown for *N. tabacum* in figure 2 be detected. Also, gene dosage did not affect the electrofocusing pattern of the polypeptides of Fraction I protein from *N. tabacum.* The polypeptide patterns for electrofocused protein from diploid *N. tabacum* cv. NC95 compared to a triploid, cv. Ky 16, and an autotetraploid, cv. Burley 21, were identical to that shown for *N. tabacum* cv. Turkish Samsun in figure 2. A fourteenth cultivar of *N. tabacum* proved to be the exception to the previous thirteen cultivars in regard to composition of its Fraction I protein. This cultivar is a male sterile Burley tobacco.

Polypeptide Composition of a Male-Sterile N. tabacum

Electrofocusing of Fraction I protein from the Burley type, male-sterile, *N. tabacum* produced a pattern whose two small subunit polypeptides had isoelectric points identical to those shown in fig. 2 for *N. tabacum* (Chen et al., 1975). However, the surprising result was that the isoelectric points of the three large subunit polypeptides were identical to those of the Australian species shown in figure 2. It was deduced that the male-sterile cultivar had been derived from an original cross where an Australian species was the female parent and *N. tabacum* the source of pollen for fertilization. Although the F_1 hybrid is self-infertile, the ova can still be fertilized by nuclei in *N. tabacum* pollen and thus by repeated backcrossing and continued introduction of *N. tabacum* nuclear genes for the small subunit of Fraction I protein, the Australian type small subunit polypeptides were completely eliminated and replaced by the two *N. tabacum* polypeptides. It was, therefore, gratifying to learn that a search of the breeding record of this particular cultivar revealed that either *N. suaveolens* or *N. megalosiphon*, both Australian species, had served as the female parent in the original cross with *N. tabacum* pollen. As shown in figure 2, the small subunit of *N. suaveolens* protein consists of three polypeptides, none corresponding in isoelectric point with the two of *N. tabacum*. The protein of *N. megalosiphon* is also composed of three polypeptides, two being identical to *N. suaveolens* protein, but none like the two of *N. tabacum*.

There are several important consequences of this analysis of the Fraction I protein in the male-sterile cultivar. First, it demonstrates that in male-sterile lines, which are of great advantage to plant breeders, the chloroplast genome cannot be altered by conventional breeding techniques. This is of particular importance with regard to the current problems with hybrid corn, where male sterility has been used extensively to facilitate hybrid seed production. The male-sterile lines have become susceptible to two fungal diseases, the susceptibility being transferred exclusively by the maternal parent. The *Nicotiana* experiments suggest that the susceptibility factors will be inaccessible to alteration by conventional breeding techniques if they are associated with the chloroplast genome.

Second, it indicates that in studies on the physiological basis of male sterility greater emphasis ought to be placed on the effects of the chloroplast genome, especially since there is no rigorous evidence for the maternal inheritance of any other extranuclear genetic material. The influence of the mitochondrial genome has traditionally been regarded as a source of the cytoplasmic factor in male sterility (see Laughnan, this volume, for a discussion of this phenomenon in maize).

POLYPEPTIDE COMPOSITION OF FRACTION I PROTEINS REPRESENTATIVE OF THE MAJOR PHYLA OF THE PLANT KINGDOM

The diagram shown in figure 3 is a compilation of the results of isoelectric focusing of Fraction I proteins obtained from photosynthetic organisms ranging from green algae to monocotyledons (K. Chen, unpublished data), and can be compared to the results of isoelectric focusing of *Nicotiana* Fraction I proteins shown in figure 2. In each instance the large subunit was found to resolve into three polypeptides. The isoelectric points of the large subunit polypeptides of some proteins are considerably different from others, although the pH difference separating the individual polypeptides making up the large subunits is nearly the same for each species. The wide difference in pH between

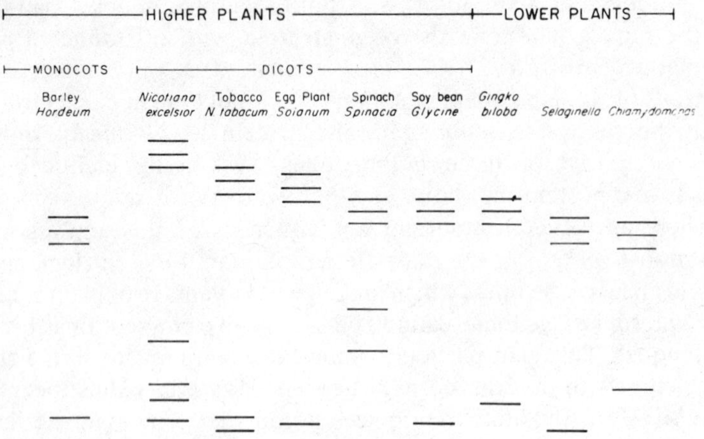

Fig. 3. Polypeptide composition of Fraction I proteins representative of the major phyla of the plant kingdom.

the *N. tabacum* and spinach large subunit polypeptides can be attributed to already known differences in sequence. Comparative fingerprinting of tryptic peptides has shown that spinach and tobacco large subunits differ by five peptides (Kawashima and Wildman, 1971b). A tantalizing question is how and why the three large subunit polypeptides evolved in the first place and have since been maintained during the evolution of Fraction I protein.

MECHANISM OF EVOLUTION OF FRACTION I PROTEIN IN *NICOTIANA*

The large subunits of Fraction I proteins from all *Nicotiana* species are very similar, and it is not possible at present to describe the evolution of the large subunits except to emphasize that the structure has been highly conserved throughout the genus. The small subunits, however, show much more variability as demonstrated both by electrofocusing and by analysis of tryptic peptides, and it is possible to describe the evolution of the small subunit of Fraction I protein in terms of the evolution of species in the genus *Nicotiana*.

Goodspeed viewed the origin of new species of *Nicotiana* as occurring by chromosome doubling following interspecific hybridization. The experimental verification of this idea was accomplished by cytogenetic studies of *N. tabacum*, the synthetic species *N. digluta*, and their parent species (Clausen, 1928). Analysis of the polypeptide composition of Fraction I proteins isolated from these species has demonstrated how thc number of small subunit polypeptides has evolved from one to four.

Origin of Two Small Subunit Polypeptides in N. tabacum

As mentioned before, the tobacco plant of commerce is not found as a wild plant, and there has been uncertainty as to whether *N. otophora* or *N. tomentosiformis* was the species with which *N. sylvestris* hybridized to give rise to *N. tabacum* (Goodspeed, 1954). Inspection of figure 2 shows that only *N. sylvestris* has the same large subunit polypeptides as *N. tabacum* and must, therefore, have been the female partner in the original hybridization. *N. sylvestris* must also have provided the genetic information for the upper of the two small subunit polypeptides of *N. tabacum*. *N. otophora* could not have served as the male partner since its single small subunit polypeptide has a isoelectric point identical to the polypeptide of *N. sylvestris*. However, *N. tomentosiformis* could have been the male partner since its single small subunit polypeptide has an isoelectric point identical to the lower of the two *N. tabacum* polypeptides. Analysis of Fraction I protein isolated from the fertile amphidiploid *N. sylvestris* × *N. otophora* showed only a single small subunit polypeptide of the same isoelectric point as *N. sylvestris*. In contrast, analysis of the protein from the fertile amphidiploid *N. sylvestris* × *N. tomentosiformis* showed the small subunit to contain two polypeptides with the same isoelectric points as those from the *N. tabacum* protein. This analysis indicates that the single small subunit polypeptides of each of the parent species have been perpetuated in *N. tabacum* to produce a Fraction I protein with two small subunit polypeptides (Gray et al., 1974).

Origin of Four Polypeptides in Small Subunit from N. digluta

In 1925 Clausen and Goodspeed described the origin of *N. digluta*, a new species of *Nicotiana* that arose under laboratory conditions. This synthetic species arose through the doubling of chromosome number in an F_1 hybrid of *N. glutinosa* × *N. tabacum*. This plant contained 36 pairs of chromosomes, 12 derived from *N. glutinosa* and 24 from *N. tabacum*. This tetraploid plant was self-fertile and produced viable seeds and, therefore, qualified as a new species of *Nicotiana*. A similar plant arose spontaneously in our greenhouse out of 1,000 otherwise self-infertile progeny of *N. glutinosa* × *N. tabacum* and, therefore, presented an opportunity to examine the Fraction I protein of this synthetic new species.

Examination of figure 2 shows that the Fraction I protein of *N. glutinosa* is composed of three large subunit polypeptides (Group C) and two small subunit polypeptides that are different from the two small subunit polypeptides of *N. tabacum*. This presents an ideal situation with which to examine the polypeptide composition of the Fraction I protein of the new self-fertile species, *N. digluta*. The new species contains a Fraction I protein whose composition could have been predicted from the composition of the parental proteins shown in figure 2. The large subunit polypeptides are the same as those of *N. glutinosa* as required by the maternal inheritance of chloroplast DNA genes. The small subunit is composed of four polypeptides, two from *N. glutinosa* and two from *N. tabacum*, indicating an equal contribution of the maternal and paternal nuclear genes (Kung et al., 1975).

These two analyses demonstrate the way in which new Fraction I proteins can arise during the evolution of new species of *Nicotiana*.

Hybridization of two species with single small subunit polypeptides could give rise to a new species with two small subunit polypeptides and a second round of interspecific hybridizations could give rise to Fraction I proteins with four small subunit polypeptides. Proteins with three polypeptides could also have evolved from two hybridizations between *Nicotiana* species, one of the species containing a single small subunit polypeptide. Evidently, where more than two small subunit polypeptides are found, the species of *Nicotiana* containing this protein must have arisen after two rounds of interspecific hybridizations. On the basis of this view, we can suspect that Australian proteins are the most highly evolved of all the proteins found in the genus *Nicotiana* because most contain three or four small subunit polypeptides. The higher degree of complexity in composition of the Australian Fraction I proteins is in keeping with Goodspeed's view that the isolation of Australia by vast reaches of water resulted in a great deal of interbreeding between *Nicotiana* species and a consequent trend toward isogenicity during evolution of these species (Goodspeed, 1954). In comparison to their Australian relatives, the Fraction I proteins of Western Hemisphere *Nicotianas* are less highly evolved. Out of eleven species analyzed, no instance of a Fraction I protein containing more than two small subunit polypeptides has been seen; six of eleven species contain only a single polypeptide. These Fraction I proteins are less complex in composition possibly because of greater opportunities for geographical isolation of the new species of *Nicotiana* as they arose and fewer subsequent opportunities for hybridization with other species.

EVOLUTION OF FERREDOXIN COMPARED TO FRACTION I PROTEIN

Among the numerous species of *Nicotiana* investigated, no instance was encountered where ferredoxin existed as more than one species of polypeptide chain when examined by isoelectric focusing. Since differences between *Nicotiana* ferredoxins can be detected by isoelectric focusing and by amino acid analysis (Kwanyuen and Wildman, 1975), the question arises as to why ferredoxin isozymes were not produced by the interspecific hybridizations that were the basis of the evolution of new species of *Nicotiana*. The conservation of a single ferredoxin polypeptide stands in stark contrast to the small subunit of Fraction I proteins whose number of polypeptides seems to be proportional to the number of interspecific hybridizations that produced the species of *Nicotiana*. Even though ferredoxin and the small subunit of Fraction I protein have roughly the same number of amino acids and are both coded by nuclear genes, the extreme conservation of the ferredoxin sequence resembles that of the large subunit of Fraction I protein that is coded by chloroplast DNA. The contrast between the evolution of ferredoxin and Fraction I protein in the *Nicotiana* can even be extended to the evolution of these proteins in spinach (*Spinacia oleracea*). This plant belongs to the family Chenopodiaceae, which is widely separated phylogenetically from the Solanaceae, to which the *Nicotiana* species belong. Ferredoxin of spinach is composed of a single polypeptide chain whose sequence is known (Matsubara and Sasaki, 1968) and which gives a single polypeptide on electrofocusing (Kwanyuen and Wildman, 1975). However, as shown in figure 3, the small subunit of spinach Fraction I protein consists of two polypeptides, which indicates that *Spinacia oleracea* is the product of interspecific hybridization. What the comparisons point to is the impossibility of predicting whether nuclear or extranuclear DNA is the probable source of the coding information for a protein by an analysis of the degree of conservation of the amino acid sequence.

Ferredoxin is more ancient in origin than Fraction I protein, but presumably at some time in evolution a primitive procaryotic photoautotrophic cell contained a single circular DNA molecule that possessed the genetic information for both proteins. With the subsequent evolution of photoautotrophic eucaryotes, the genetic information for the large subunit of the photosynthetic carboxylation enzyme became part of the DNA confined to the chloroplasts, whereas the code for the small subunit and ferredoxin became a part of the nuclear DNA. As the evolution of non-photosynthetic eucaryotes ensued, the nuclear DNA coding for ferredoxin was continually perpetuated so that this protein could become a vital constituent of both plants and animals, as well as microorganisms. The information for ferredoxin could always be contributed by both parents during alternation of generations, whereas the transmission of information for the enzymatic function of Fraction I protein was restricted to just one of the parents. Perhaps this information is a clue to the identity of the other proteins coded by chloroplast DNA. We suspect that any chloroplast protein found in both non-photosynthetic and photosynthetic cells will be coded by nuclear DNA as in the case of ferredoxin. Only proteins uniquely found in photosynthetic cells, as in the case of Fraction I protein, seem likely to be coded by chloroplast DNA.

ACKNOWLEDGEMENT

Supported by U.S. Atomic Energy Commission, Division of Biology and Medicine, Research Contract AT(04-3)-34, Project 8, and by U.S. Public Health Service Grant AI-00536. Dr. Kung's present address is Department of Biological Sciences, University of Maryland/Baltimore County, Baltimore, Maryland; Dr. Sakano's present address is Department of Botany, Faculty of Science, University of Tokyo, Hongo, Tokyo, Japan.

REFERENCES

Bourque, D. P., and S. G. Wildman. 1972. Evidence that nuclear genes code for several chloroplast ribosomal proteins. Biochem. Biophys. Res. Commun. 50:532–37.

Burbidge. N. T. 1960. The Australian species of *Nicotiana* L (Solanaceae). Austr. J. Bot. 8:342–80.

Chan, P. H., K. Sakano, S. Singh, and S. G. Wildman. 1972. Crystalline Fraction I protein: Preparation in large yield. Science. 176:1145–46.

Chan, P. H., and S. G. Wildman. 1972. Chloroplast DNA codes for the primary structure of the large subunit of Fraction I protein. Biochim. Biophys. Acta. 277:677–80.

Chen, K., S. D. Kung, J. C. Gray, and S. G. Wildman. 1975. Polypeptide composition of Fraction I protein from *Nicotiana glauca* and from cultivars of *Nicotiana tabacum*, including a male sterile line. Biochem. Genetics (in press).

Clausen, R. E. 1928. Interspecific hybridization and the origin of species in *Nicotiana*. Zeits. Induktive Abst. Verer. 46:25.

Clausen, R. E., and T. H. Goodspeed. 1925. Interspecific hybridization in *Nicotiana*. II. A tetraploid *glutinosa-tabacum* hybrid, an experimental verification of Winge's hypothesis. Genetics 10:278–84.

Dawid, I. B., and A. W. Blackler. 1972. Maternal and cytoplasmic inheritance of mitochondrial DNA in *Xenopus*. Develop. Biol. 29:152–61.

Goodspeed, T. H. 1954. The genus *Nicotiana*. Chronica Botanica, Waltham, Mass. 536 pp.

Gray, J. C., S. D. Kung, S. G. Wildman, and S. J. Sheen. 1974. Origin of *Nicotiana tabacum* L. detected by polypeptide composition of Fraction I protein. Nature. 252:226–27.

Hall, D. O., R. Cammack, and K. K. Rao. 1973. The plant ferredoxins and their relationship to the evolution of ferredoxins from primitive lifc. Pure and Applied Chem. 34:553–77.

Kawashima, N. 1969. Comparative studies on Fraction I protein from spinach and tobacco leaves. Plant Cell Physiol. 10:31–40.

Kawashima, N., S. Y. Kwok, and S. G. Wildman. 1971. Studies on Fraction I protein. III. Comparison of the primary structure of the large and small subunits obtained from five species of *Nicotiana*. Biochim. Biophys. Acta. 236:578–86.

Kawashima, N., Y. Tanabe, and S. Iwai. 1974. Similarities and differences in the primary structure of Fraction I proteins in the genus *Nicotiana*. Biochim. Biophys. Acta, 371:417-31.

Kawashima, N., and S. G. Wildman. 1970. A model of the subunit structure of Fraction I protein. Biochem. Biophys. Res. Commun. 41:1463-68.

Kawashima, N., and S. G. Wildman. 1971a. Studies on Fraction I protein. I. Effect of crystallisation of Fraction I protein from tobacco leaves on ribulose diphosphate carboxylase activity. Biochim. Biophys. Acta. 229:240-49.

Kawashima, N., and S. G. Wildman. 1971b. Studies on Fraction I protein. II. Comparison of physical, chemical, immunological and enzymatic properties between spinach and tobacco Fraction I proteins. Biochim. Biophys. Acta. 229:749-60.

Kawashima, N., and S. G. Wildman. 1972. Studies on Fraction I protein. IV. Mode of inheritance of primary structure in relation to whether chloroplast or nuclear DNA contains the code for a chloroplast protein. Biochim. Biophys. Acta. 262:42-49.

Kung, S. D., K. Sakano, and S. G. Wildman. 1974. Multiple peptide composition of the large and small subunits of *Nicotiana tabacum* Fraction I protein ascertained by fingerprinting and electrofocusing. Biochim. Biophys. Acta. 365:138-47.

Kung, S. D., K. Sakano, J. C. Gray, and S. G. Wildman. 1975. The evolution of Fraction I protein during the origin of a new species of *Nicotiana*. J. Mol. Evol., in press.

Kung, S. D., J. P. Thornber, and S. G. Wildman. 1972. Nuclear DNA codes for the photosystem II chlorophyll-protein of chloroplast membranes. FEBS Letters, 24:185-88.

Kwanyuen, P., and S. G. Wildman. 1975. Nuclear DNA codes for the primary structure of plant ferredoxin. Biochim. Biophys. Acta., in press.

Matsubara, H., and R. M. Sasaki. 1968. Spinach ferredoxin. II. Tryptic, chymotrypic and thermolytic peptides and complete amino acid sequence. J. Biol. Chem. 243:1732-57.

Rosen, W. G., S. R. Gawlik, W. V. Dashek, and K. A. Siegesmund. 1964. Fine structure and cytochemistry of *Lilium* pollen tubes. Amer. J. Bot. 51:61-71.

Rutner, A. C. 1970. Estimation of the molecular weight of ribulose diphosphate carboxylase subunits. Biochem. Biophys. Res. Commun. 39:923-29.

Rutner, A. C., and M. D. Lane. 1967. Nonidentical subunits of ribulose diphosphate carboxylase. Biochem. Biophys. Res. Commun. 28:531-37.

Sakano, K., S. D. Kung, and S. G. Wildman. 1974. Identification of several chloroplast DNA genes which code for the large subunit of *Nicotiana* Fraction I proteins. Molec. Gen. Genet. 130:91-97.

Sakano, K., J. E. Partridge, and L. M. Shannon. 1973. Absence of carbohydrate in crystalline Fraction I protein isolated from tobacco leaves. Biochim. Biophys. Acta. 329:339-41.

Singh, S., and S. G. Wildman. 1973. Chloroplast DNA codes for the ribulose diphosphate carboxylase catalytic site on Fraction I proteins of *Nicotiana* species. Molec. Gen. Genet. 124:187-96.

Wildman, S. G., C. Lu-Liao, and F. Wong-Staal. 1973. Maternal inheritance, cytology, and macromolecular composition of defective chloroplasts in a variegated mutant of *Nicotiana tabacum*. Planta. 113:293-312.

Wong-Staal, F., and S. G. Wildman. 1973. Identification of a mutation in chloroplast DNA correlated with formation of defective chloroplasts in a variegated mutant of *Nicotiana tabacum*. Planta. 113:313–26.

JOHN R. LAUGHNAN AND SUSAN J. GABAY

An Episomal Basis for Instability of S Male Sterility in Maize and Some Implications for Plant Breeding

10

For sixteen years a type of male-sterile cytoplasm called T (Texas origin) was used to avoid hand detasseling of corn plants in the production of hybrid seed. The corn blight epiphytotic of 1970 led to abandonment of the use of Texas male-sterile cytoplasm (*cms-T*) for this purpose. (The reader is referred to Ullstrup [1972] for a review of the procedures employed in production of hybrid seed corn and of the southern corn-leaf blight epiphytotic.) Postmortem analyses indicated that plants with T cytoplasm are highly susceptible to race T of the ascomycete *Helminthosporium maydis,* and led to the consideration of alternative methods to avoid detasseling in production fields. Chief among these are the conversion of currently popular inbred lines to other male-sterile cytoplasms, notably S (USDA) and C (Charrua), and the use of genic male sterility in conjunction with chromosome aberrations according to the method developed by Patterson (1973). It is reassuring to note that these efforts will almost certainly lead to greater diversification of cytoplasmic genotypes among commercially grown hybrids, but there is reason to believe that they will not ease the problem of narrowing germ plasm at the nuclear level since there is a great reluctance in the hybrid seed corn industry to abandon the relatively few inbred line genotypic combinations that have proven to be superior in yield and other agronomic characteristics.

Provisional Department of Genetics and Development, 515 Morrill Hall, University of Illinois, Urbana, Ill. 61801.

There are other possible approaches to the problem, but, because of our lack of understanding concerning the occurrence, transmission, and mutability of cytoplasmic factors in general, and in corn in particular, these can hardly be regarded as short-term solutions. One of these involves the attempt, now being made in a number of laboratories, to induce mutations in T cytoplasm that would alter susceptibility to race T of *H. maydis* without impairing its male-sterile characteristics. On the assumption that the various male-sterile cytoplasms in corn originated by missense mutations from strains with normal cytoplasm, some attempts are also being made to obtain new male-sterile strains that are resistant to *H. maydis* race T by applying mutagens to strains with normal cytoplasm. In addition, a search is under way to identify other male-sterile, resistant cytoplasms among existing varieties of corn; so far, this approach has not been successful.

Unfortunately, the circumstantial evidence now available strongly suggests that the male-sterile and race T susceptibility attributes of T cytoplasm are inseparable, that is, that they are expressions of the same cytoplasmic defect. This conclusion is based on the finding that all of a number of independently isolated strains, whose male-sterile cytoplasms fit into the *cms-T* category, have also been shown to be susceptible to *H. maydis* race T. It is conceivable that these two characteristics are based on different defects or mutations in T cytoplasm but have, for many generations, been transmitted together in the propagation of *cms-T* strains. If we knew enough about the characteristics of cytoplasmic inheritance in corn, we might arrange for recombination between them to isolate one trait without the other; but this, unfortunately, is beyond our reach at the present time. However, if the male sterility and race T susceptibility of T cytoplasm result from mutations in different cytoplasmic genes, efforts to back-mutate one but not the other of these factors in T cytoplasm, or to secure one but not the other by mutation in normal cytoplasm, may well be successful. Our knowledge concerning the mutability of cytoplasmic factors in corn is so meager that it leaves us without a basis for predicting the success of these experiments.

We will return at a later point to a more general consideration of the practical need for a better understanding of a number of aspects of cytoplasmic inheritance in corn but at this point will deal with only one of these, the matter of mutation and mutability. It is widely recognized that one of the major concerns in adapting the use of cytosteriles to the production of hybrids in corn is the matter of a stable sterile phenotype in male-sterile strains, and there is general recognition that the external environment, especially in the case of certain genotypes, can greatly influence this expression and may sometimes lead to serious production problems. It is not generally appreciated, however, that "mutations" may occur in male-sterile strains, at both the cytoplasmic and nuclear levels, and that these, unlike the environmentally induced fertility, will lead to inherited loss of the male-sterile phenotype. The main reason for lack of concern over this possibility is that the cytoplasmic factor causing male-sterility in *cms-T* plants appears to be highly stable in a wide range of nuclear genotypes. We are not aware of any cases of spontaneous mutations from male-sterile to male-fertile condition in plants with T cytoplasm. This is one problem, therefore, that did not require the serious attention of corn breeders in their work with *cms-T.* The same can not be said for *cms-S* and, in view of the current tendency to shift to use of other male-sterile cytoplasms in corn-breeding efforts, it would be wise to take this into account.

Our research over the past several years has dealt with male-fertile plants that appear from time to time in male-sterile strains with S cytoplasm (*cms-S*). Such occurrences are often associated with the environment, however, and are not inherited. We have been particularly interested in those male-fertile exceptions that represent inherited changes in either cytoplasm or nucleus. Before describing these changes in greater detail it will be helpful to review some of the basic characteristics of male-sterile cytoplasms, and of their restoration, in corn.

Since it is not possible at this time to deal with segregation, assortment, and recombination at the cytoplasmic level in corn, these techniques cannot be used to categorize the various types of male-sterile cytoplasms. Nevertheless, the existence of restorer genes, usually dominant in action, that circumvent the effect of sterilizing cytoplasms has provided the opportunity to characterize and group the rather large number of cytoplasmic male-sterile strains that have so far been discovered. The usual method is to cross the cytoplasmic male-sterile individuals with pollen from a series of established inbred line testers and note the pattern of restoration among the offspring. In an alternative and more reliable procedure, the one employed by Beckett (1971), the various *cms* strains to be classified are crossed, and backcrossed repeatedly, with a series of inbred lines used as pollinators. As a result of such studies most *cms* strains can be assigned to the S group, relatively few to the T group, and only several to the recently identified group known as C. Within each group the pattern of restoration, based on crosses with inbred line testers, is similar, but evidence is accumulating that the male-sterile cytoplasms within both the S and T groups are not identical. These differences are apparently minor, however, and in this regard it is of interest that all male-sterile cytoplasms assignable to the T group, for example, T, P, HA, Q, RS, and others tested but as yet unnamed are susceptible to *H. maydis* race T, and that all *cms-S* sources so far tested are not susceptible. The possibility cannot be excluded that *cms-T* strains resistant to race T of *H. maydis* might be discovered at some time in the future, but the data now available indicate that a search for these would be unrewarding.

We are accustomed to classify plants as male-fertile or male-sterile according to whether or not they exsert anthers (fig. 1) and shed pollen. On these criteria one might well overlook the difference between restored *cms-T* and *cms-S* plants. Actually, two gene loci, Rf_1 and Rf_2, have been identified as restorer factors for T male-sterile cytoplasm (see Duvick, 1965, for review). In the *cms-T* system, restoration is determined by the genotype of restorers at the diploid or sporophytic level. This means that $Rf_1\ rf_1\ Rf_2\ rf_2$ plants with T cytoplasm produce all normal pollen in spite of the fact that only one-fourth of the pollen grains produced by these plants carry the $Rf_1\ Rf_2$ genotype. On the other hand, restoration of fertility to plants carrying S cytoplasm is determined by the genotype of the pollen grain, so we say it is gametophytic. A plant with S cytoplasm that is heterozygous for the appropriate restorer locus, in this case Rf_3 rf_3, is semifertile in that about half of the pollen grains, those carrying Rf_3, are normal and functional, and the other half, those carrying rf_3, are aborted and nonfunctional. When such plants are crossed with S male-sterile females, the only functioning pollen grains are those carrying the restorer allele Rf_3, so that all offspring have the genotype $Rf_3\ rf_3$, and are again semifertile, exhibiting 50% pollen abortion (Buchert, 1961). Since male-sterile cytoplasms do not similarly affect the female gametophyte, when $Rf_3\ rf_3$ plants with S cytoplasm are crossed as females with $rf_3\ rf_3$ maintainers, defined as tester plants with normal cytoplasm and nonrestoring genotype, half of the offspring are male-sterile ($rf_3\ rf_3$), and half are semifertile ($Rf_3\ rf_3$), exhibiting 50% pollen abortion. When plants with S cytoplasm and the genotype $Rf_3\ rf_3$ are self-pollinated, they produce equal numbers of two kinds of offspring, those with all normal pollen, whose crosses with maintainer plants indicate that they are $Rf_3\ Rf_3$ in genotype, and those, with 50% aborted pollen, whose crosses with maintainer plants indicate

Fig. 1. Tassels of maize plants illustrating changes from male-sterile to male-fertile condition: (*upper left*) sterile tassel on plant with S cytoplasm and nonrestoring rf_3 rf_3 genotype; (*upper right*) entirely fertile tassel; (*lower left*) fertile-sterile tassel chimera showing exsertion of normal anthers at left and sterile portion of tassel at right; (*lower right*) fertile-sterile tassel chimera with two separate fertile sectors.

that they carry the $Rf_3\ rf_3$ genotype. Thus, restorer genes for S male-sterile cytoplasm act late, at the gametophyte or pollen level, whereas the restorers of T male-sterile cytoplasm produce their effect at an earlier developmental stage in the mother plant or sporophyte.

A number of years ago we commenced a backcrossing program designed to introduce male-sterile cytoplasm into seven sweet corn lines carrying the *shrunken-2* allele. The male-sterile cytoplasm incorporated into these inbred lines traces to a *Vg* (dominant *vestigial glume*) source whose pattern of restoration places it in the *cms-S* group. We therefore have male-fertile maintainer and S male-sterile versions of seven *shrunken-2* inbred lines, and these have been maintained through fifteen to twenty generations. The inbred lines involved are R839, R851, R853, R853N, R825, M825, and E1. Two sublines of M825, K and L, were established in 1964 and have been maintained since that time.

Two years ago we reported (Singh and Laughnan, 1972) a number of cases of instability of S male-sterile cytoplasm. The male-fertile plants occurred among otherwise male-sterile progenies of S male-sterile M825 female parents crossed with inbred line R138-TR, a maintainer for S cytoplasm. The unexpected male-fertile plants were crossed onto S male-sterile testers, and either crossed with S maintainer pollen or successively self-pollinated to produce F_2, F_3, and F_4 progenies. These tests indicated that the newly arisen male fertility was propagated through the female but was not transmitted through male germ cells to progeny of test crosses onto S male-sterile testers. It was not possible to explain the newly arisen male fertility on the basis of dominant or recessive nuclear restorer genes. We concluded, therefore, that it resulted from a change from male-sterile to male-fertile condition in the cytoplasm of the male-sterile M825 plant involved as the female parent in the above cross. It appears that this plant bore an ear in which there was a relatively early "mutational" event at the cytoplasmic level resulting in a chimera involving some kernels that carried S male-sterile cytoplasm, and others that carried the mutated fertile cytoplasmic condition.

Several hundred instances of inherited, cytoplasmic changes from S male-sterile to normal male-fertile condition have since been identified among the otherwise male-sterile progenies involving the *shrunken-2* inbred lines described above. These have occurred most frequently in the M825 inbred line background, particularly in the L version, but such cases have been found in each of the other six inbred lines as well. The tassels of the newly arisen male-fertile plants may be entirely fertile or they may occur as chimeras having both fertile and sterile portions (fig. 1). We have also identified a number of instances of fertile-sterile ear chimeras, thus indicating that the cytoplasmic change may register initially in both male and female reproductive tissues.

We have used the term "mutation" to describe the event that leads to the unexpected appearance of male-fertile plants in otherwise S male-sterile progenies. It is clear that the event occurs most frequently at the somatic level in S male-sterile plants, and genetic analysis indicates that a change has occurred at the cytoplasmic level. It is not established, however, that the causal event involves a qualitative change in a cytoplasmic factor governing the expression of male fertility; it is equally plausible that the male-fertiles we have encountered result from occasional transmission of normal cytoplasm by the male-fertile maintainer parents that are employed routinely as pollinators in crosses with male-sterile plants to produce the progenies among which the male-fertile exceptions have been identified. We can illustrate the difference between these hypotheses by considering the cross between M825 *cms-S* male-sterile plants of the L version and the isogenic M825 maintainer. Among the progeny of this cross, cases of male fertility are encountered with relatively high frequency. On the first hypothesis these would be regarded as the result of mutation of the S male-sterile cytoplasmic factor to male-fertile condition in F_1 offspring; according to the second hypothesis, these male-fertile offspring result from transmission of normal cytoplasm through the male germ cells of the male-parent maintainer. We are continuing our efforts to distinguish between these two alternatives but at this time have no reasons to favor one or the other.

When we initially encountered cases of male-fertile individuals in S male-sterile progenies, our first concern in analyzing them was to determine whether the fertility was based on a change at the cytoplasmic or nuclear-gene level. After all, if mutation was involved, it could occur in the cytoplasm or in the nucleus. We expected the newly isolated male-fertile strain to behave as a maintainer in the former case, and as a restorer in the latter. We were surprised, therefore, to find that all the several hundred newly arisen male-fertile strains that we tested, the vast majority of these occurring in inbred line M825, were in the former category. However, we have since encountered a number of cases of the latter type. In general these strains behave as if a mutation had occurred at a restorer locus in the nucleus. We now turn to a more detailed account of the behavior of these mutant strains.

In the summer of 1971 over 300 cases of male fertility in the S male-sterile *shrunken-2* inbred lines were progeny tested to determine whether the fertility they exhibited occurred at the cytoplasmic or nuclear level. Included in this group were individuals with entirely fertile tassels and larger numbers of plants with fertile sectors in the tassel. In all but five of these cases the progenies of testcrosses with S male-sterile plants indicated that the male-fertile character of exceptional plants was not transmitted through the pollen. These results were consistent, therefore, with those of earlier tested cases in indicating that the sudden appearance of male fertility in otherwise S male-sterile progenies is based on a change in the cytoplasm. However, the five cases referred to above behaved differently; in each instance, testcrosses of these male-fertile individuals with S male-sterile testers produced male-fertile offspring, suggesting a Mendelian, or nuclear, basis for the fertility. Results of studies dealing with four of these strains have been published (Laughnan and Gabay, 1973a). Meanwhile, several additional cases of mutations leading to nuclear restoration in S cytoplasm have been encountered. At this time we have a considerable amount of data on ten newly arisen restorer strains that have so far been identified.

Table 1 provides information on the origin and other characteristics of each of the ten newly arisen restorer strains. These have been assigned Roman numeral designations until such time as their relationships, if any, with the already described restorer loci Rf_1, Rf_2, and Rf_3 have been clearly defined. Seven of the cases arose in inbred line backgrounds, two in F_1 populations and one in an F_2 progeny. In all but one, the initial event leading to male fertility occurred in *cms-S* plants. Restorer IV is the single exception; here the mutation occurred in a maintainer plant of inbred line R853N. Apparently this has special significance, for it now appears that restorer IV has characteristics not shared by the others, a point to which we will later return.

As indicated in table 1, restorers III, V, VII, VIII, and IX first made their appearance in plants with entirely fertile tassels, whereas restorers I, II, VI, and X derive from fertile cases that first appeared in fertile-sterile tassel chimeras. Restorers III and IX originated in plants with entirely fertile tassels. When testcrossed with maintainer pollen, both plants gave male-fertile and male-sterile progeny as would be expected if the tested plants were heterozygous for a nuclear restorer. However, the ratio observed for these classes in the case of restorer III was 13:36, a highly significant deviation from the expected 1:1 ratio for male-fertile and male-sterile progeny. Progeny

TABLE 1

CHARACTERISTICS OF THE TEN NEW S RESTORERS AT THE TIME OF FIRST APPEARANCE

Restorer	Source line, F_1 or F_2	Cytoplasm*	Tassel character of exceptional plant	Female transmission: fertile exception × maintainer male
I	M825K	S	Chimera	All ms
II	R853	S	Chimera	19 fert: 32 ms
III	R853	S	Entirely fertile	13 fert: 36 ms
IV	R853N	N†	—	—
V	F_1: M825L/R853	S	Entirely fertile	No test
VI	F_1: M825L/Oh07	S	Chimera	All ms
VII	R825	S	Entirely fertile	No test
VIII	M825L	S	Entirely fertile	All ms
IX	M825L	S	Entirely fertile	10 fert: 14 ms
X	F_2: M825L/Oh07	S	Chimera	No test

*We have used the generic S symbol here; actually, all were *cms-Vg*, a subgroup of S.

† Since, as noted in the text, Restorer IV occurred in a maintainer plant with normal cytoplasm, the following two entries in this row are not relevant.

tests in advanced generations of restorers III, IX, and others indicate that these are not chance deviations, and we will return at a later point to discuss the significance of this lowered transmission of female gametes carrying the newly arisen restorer alleles. Restorer VIII also originated in a plant with an entirely fertile tassel, but, as noted in table 1, only male-sterile progeny were obtained from the testcross of this plant with maintainer pollen. We have previously encountered a number of cases in which an exceptional plant with an entirely fertile tassel, whose testcross progeny indicated that the exceptional event had occurred in the cytoplasm, failed to transmit the male-fertile trait through its female gametes. It must be concluded that whether the exceptional event leading to the male-fertile phenotype occurs at the nuclear or cytoplasmic level, it can occur sufficiently early in the development of initials to involve the entire tassel, yet none of the ear elements of such a plant.

Of the four cases appearing initially as fertile tassel chimeras, three were tested for female transmission of male fertility. Of these (table 1), only restorer II showed female transmission of male fertility. It is most reasonable to conclude that in this case the ear was included in the fertile portion of this chimera whereas, in the cases of restorers I and VI, it was not. In this connection it should be noted that the restorer II chimera included male-fertile elements in the main rachis of the tassel, whereas in the cases of restorers I and VI only lateral branches of the tassel were fertile.

It goes without saying that each of the ten newly arisen restorer strains transmitted the male fertility trait through the pollen in crosses with *cms-S* testers, and has continued to do so in subsequent generations. In addition, the mode of restoration observed for the ten newly arisen nuclear restorers is gametophytic, as it is with the standard S restorer, Rf_3, rather than sporophytic, which is characteristic of the *cms-T* system and restorers Rf_1 and Rf_2.

The simplest explanation for the occurrence of these new restorers, the one we considered most seriously at first, involves the assumption that they arose from mutation at the standard Rf_3 locus, and that they have identical or at least very similar characteristics. When we discovered, however, that two of the four restorer strains that we first tested, when crossed with S male-sterile plants of the R839 inbred line, gave all male-sterile offspring, whereas the other two produced fully restored F_1 progeny, we were obliged to seriously question this assumption. Further indication of differences among the new restorer genes is apparent from a study of their patterns of fertility restoration in F_1 hybrids with various inbred line female parents carrying S-type male-sterile cytoplasm. The results are presented in table 2, where it is apparent that restorers I and II give responses different from most other restorers and are distinguished from each other on the basis of their fertility restoration patterns in crosses with R853, WF9, and I153 male-sterile inbred lines. In fact, the data in table 2 suggest that there may be as many as four or more unique patterns of restoration among the ten newly arisen restorer strains.

It is conceivable that the differences in patterns of restoration noted above are attributable not to differences in the restorer strains but rather to the influence of background nuclear genotypes, which differ in most of these testcross progenies since the restorers themselves arose in different line backgrounds. But this can not be argued to explain the different responses of restorers II and III since both of these arose in the same *cms-S* inbred line, namely R853, or of restorers VIII and IX, both of which arose in inbred line M825L. The data must therefore be interpreted to indicate that there are real differences between the newly arisen restorers in regard to their restoring capabilities.

TABLE 2

RESTORATION PATTERNS OF TEN NEW S RESTORERS. TESTCROSS PROGENY INDICATED AS + (MALE-FERTILE) OR − (MALE-STERILE)

FEMALE PARENT		MALE PARENT RESTORER STRAIN									
Inbred line	Cytoplasm	I	II	III	IV	V	VI	VII	VIII	IX	X
R839	Vg	−	−	+	+	+	+				
M825L	Vg	+	−	+	+	+	+		+	+	+
R853	Vg	−	+	+	+						
WF9	S	+	−	+	+	+	+				
K55	Vg	−	−	+	+	+	+	−	−	+	+
M14	M1	−	−	+	+	+	+	−	−	−	+
IllA	S	−	−	+	+	+	+	−		+	+
N6	S	−	−	+	+	+	+	−	−	−	+
I153	Vg	−	+	+	+	+	+	+		−	

There is additional evidence indicating that the new restorers are not alike. We were first alerted to this when we discovered that plants carrying S cytoplasm, and heterozygous for either restorer II or III, when crossed with maintainer pollen frequently produce a significantly greater number of male-sterile than male-fertile offspring. At first we were inclined to attribute this to chance variation, but our classifications of larger numbers of testcross progenies confirmed the observation. In the course of preparing seed for the 1973 and 1974 summer plantings, reduced kernels were observed on many of the ears that also segregated for a new restorer gene (fig. 2). In fact, reduced kernels were found on ears segregating for each of the ten new restorers except restorer IV. This is of special interest because, of the ten cases under consideration here, restorer IV is the only one in which the restorer mutation occurred in nonsterile cytoplasm. Reduced kernels from some of the ears were planted separately from the normal ones to determine whether the reduced-kernel phenotype could be correlated with the presence of a new restorer gene. Surprisingly these progenies indicated that a very high proportion of reduced kernels produced male-fertile plants, indicating the presence of the restorer allele, while kernels of normal size yielded principally male-sterile plants. This association has been firmly established for restorers II, III, V, VI, and VII. Preliminary results suggest a similar association for restorers I and VIII, but restorers IX and X have not yet been progeny tested.

Having found an association between kernel size and restorer genotype, we were inclined to consider that the aberrant ratios favoring

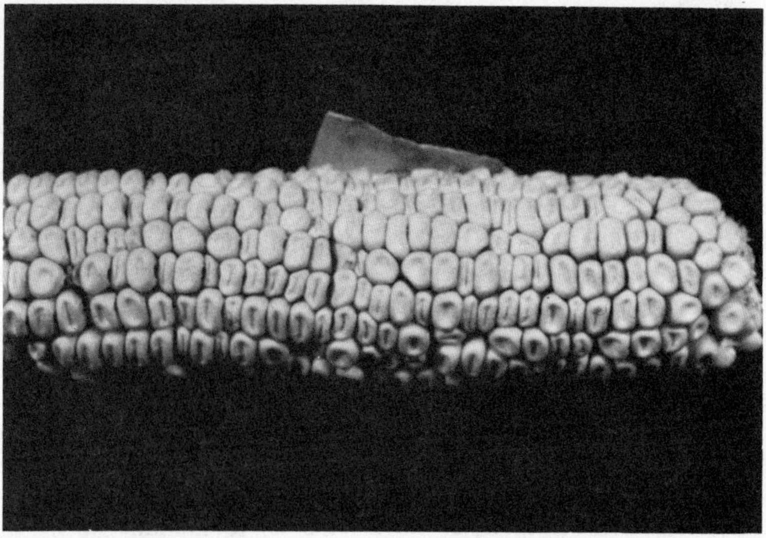

Fig. 2. Normal and reduced kernels on an ear of maize segregating for new restorer VI.

male-sterile over restored male-fertile individuals in testcross progenies of *Rf rf cms-S* plants was attributable to the inadvertant selection of larger kernels from these ears in preparation for planting. However, we have noted that not all such ears exhibit obvious differences in kernel size and that, in fact, discrepant ratios for male-sterile and male-fertile offspring have frequently been encountered among plants from testcross ears that showed no apparent size differences among kernels. Where reduced kernels are evident on testcross ears we have noted that they exhibit a considerable range in sizes. In testing for the association of kernel size with restorer genotype, we have tended to select for planting the extremes in the respective size classes. In a few such cases all reduced kernels gave male-fertile plants, and all the plants from large kernels were sterile. More commonly, however, the separation was not complete; in particular, it was found that large kernels frequently carry the restorer genotype. Therefore, the deleterious effect of these newly arisen restorers on development of the caryopsis is not absolute and may fail to be expressed under some circumstances. This would explain why some ears on *Rf rf cms-S* plants crossed with maintainer pollen exhibit no small kernels and why, among those which do, some kernels carrying the restorer allele are not reduced. It is conceivable that, at the other extreme, some ovules carrying the restorer genotype are aborted; this possibility has not yet received an adequate test.

In the course of these investigations attempts have been made, through analyses of progenies of self-pollinations, to obtain plants homozygous for each of the new restorer genes in S male-sterile cytoplasm. These would be routinely identified as F_2 plants, with all normal pollen, whose crosses with maintainer pollen yield all male-fertile progeny. Since the self-pollinated *cms-S* F_1 plant has the genotype *Rf rf*, and since *rf* pollen from such plants is aborted and nonfunctional, the F_2 offspring should consist of equal numbers of *Rf Rf* and *Rf rf* offspring. Again, in this regard the restorer IV pedigree exhibits normal behavior; that is, homozygotes and heterozygotes occur with equal frequency among the F_2 progeny. The *Rf Rf* homozygote is also encountered in F_2 progenies involving restorer V where, however, it is commonly found with much lower than the expected frequency. In other words, the *Rf Rf* homozygote in the restorer V pedigree is semi-lethal. The remaining restorers have a more drastic effect. In spite of the examination of large numbers of F_2 plants in the pedigrees of restorers II, III, and VI, no plants with all normal pollen have been identified, and appropriate test crosses of F_2 plants have produced no progenies with all male-fertile plants. Testcross progenies of F_2 plants have not yet been observed in the cases of restorers I, VII, VIII, IX, and X, but pollen examination of F_2 plants in all these pedigrees have failed to identify any plants with all normal pollen. A final decision in the matter must await the results of confirming test crosses, but the data now available strongly indicate that the *Rf Rf* homozygote, in S cytoplasm, is lethal for the vast majority of the newly identified restorers.

As indicated above, we have not been able to distinguish between two alternatives for the origin of male-fertile exceptions in plants with S-type male-sterile ctyoplasm. We assume they arise either as a result of a qualitative change in an element of S male-sterile cytoplasm, or as the result of occasional transfer of normal cytoplasm through the male germ cells of maintainer pollen parents. Leaving aside the question of the mechanism by which the fertile condition arises, it is clear, from what we now know, that the male-fertile exceptions involve either a change at the cytoplasmic level, which is most often the case, or a change in the nucleus. Because the two kinds of male-fertile exceptions have arisen in the same strains, and in both cases are expressed initially as either entirely male-fertile plants or as fertile-sterile tassel chimeras, we consider them to have a common origin. According to this scheme, the male-fertile element has the characteristics of an episome. If the latter is fixed in the cytoplasm, the newly arisen male-fertile behaves as a maintainer strain; if it is fixed in the nucleus, it behaves as a restorer strain.

One of the characteristics of episomes, as defined in bacterial systems, is their capacity to transpose, that is, to move from one state or intracellular locale to another, or to be lost altogether. There are several ways in which we might test for transposition of S restorers in the sytem described above. We have commenced such experiments but have no significant data to report at this time.

Another approach to the problem is to carry out tests for allelism involving Rf_3, the standard S restorer allele, and the newly arisen restorer strains. If the new restorers resulted from mutations of a nonrestoring allele, it might be expected that all the new restorer mutations we have encountered involve only a single gene locus, possibly rf_3 itself. If the phenomenon is episomal in nature, however, this constraint would not necessarily apply; and it could well be that each of the new restorer events involves fixation of the fertility element, or episome, at a different chromosomal site. Accordingly, we have examined the progeny of crosses designed to determine whether the newly arisen nuclear restorers are allelic with the standard Rf_3 restorer carried by inbred line CE1. If a particular new restorer gene in question is allelic with Rf_3, crosses of plants carrying the new restorer in S cytoplasm with pollen from inbred line CE1 should produce some progeny with all normal pollen. Nonallelism, on the other hand, would be indicated by the presence in such progenies of plants with 25% aborted pollen. On the basis of pollen analyses of these progeny, none of the restorers I through VI is allelic with Rf_3. Testcross progeny from plants with 25% pollen abortion, crossed by maintainer pollen parents, are now being grown to confirm this observation. If the two restorers being tested are not allelic, such crosses should produce progenies in which 25% of plants are male-sterile. Such an analysis has been completed for restorer IV; classifications of eight testcross progenies gave 240 male-fertile, 84 male-sterile, and 32 partially male-fertile plants, a result that is consistent with the pollen evidence indicating that restorer IV is not allelic with Rf_3. Restorers VII through X remain to be tested for allelism with the standard restorer Rf_3 carried by inbred line CE1.

Studies have also been carried out to determine whether the newly arisen nuclear restorers resulted from mutation at the same gene locus, that is, whether they are allelic with each other. Here again, nonallelism would be indicated by the occurrence of plants with 25% aborted pollen grains among the progeny of crosses between plants carrying the different new restorer genes in S cytoplasm. Of the 15 possible F_1 combinations of crosses involving restorers I through VI, the ones so far tested all indicated the presence of plants with 25% aborted pollen, and thus that the restorer genes involved in these crosses are not allelic. Restorers VII through X have not yet been tested in this system.

The preliminary evidence now in hand from pollen classifications indicates that six of the ten new restorers are neither allelic with the standard restorer Rf_3 nor with each other. It appears, therefore, that the newly arisen restorers occupy unique chromosomal sites. This is consistent with the view that, in the case of a new restorer, the fertile element originates in the cytoplasm, either by mutation of the sterile S element to the normal or fertile condition, or by transmission of the normal fertile element through germ cells of the male parent, and is subsequently transposed to, and stabilized in, a chromosome.

Following this preliminary indication that the new restorers are not allelic, we have recently attempted, through the use of a series of *waxy* translocation testers, to establish their linkage groups. This new technique, whose details will be described elsewhere, makes it possible to assign the linkage group of a particular restorer merely through the identification of distorted ratios for blue- and red-staining pollen grains from appropriately synthesized heterozygotes. There is now good reason to believe that what would ordinarily be a laborious task may be greatly simplified by this procedure. By this means, restorer IV has been assigned to chromosome 3, and restorer I to chromosome 8. These assignments have been confirmed recently by data from testcross progenies. In addition, restorers VII and VIII have tentatively been assigned to chromosomes 3 and 8, respectively, and both restorers IX and X appear to reside in chromosome 1. It remains for testcross progenies to confirm these locations, and for further analyses to identify the linkage groups of restorers II, III, V, and VI. There is already strong indication, however, that the newly arisen nuclear restorers occupy different chromosomal sites, and this in turn supports the episomal nature of these elements.

We have suggested that each of the new restorers arose following the first appearance in the cytoplasm of a fertile element that later, but not much later, established residence at one or another site in a chromosome. This is a working hypothesis that has not been substantiated fully but has already proved its usefulness. If it is correct, we might expect to obtain evidence of secondary transpositions of the male-fertile element from chromosome to cytoplasm in one or more of the new restorer strains, and from cytoplasm to chromosome in fertile strains that originated through changes at the cytoplasmic level. The above indicated assignment of chromosomal sites for at least a number of the new restorers will afford the opportunity to search for such transpositions.

We are also undertaking a detailed study of fertile chimeras when they first make their appearance in tassels of S male-sterile plants. There is evidence that both cytoplasmic and nuclear fixations of the male-fertile element are represented in individual fertile sectors. If this is confirmed in more extensive studies, it will add strong support for the episome hypothesis.

At this point let us reconsider briefly the evidence that a number of the new restorers so far analyzed, perhaps all but restorer IV, have associated phenotypic effects. How can we explain the finding that, in the cases of most of the restorers, kernels carrying the restorer gene are strikingly reduced in size and that in apparently eight of the ten cases the homozygous restorer genotype is lethal? And why do the restorers exhibit different patterns of restoration? At the present time we can only speculate about this. The differential behavior of the new restorers may relate to differences in integration sites (position effect) in the chromosomes. Or it could be the result of qualitative differences in the fertile elements that are determined prior to chromosome integration. Are these fertile elements rendered defective through some kind of enforced recombination with defective S elements in the cytoplasm before integration in the chromosome occurs? We are at present much impressed by the latter possibility since restorer IV, the only one of the ten new restorers that has been shown not to be associated with deleterious side effects, is also the only one that arose in fertile (maintainer) cytoplasm, in which there would be no opportunity for recombination with an aberrant S male-sterile element prior to integration. We anticipate that further investigation of this system will answer at least some of these questions.

It is hoped that the studies reported here will help focus attention on another category of concern for corn breeders, and for plant breeders in general. This has to do with the likely practical application, at some time in the future, of basic information concerning the inheritance of cytoplasmic traits. The production of hybrid seed corn by the male-sterile method represents the most extensive practical application of cytoplasmic inheritance in any species. Cytoplasmic male sterility is being used to exploit heterosis in a number of plant species, and apparently only technical difficulties stand in the way of adapting it for that use in many others. In this procedure advantage is taken of cytoplasmic sterility merely as a technique to facilitate the mass production of hybrid *nuclear* genomes. But genes are also found in cytoplasmic organelles, and these genes, in the most thoroughly studied species, *Chlamydomonas reinhardi,* a unicellular green alga, and *Saccharomyces cerevisiae* (yeast), are known to mutate, segregate, and recombine (see Sager, 1972; Coen et al., 1970). In short, they exhibit the characteristics of nuclear genes. Should we not expect, therefore, that if technical difficulties were surmounted and a higher plant system were to afford the opportunity, selection for recombinant *cytoplasmic* genomes, and the development of *cytoplasmic* hybrids would lead to corresponding advantages in yield and other desirable agronomic attributes? We are convinced that this is an achieveable goal, and a considerable part of our current research effort is oriented in that direction.

The technical difficulties referred to above, however, are formidable. Most of the available basic information on inheritance of cytoplasmic traits in plants comes from *Chlamydomonas,* an organism without great economic significance. Nevertheless, the extensive studies on cytoplasmic inheritance carried out with this species, and with yeast, have called attention to the kinds of information we need to develop in corn, and other higher plants, if we hope to gain advantages from manipulating cytoplasmic genomes in these species. First, we need to extend the list of traits known to be cytoplasmically inherited; at the present time relatively few such traits have been identified in corn, and to our knowledge no single strain has been found to carry more than one such trait unless, as appears unlikely, the male-sterile phenotype and susceptibility to *H. maydis* race T are to be regarded as separable attributes of Texas cytoplasm. Second, we need to develop procedures that will afford routine hybridization of cytoplasmic genomes, an indispensable prerequisite for studies on segregation and recombination of cytoplasmic factors and for the selection of favorable combinations.

In pursuing the identification of additional cytoplasmic mutants, it is obvious that we cannot rely on the occurrence of spontaneous mutations since, in spite of the extensive genetic and plant-breeding research with corn, relatively few naturally occurring cytoplasmic mutations have been identified. In *Chlamydomonas* (Sager, 1972) the antibiotic streptomycin has been highly effective in inducing cytoplasmic mutations; in fact, practically all such mutants have come from treatments with this mutagen, and to date it has not been found to produce nuclear gene mutations in this species. Ethidium bromide appears to have a selective effect in the induction of mitochondrial mutations in yeast. We are currently using these and other potential mutagenic agents in attempts to obtain cytoplasmic mutations in corn. The finding (Laughnan and Gabay, 1973b) that the pathotoxin complex produced by race T of *H. maydis* inhibits the *in vitro* germination of pollen grains carrying T cytoplasm makes it possible to adapt this procedure to search at the pollen level for mutations conferring resistance to the toxin; we are pursuing these studies. Kermicle (1969; 1971) discovered a mutant called *ig* (indeterminate gametophyte) that, when carried in the female parent of crosses, yields relatively high frequencies of paternal monoploids with cytoplasmic contribution from the female parent. As he suggests, *ig* affords a convenient and rapid procedure for the isolation of any particular nuclear genome in a variety of cytoplasms. It is also possible that this technique may be employed to quickly identify cytoplasmic changes that arise through placement of nuclear genomes of related species in maize cytoplasm.

We are accustomed to think of cytoplasmic inheritance as uniparental inheritance, that is, as involving transmission of gene-like elements through the female but not the pollen parent. In fact, some have defined it as the inheritance of traits exclusively through the female parent. If, in fact, there were no exceptions to the rule that cytoplasmic factors are transmitted solely through the female parent, it would mean that there is no sexual mechanism in higher plants to bring together in a common cytoplasm the varying cytoplasmic genomes of different individuals. In this regard it is encouraging to note that there are recognized instances of transmission of cytoplasmic factors through the pollen of several genera of higher plants, notably *Pelargonium* and *Oenothera*. Such instances of biparental transmission raise the hope that ways may be found to induce it in species, such as corn, in which only uniparental transmission of cytoplasmic factors has so far been observed. Recent studies in *Chlamydomonas* by Sager and Ramanis (1973) are particularly encouraging. In this organism the transmission of cytoplasmic factors is strongly uniparental. In most matings the cytoplasmic gene contribution is from the mt^+ parent only, but about 1% of matings exhibit biparental transmission in which both parents make equal contributions of cytogenes. It was found that the frequency of biparental cytoplasmic inheritance could be enhanced, in some cases dramatically, by exposing mt^+ gametes, prior to fusion, to ultraviolet radiation or to ethidium bromide or other chemical items. These studies indicate that the strong natural tendency toward uniparental inheritance of cytoplasmic factors in *Chlamydomonas* is not based on physical exclusion of these factors from the mt^- parent but rather is the result of their destruction in the zygote. They therefore invite us (Sager, 1972) to consider that uniparental inheritance in higher plants may have a similar physiological basis for exclusion of the cytoplasmic contribution from the pollen parent. If the exclusion in higher plants is physiological, rather than purely physical as has long been considered to be the case, we should expect, based on the results with *Chlamydomonas,* that ways will be found to artificially induce biparental inheritance in corn and other higher plant species.

There are strong indications that hybridization of cytoplasmic factors in corn and numbers of other higher plants need not await the development of techniques designed to circumvent uniparental inheritance in the sexual cycle. We refer to the recent remarkable investigations dealing with isolated protoplasts of cells of several higher plant species. Carlson (1973), in particular, has provided convincing demonstration of the use of protoplasts, derived from tobacco species, in cell hybridization, organelle transfer and selection for resistance at the cell-culture level. At the present time these techniques have not been applied successfully in corn and most other higher plant species. There is reason to believe, however, that only technical difficulties remain to be resolved in the use of the parasexual cycle to accomplish protoplast fusion and the accompanying hybridization of cytoplasms in these species. This would surely lead to productive genetic investigations of cytoplasmic genes in these species, and hopefully to the improvement of crop plants through the genetic manipulation of cytoplasmic genomes.

ACKNOWLEDGMENTS

The authors' research is supported by the Departments of Agronomy and Botany, and the Provisional Department of Genetics and Development, University of Illinois, Urbana, and by CSRS grant 177-15-04, PL 89-106.

LITERATURE CITED

Beckett, J. B. 1971. Classification of male-sterile cytoplasms in maize (*Zea mays* L.). Crop Sci. 11:724–27.

Buchert, J. G. 1961. The stage of genome-plasmon interaction in the restoration of fertility to cytoplasmically pollen-sterile maize. Proc. Nat. Acad. Sci. USA. 47:1436–40.

Carlson, P. S. 1973. The use of protoplasts for genetic research. Proc. Nat. Acad. Sci. USA. 70:598–602.

Coen, D., J. Deutsch, P. Netter, E. Petrochilo, and P. P. Slonimski. 1970. Mitochondrial genetics. I. Methodology and phenomenology. Symp. Soc. Exp. Biol. 24:449–96.

Duvick, D. N. 1965. Cytoplasmic pollen sterility in corn. Advan. Genet. 13:1–56.

Kermicle, J. L. 1969. Androgenesis conditioned by a mutation in maize. Science 166:1422–24.

Kermicle, J. L. 1971. Pleiotropic effects on seed development of the indeterminate gametophyte gene in maize. Amer. J. Bot. 58:1–7.

Laughnan, J. R., and S. J. Gabay. 1973a. Mutations leading to nuclear restoration of fertility in S male-sterile cytoplasm in maize. Theor. Appl. Gen. 43:109–16.

Laughnan, J. R., and S. J. Gabay. 1973b. Reaction of germinating maize pollen to *Helminthosporium maydis* pathotoxins. Crop Science 13:681–84.

Patterson, E. B. 1973. Genic male sterility and hybrid maize production. Proceedings of the Seventh Meeting of the Maize and Sorghum Section of Eucarpia (European Association for Research on Plant Breeding).

Sager, R. 1972. Cytoplasmic genes and organelles. Academic Press, New York.

Sager, R., and Z. Ramanis. 1973. The mechanism of maternal inheritance in *Chlamydomonas:* Biochemical and genetic studies. Theoret. Appl. Genet. 43:101–8.

Singh, A., and J. R. Laughnan. 1972. Instability of S male-sterile cytoplasm in maize. Genetics 71:607–20.

Ullstrup, A. J. 1972. The impacts of the southern corn leaf blight epidemics of 1970–71. Ann. Rev. Phytopath. 10:37–50.

Concluding Remarks

A major purpose of the colloquium and this volume was to provide a forum for workers in this area to interact, and we were pleased, therefore, to find that the participants presented a clear view of work in each area and that new ideas and insights surfaced. It is the purpose of this brief section to mention some of these points since, in several instances, they are not fully apparent from reading any one chapter in this volume.

Mahler and Attardi documented the ever growing body of knowledge suggesting that the roles of mitochondrial genes are virtually the same in yeast and HeLa cells despite the fivefold difference in mitochondrial genome sizes. It is clear that most of the major proteins that are mitochondrial gene products involved in the specification of the respiratory chain have been identified in fungi, whereas in HeLa cells much progress has been made in the identification of mitochondrial mRNAs; it appears that the linear organization of genes on animal mtDNA will be known within the next few years, but it may take longer to sequence the yeast genes.

Much of the research on organelle biogenesis of the last fifteen years has focused on establishing the degree of autonomy of the organelle genetic systems; only recently have researchers sought evidence relating to regulatory interactions between nucleus and cytoplasm and in particular the interaction of mitochondria and chloroplasts with other components of the cell. It is clear from Attardi's comments that this latter subject will be a major theme of research in the next decade, and his studies of the cell cycle dependence of mitochondrial macromolecular synthesis and division constitutes an important first step.

This general theme was also reflected in other ways by the description by several participants of specific cases in which nuclear and organelle genes interact. Mahler and Wildman showed that specific organelle enzymes are composed of polypeptide chains specified by nuclear *and* organelle genes; in the case of cytochrome oxidase the organelle products do not appear to be required for activity of the isolated enzyme, while in the case of RUDP carboxylase, the organelle gene product seems to contain the active site of the enzyme. Hoober described a model for the expression of certain nucleus-specified membrane proteins that involves a chloroplast gene product as a regulatory element. Sager and Tilney-Bassett described several nuclear mutations that appear to control plastid inheritance while Birky and Perlman presented molecular models for polarity and suppressiveness that involve putative nucleus-specified nucleases.

The contributions on organelle genetics by Birky, Sager, and Tilney-Bassett discussed the situation in diverse kinds of organisms. It is clear that all have certain elements at least superficially in common, including vegetative segregation and uniparental inheritance. As noted by Birky, these similarities warrant a serious attempt to deduce general properties of organelle genetic systems, and that is clearly his and Sager's intent for future work.

At the present time the most attractive hypotheses for the mechanism of uniparental inheritance are ones that postulate (1) preferential replication of organelle genomes from one parent in zygotes or (2) the preferential destruction of organelle DNA from one parent in the zygote. In at least some organisms (yeast, *Chlamydomonas,* some higher plants) the physical exclusion of paternal organelle DNA from the zygote is not operative. Sager described work that suggests that mechanism (2) occurs in *Chlamydomonas;* she presented a specific model for the process that is broadly analogous to restriction and modification phenomena in bacteria and extended the model to the similar situation in yeast known as polarity. Birky, on the other hand, pointed out differences between polarity and uniparental inheritance in *Chlamydomonas* and presented a new model for polarity that is based on preferential gene conversion at specific heteroduplex regions of mtDNA. Perlman presented a version of the gene conversion model as a possible mechanism for suppressiveness of certain petite mutants of yeast.

It is clear from the content of all of the genetic contributions that the study of organelle genetics is beginning to advance beyond the phenomenological stage and will be concerned in the future primarily with the elucidation of the genetic mechanisms responsible for the existing phenomena.

Although much of the research in this area is being pursued for its own sake, there are several instances in which these studies may provide important new insights to problems of quite general interest. Wildman's analysis of the distribution of isoelectric-focusing variants of specific chloroplast proteins provides new understanding of the evolutionary relatedness of many species of *Nicotiana*. Perlman reviewed the current status of studies of the petite mutation in baker's yeast and highlighted the utility of that system for studying mutagenic and repair mechanisms and more generally the genetic properties of reiterated sequences in eucaryotes. In particular, since petite mutants have mitochondrial DNA that is extensively reiterated but derived from unique wild-type mitochondrial DNA, he noted that it may be possible to gain insight to mechanisms capable of forming repeated genetic information by a careful study of the molecular aspects of that mutagenic process. Griffiths discussed his progress in the genetic and biochemical dissection of the oxidative phosphorylation machinery of yeast mitochondria; these studies have great promise of providing further understanding of the major energy-producing reactions in mitochrondria.

Two contributors reported new, unexpected cytoplasmic genetic elements, both of which resemble episomes. Griffiths described a genetic element that affects mitochondrial drug sensitivity in yeast and that is readily distinguishable from mitochondrial DNA. We may anticipate further active work aimed at characterizing that element, at defining its role in the cell, and at determining whether such elements exist in organisms other than baker's yeast. Laughnan, in his discussion of cytoplasmic pollen sterility in maize, found little reason to connect that phenomenon with mitochondrial or chloroplast genes and instead concluded that a new episome-like factor may best explain his observations.

In summary, the contributors to this volume provide us with new data and insights to problems that have been under investigation for some years. In addition, they have indicated new directions and themes that will be reflected in future investigations.

Index